Science After the Practice the Philosophy, History, ai Studies of Science

T0227821

"This is a terrific collection, which has brought together contributions by the people who have thought hardest about practices in science. It is essential reading for anyone who employs this problematic concept in the study of science. Its special virtue is that it brings social studies of science and the philosophy of science back together, and in a fruitful and interesting way".
—**Stephen Turner**, *University of South Florida, USA*

In the 1980s, philosophical, historical and social studies of science underwent a change which later evolved into a turn to practice. Analysts of science were asked to pay attention to scientific practices in meticulous detail and along multiple dimensions, including the material, social and psychological. Following this turn, the interest in scientific practices continued to increase and had an indelible influence in the various fields of science studies. No doubt, the practice turn changed our conceptions and approaches of science, but what did it really teach us? What does it mean to study scientific practices? What are the general lessons, implications, and new challenges?

This volume explores questions about the practice turn using both case studies and theoretical analysis. The case studies examine empirical and mathematical sciences, including the engineering sciences. The volume promotes interactions between acknowledged experts from different, often thought of as conflicting, orientations. It presents contributions in conjunction with critical commentaries that put the theses and assumptions of the former in perspective. Overall, the book offers a unique and diverse range of perspectives on the meanings, methods, lessons, and challenges associated with the practice turn.

Léna Soler is Maître de conférences in the Laboratoire d'Histoire des Sciences et de Philosophie—Archives Henri Poincaré at the Université de Lorraine, France.

Sjoerd Zwart is assistant professor at the Universities of Technology in Delft and Eindhoven, The Netherlands.

Michael Lynch is professor in the Department of Science and Technology Studies at Cornell University, USA.

Vincent Israel-Jost is a post-doctoral researcher in the Laboratoire d'Histoire des Sciences et de Philosophie—Archives Henri Poincaré at the Université de Lorraine, France.

Routledge Studies in the Philosophy of Science

Science After the Practice Turn in the Philosophy, History, and Social Studies of Science

Edited by Léna Soler, Sjoerd Zwart,
Michael Lynch, and Vincent Israel-Jost

Routledge
Taylor & Francis Group

NEW YORK AND LONDON

First published 2014
by Routledge
711 Third Avenue, New York, NY 10017, USA

and by Routledge
2 Park Square, Milton Park, Abingdon, Oxfordshire OX14 4RN

First issued in paperback 2017

Routledge is an imprint of the Taylor & Francis Group,
an informa business

Library of Congress Cataloging-in-Publication Data

Science after the practice turn in the philosophy, history, and social studies
  of science / edited by Léna Soler, Sjoerd Zwart, Vincent Israel-Jost, and
  Michael Lynch. — 1 [edition].
    pages cm. — (Routledge studies in the philosophy of science ; 14)
  Includes bibliographical references and index.
  1. Science—Philosophy.   2. Science—Social aspects.   I. Soler, Lena,
1966– editor of compilation.
  Q175.S4157 2014
  501—dc23
  2013037522

ISBN 13: 978-1-138-06273-3 (pbk)
ISBN 13: 978-0-415-72295-7 (hbk)

Typeset in Sabon
by Apex CoVantage, LLC

Contents

the implicit, hidden, intuitive, and often partly unconscious dimensions of science.

A third way to portray science more realistically has been to take into account the fact that science, as it is actually practiced, is an *inherently collective* activity. If we think of science "in practice" and not just "in principle", it becomes obvious that no single scientist can possess, least of all produce, all present-day science, or even all what goes into most contemporary scientific projects. Even at a small scale, a scientist is far from independent from other scientists, or from previously established techniques and theories that are taken up in current projects (see the contribution of Hanne Andersen in Chapter 5 of this volume). The recognition of this fact led to the development of "social epistemology": an attempt to address the production of knowledge at a collective rather than individual level.[41] As Woody analyzes in Chapter 4 of this volume, this move "from the perspective of the individual scientist to that of a particular disciplinary community (and its interaction with other such communities)" introduced "a variety of new issues surrounding how knowledge is generated and transmitted, how knowledge is intertwined with issues of authority, expertise, trust, and divisions of labor, and how social structure is established and perpetuated to coordinate community deliberation and action" (p. 124).

Finally, a fourth way to consider science in a less idealized perspective—which could be viewed as encompassing the third way as a particular case—has been to move from "purely cognitive" and "internal" to "psycho-social" and "external" conceptions of science. (Eventually, this move led some analysts to question the applicability of these distinctions themselves.) According to this move, an empirically based study of science should pay attention to the fact that scientists are *singular* human beings whose actions are embedded in *particular* social contexts; this implies to leave idealizations, such as '*the subject of knowledge*' treated as an abstract universal rational function, or 'cognition' as a trans-historical, immutable reality, and, instead, to take into account the *specificities* of the individuals and the societies under study. Beyond this minimal message, however, we find variations and disagreements at many levels, and we meet the highly contentious issue of social constructivism.

Admittedly, nobody would deny that the practice turn helped to recognize that science and society are not two separated spheres, but are intertwined in multiple ways. The practice turn explicitly placed epistemic politics on the agenda, and the importance of this point was here to stay. For instance, Chang, in his 2012 book, still insists on "the inevitable political dimension of knowledge, and the ineliminable link between knowledge and politics, between science and policy" (Chang, 2012, p. 264). To take another representative example, Rouse (2002), reflecting on the lessons of the practice turn, speaks of "the intimate entanglement of the sciences with other practices". The "traffic in all directions" is "too extensive and wide-ranging", he writes, to allow for any significant autonomy of a domain of scientific practices. The sciences are contextually integrated in multiple ways: in acquiring funding and material resources; in recruiting and training scientists and auxiliary personnel; embedded in institutions such as universities, academies, bureaucracies, or foundations; in the articulation and application of social norms, cultural forms, and bodily disciplines; but also in acquiring and answering to much of what is ultimately at issue and at stake in scientific practices. (p. 165)

He pursues: "The reverse is also true, that is, what is left of the social or cultural 'world' apart from scientific practices is not autonomously intelligible" (p. 165). Given such conditions, Rouse views "the construal of scientific practices as clearly bounded and distinct from extrascientific 'context'" as a "danger to avoid" (p. 164).

Beyond the widespread acceptance of the general idea that science is not an autonomous sphere separated from the rest of society, however, the issue of the specific nature of the influence of social factors on scientific knowledge, or even of the possibility of separating "social factors" from "scientific knowledge", continues to be a subject of hot debate. Schematically, two general positions can be distinguished. According to the first, 'weak' one, social factors only influence the topics that scientists investigate (for example, by giving national priority and funding to specific problems or problem areas), and they possibly influence the dynamics of scientific research, but accepted results ultimately are determined by the object under investigation (typically on the basis of experimental data, elucidated through correct procedure, testing, and replication of results). This is, for instance, the position of Franklin (1998/2012) in relation to the natural sciences, and seems to be the position of Van Bendegem, in the present book, in relation to mathematics.[42] The second, 'stronger' position is often equated with "social constructionism" (although different kinds of social constructionism can be differentiated; see Hacking, 1999). Historically, it has been closely associated with the Strong programme in the sociology of scientific knowledge, and in philosophy of science, Latour is almost invariably mentioned as an exemplar (despite his explicit rejection of "social" constructionism, and his move away from the sociology of scientific knowledge). Roughly put, the claim is that any scientific product is *relative to* a social context, in the sense that its construction and establishment are *essentially* dependent of such context, so that social factors cannot be eliminated from the reasons why scientific achievements are accepted or rejected. As Lynch sums up the outcome of the practice turn at this level: "like other forms of knowledge, science was said to be inflected by vested interests, culturally located, and open to political influence, so that controversies are closed, not because of crucial experiments that falsify contending theories, but because persistent dissenters are discredited on 'social' grounds and marginalized from the core groups that dominate particular research fields at particular times" (this volume, Chapter 3, p. 102).

Acknowledgments

We would like to offer our greatest thanks to many people and institutions for their indispensible roles in the realization of this edited volume. First of all, let us express our great appreciation to all the authors. The time and effort they investigated made the present project so successfully achieve its end. We are grateful for their important contributions to the present-day research in philosophy, history, and social studies of science and technology after the practice turn. Second, we would like to thank all the members of the PratiScienS group, who provided the starting point of the present volume: the conference *Rethinking Science After the Practice Turn,* held at the Nancy University in June 2012. They helped to turn the conference into a great success, and unrelentingly participated in the various demanding and inspirational collective discussions of all the contributions of this book. In addition, our special thanks are extended to the staff of Routledge for their warm support and collaboration in the publishing process. Finally, we would like to acknowledge the institutions that support or have been supporting the activities of the PratiScienS team, viz., the:

Agence Nationale de la Recherche (ANR)
Maison des Sciences Humaines de Lorraine (MSHL);
Région Lorraine;
Laboratoire d'Histoire des Sciences et de Philosophie—Archives Henri Poincaré;
Université de Lorraine.

Introduction

*Léna Soler, Sjoerd Zwart, Vincent
Israel-Jost, and Michael Lynch*

I. AIM, SCOPE, AND GENESIS OF THE BOOK

The aim of the book is to take stock of the turn to practice in science studies, including the natural sciences as well as mathematics and engineering. "Science studies", as used here, refers to all studies that take science as their object, whatever the perspective—philosophical, historical, sociological, or other.

It is generally acknowledged that science studies, defined in this broad sense, underwent a change that began in the 1970s and was later often called a "practice turn".[1] Researchers in science studies, who we will refer to as "analysts of science", were progressively more encouraged to pay attention to scientific practices in meticulous detail and along multiple dimensions, including the material, tacit, and psycho-social ones. This turn to practice emerged from criticisms of traditional philosophy of science. The latter was deemed to be too idealized and too disconnected from how science actually is performed in laboratories and other research settings. In particular, ethnographic studies conducted in scientific laboratories showed that science was a complex activity that involved much more than experiments and logical inferences. As a result, in the 1980s and even more in the 1990s, analysts of science developed a broad interest in scientific practices.

This interest continued to increase in all the fields of science studies, to the point that today, the reference to practice is a pervasive feature in diverse studies devoted to science. It is present in philosophy of science and, to some extent, in epistemology, including the analytic tradition. It remains very important in science and technology studies (STS), particularly in sociological and anthropological investigations that use ethnographic and ethnomethodological approaches. It also is prominent in history of science, and in mixed forms of research, such as social history of science and historically inspired philosophy of science.

The practice turn is thus deeply entrenched in science studies today, and there is little doubt that it changed our conceptions of and approaches to science. But what did it actually teach us? What does it mean to study scientific practices? What are the general lessons, implications, and new challenges?

Surprisingly, few works have attempted to explore such questions systematically. Admittedly, the task is not easy, given the diversity of the fields and orientations in studies of science in practice; the practice turn has been associated with diverse inquiries that have heterogeneous aims, ideals, and methods, often leading to conflicting arguments and conclusions.

By extending beyond the natural sciences to address mathematics and engineering, the proposed volume attempts to cover different scientific fields, time periods, and regions. While not claiming to be exhaustive, it promotes interactions between acknowledged experts from different fields (including historians, sociologists, and philosophers of science from different traditions). The contributors have been encouraged to make explicit their conceptions of practice when expounding upon their interpretations of the practice turn.

Each of the nine chapters consists of a "main paper" that is followed by a critical commentary, usually written by more than one author. Commentaries are designed to put the theses and assumptions of the paper in perspective, and generally to provide readers with a critical appraisal of the paper. Overall, the ambition is to offer a diverse and up-to-date range of perspectives on the meanings, methods, lessons, and challenges associated with the practice turn.

The main papers and commentaries in the book are fully revised versions of papers originally presented at a recent international conference, "Rethinking Science After the Practice Turn", organized by Léna Soler and the PratiScienS group at the University of Nancy, France, in June 2012. This was the final conference of a research project entitled "Rethinking Sciences from the Standpoint of Scientific Practices" (PratiScienS[2] for short). Most members of the PratiScienS core team are authors of commentaries in this book, and one member, Jean-Michel Salanskis, is the author of a main paper (Chapter 1). In addition to the members of the PratiScienS core group who are based in France, other researchers from many different countries are associated with the work of PratiScienS,[3] though on a more occasional basis. Many of these partners of PratiScienS wrote main papers in this book.

The book came about in an original way. All of the first drafts of the main papers, as well as all first drafts of the commentaries, were discussed within the PratiScienS core team prior to the June 2012 conference. This process permitted to the PratiScienS team to give abundant feedback to the authors before, during, and after the conference. We hope that this inclusive effort to share and criticize each paper and commentary will serve the clarity, unity, and interest of the book. In spite of the highly collaborative nature of the process that led to this book, each author is, of course, responsible for his or her own contribution.

II.   ONE SINGLE TURN TO PRACTICE, OR MULTIPLE PRACTICE TRENDS?

We have retained the phrase "practice turn" for the title of the present book, although we recognize that the studies we cover with that term are quite diverse, and that other terms are in play for addressing practical, practice-based, and pragmatic tendencies associated with "science in practice". We also recognize that there has been an inflation of the announcement of "turns" in science studies. We have sometimes joked during the PratiScienS meetings that so many "turns" have been asserted that following them could easily make us dizzy. Talk of "turns" is, of course, a rhetorical artifice that serves to provoke interest in a given project and should not be taken at face value. In particular, to talk about a "practice turn" suggests, on the one hand, a sufficiently well-defined origin that enables "before" and "after" phases to be delimited, as well as a sufficiently coherent project. However, neither the origin nor the coherence of the so-called "practice turn" is unproblematic, as they both depend upon a preconception of what it means to study science in practice. This, of course, is not so straightforward, as scientific practices can be understood in various overlapping, but not equivalent, ways.[4]

Clearly, practice-based studies of science are concerned with "what scientists actually do", but scientists do many different things, individually and collectively, including technical work at the laboratory bench, planning and administering projects, writing and reviewing scientific papers, discussing research with their colleagues and outside audiences, and writing reports to obtain funding. What is relevant, among the range and variety of things that scientists and groups of scientists do? How can analysts of science access the relevant settings and actions? In which frame can they interpret the relevant materials? At what level of detail or generality can they elaborate the resulting accounts? The answers to such questions prove to be quite diverse.

Given such diversity, it might be preferable to talk about "practice trends" rather than a single "turn" in studies of science: trends that pay attention to scientific practices with different orientations to, and definitions of, "practices", with possibly different aims and methodologies, as well as trends that converge on the theme of practice from diverse points of view and to different purposes. In the remainder of this introduction, however, we shall stick with the term "practice turn".

III.   LANDMARKS IN THE PRACTICE TURN

As for any movement constituted of diversified internal trends, the issue of the origin is not straightforward, and may lead to disagreements. Below, we only provide some landmarks.

Early works that, in retrospect, can be associated with the practice turn in science studies can be dated back to the mid-1970s. The turn culminated in the 1990s, as indicated by exemplary publications, such as Andrew Pickering's 1992 edited volume, *Science as Practice and Culture*; Michael Lynch's 1993 book, *Scientific Practice and Ordinary Action: Ethnomethodology and Social Studies of Science*; Stephen Turner's 1994 book, *The Social Theory of Practices: Tradition, Tacit Knowledge and Presuppositions*; Jed Buchwald's 1995 edited volume, *Scientific Practice: Theories and Stories of*

Doing Physics; Theodore Schatzki, Karin Knorr Cetina, and Eike Von Savigny's 2001 edited volume, *The Practice Turn in Contemporary Theory;* and Joseph Rouse's 2002 book, *How Scientific Practices Matter.*

Pioneering works have two relatively distinct sources. One is sociology of science—more specifically, sociology of scientific knowledge and laboratory studies of science in action based on ethnographic and related methodologies. In his contribution to the present volume, Michael Lynch, for example, places stress on this first source, writing that the "turn toward practice was self-consciously initiated by sociologists of science in the 1970s (particularly, Barry Barnes, David Bloor and Harry Collins)" (see Chapter 3). The other source is philosophy of science—more specifically, a renewed historically based philosophy *of experimentation,* later labeled "New Experimentalism". Andrea Woody, for example, puts forward this second source (without denying the first), when she writes: "Arguably the first work in philosophy of science to advocate explicitly and self-consciously for a turn to practice was the self-labeled 'New Experimentalism' led by Ian Hacking and Allan Franklin, and in conjunction with scholarship in science studies by Pickering, Latour, Woolgar, Collins, Galison, and others" (this volume, Chapter 4).

Before dealing with these two origins of the practice turn in some more detail, we would like to note that the emphasis on scientific practices was congruent with trends that became prominent in the 1960s in philosophy, history, and social sciences that were not particularly focused on *the sciences.* Many 'towering figures' and 'emblematic works' could be mentioned in this respect, but let us simply sketch some landmarks, without any pretension to be exhaustive. One of them is Ludwig Wittgenstein (1889–1951) and his influential *Philosophical Investigations* (1958), which develops practice-based ideas about "meaning is use", "forms of life", and rule-following. Another is Martin Heidegger (1889–1976) and the notion of "being-in-the-world" he elaborated in *Being and Time* (1927/1962).[5] Others are Charles Taylor (1971, 1985) and Hubert Dreyfus (1991), who developed conceptions of cognition as strongly context- and practice-bound, and Richard Rorty (1979), who, taking up Quine's and Sellars' critiques of logical empiricism, drew upon phenomenology, Wittgenstein's later philosophy, and American pragmatism, weaving threads from 20th-century philosophy into a coherent focus on practice. Michel Foucault (1926–1984), who continued the line of the French "historical epistemology" started by Gaston Bachelard (1884–1962) and pursued by Georges Canguilhem (1904–1995), also deserves mention. Foucault studied the history of the human sciences by concentrating on "regimes of practice" rather than social institutions,[6] and developed a notion of "episteme" that inspired countless studies and theoretical discussions throughout the humanities and social sciences. Relevant sociologists are, among others: Pierre Bourdieu (1930–2002) and his *Outline of a Theory of Practice* (1977), in which, inspired by Husserl, Heidegger, and Wittgenstein, he elaborates his notion of "practical dispositions"; Harold Garfinkel (1917–2011), who meticulously studied social actions in observable situations of practical action, with a focus on the phenomenological "point of view" of persons performing the actions, and who coined the term "ethnomethodology" (1967) for his program of sociological investigation; and Anthony Giddens (1938–) and his basic "structuration theory".[7] We also could add many pragmatist and Marxist philosophers, but we will leave off at this point, in order to focus on the practice turn in studies specifically dedicated *to science*.

Most early works associated within the turn to *scientific* practices, either in sociology of science or in the New Experimentalism, have been strongly influenced by Wittgenstein (1958) and Thomas Kuhn (1970). Two other pioneering authors who have been an important source of inspiration are Ludwik Fleck (1935/1979) and Michael Polanyi (1958/1962, 1967).

III.1. Origin of the Practice Turn in Sociology of Science

In sociology of science, the first initiatives to treat scientific practices as an explicit object of study were published in the mid-1970s, when sociologists started to turn from studying the social structures in which scientists worked towards the specific daily practices, present and past, that produced scientific knowledge. Several "schools" began to engage this type of enterprise, but they did so in different ways.

In Edinburgh, the approach to practice taken by Barnes, Bloor, Shapin, and others[8] was "macro-sociological" and "socio-historical". Their aim was to discuss causal connections between the "contents" of scientific knowledge and sociological variables, such as "interests" of the groups under scrutiny. Bloor did not carry out ethnographic studies, as he preferred to use socio-historical case studies, especially of mathematics, to exemplify the "Strong programme" that he and Barry Barnes worked out. Their program was very influential in what came to be called the sociology of scientific knowledge (SSK).

Harry Collins and the "Bath school" (including Collins' students David Travis and Trevor Pinch) conducted more micro-sociological "controversy studies", with the intention to show how "closure" is produced and consensus reached through "negotiations" among members of "core sets" of scientific research groups. As early as 1974, Collins began to report on his interviews and other findings about laboratories engaged in building or trying to build TEA lasers[9] and experimenting on gravitational waves.[10]

Outside the UK, ethnomethodologists, namely Harold Garfinkel and some of his students (particularly Michael Lynch and Eric Livingston), focused on scientific and mathematical practices such as performing experiments, using instruments, preparing and interpreting visual displays of data to distinguish between artifacts and presumed natural phenomena (a topic convergent with the new experimentalists discussed below), and performing mathematical proofs. The aim of this group was to examine natural sciences and mathematics as embodied performances with instruments and graphic materials.[11]

Inspired by the Strong programme, ethnomethodology, and other late-1970s initiatives in sociology and anthropology, several investigators began detailed studies of scientific practices in laboratory settings. *Laboratory life,* by Bruno Latour and Steve Woolgar, was the first laboratory ethnography published as a monograph (in 1979), and is certainly the best-known ethnographic study of the "construction" of a Nobel Prize-winning "fact" at Salk Institute in San Diego. However, other studies of day-to-day laboratory practices also were conducted in the 1970s, such as a study of a major plant protein research institution in Berkeley, California, by Karin Knorr-Cetina (Knorr, 1977; Knorr-Cetina, 1979, 1981); dissertation research on high-energy physicists in California and Japan by anthropologist Sharon Traweek (eventuating in the 1988 publication of her book); and dissertation research on a neuroscience laboratory by Lynch at the University of California, Irvine (published as a book in 1985).

Many other early sociological studies also could be mentioned that, in the 1980s[12] through 1990s, exemplified and reinforced the interest in scientific practices. Pickering's 1992 book, with its emblematic title, *Science as Practice and Culture,* which emphatically shifted attention to "science-as-practice" in opposition to "science-as-knowledge", played an important role in the widespread tendency to talk of a "turn to practice" in the science studies.

Although there is much more to say about the differences between these studies and their import for the practice turn, we will only briefly consider two points. The first concerns the relation of these studies to social constructivism (or, as it is sometimes called, "constructionism").[13] Most of the work mentioned so far explicitly coupled anthropological or ethnographic methods to a strong social constructivist orientation.[14] However, this was not the case for ethnomethodologists. Lynch, in particular, refrains from attaching the ethnomethodological approach and his descriptions of scientific practices to social constructivist conclusions.[15]

The second point is related with differences in method concerned with studies of *past* scientific practices as opposed to descriptions of the *present unfolding* of scientific practices in *contemporaneous* settings. Regarding the latter, researchers can enter laboratories and observe or participate in the research as it goes on. Applying ethnographic methodologies, they can gather all kinds of information through interviews, tape recording, and participant-observation of formal and informal behaviour during the processes of science-in-the-making. Various authors have emphasized the special interest of this kind of configuration. For example Collins, in his 1975 study of gravity wave research, explains his preference for "the contemporaneous study of contemporary scientific developments". He argues that such studies pose fewer "philosophical and methodological problems", because they focus on ideas and facts "while they are being formed, before they have become 'set' as part of anyone's natural (scientific) world", and thus help to escape the "ethnocentricism of now" and the "cultural determinism of current knowledge" (p. 205). Latour and Woolgar warn that historical accounts miss the "process of solidification" through which laboratory statements become scientific facts, and insist that scientists, when talking among themselves, provide very different accounts of what is going on in laboratories than when talking to outsiders (1979, pp. 28–29). Claims of this kind suggest that the best way to gain access to actual scientific practices is to enter the laboratory.

Concerning the study of *past* scientific practices, however, analysts of science must proceed differently. They have to consider all relevant available material witnesses and documents—or possibly conduct oral-history interviews when the research was not carried out very long ago (see, e.g., Pickering's (1984a) study of high-energy physics in the second half of the 20th century). Moreover, they have to follow certain methodological principles for interpreting such historical sources. The tenets of the Strong programme elaborated by Bloor and Barnes—the well-known principles of impartiality regarding truth and falsity, symmetry in explanations of true and false beliefs, and so on (Bloor, 1976/1991, p. 7)—can be understood as an attempt to elaborate a systematic historiographic methodology that enables the historian to reconstitute, in an empirically adequate way, the past scientific practices that led to the acceptance or rejection of past scientific contents.[16]

III.2. Origin of the Practice Turn in Philosophy of Science

On the side of historians and philosophers of science, the origin of the practice turn is almost always associated with New Experimentalism. Under this label[17] are gathered a set of authors—such as Ian Hacking, Allan Franklin, Peter Galison, Nancy Cartwright, Robert Ackermann, and others— who, from the early 1980s, have taken a fresh look at experimentation and instrumentation by considering them from the standpoint of practice.[18] Beyond differences of agendas, these historians and philosophers "suggested that we clear away the obstacles created by old-style accounts of how observation provides a basis for appraisal (via confirmation theory or inductive logic) and repave the way with an account *rooted in the actual procedures for arriving at experimental data and experimental knowledge*" (Mayo, 1994, p. 270, italics added).[19]

New experimentalists simultaneously criticized four closely related components of the previously dominant philosophy of science. The first was theory-centrism, which took scientific theory as the primary, if not the exclusive, object of interest, with the result that the key philosophical challenge was to characterize how to choose between competing theories in light of experimental evidence. Second was the idea that the main, if not the only, aim of experiment was to test theoretical hypotheses, so that theory came first, and experiment entered into play subsequently. Third was the reduction of experiment to observational reports and data that either would confirm or refute theoretical hypotheses. And fourth was the framework of the "spectator theory of knowledge" (Hacking, 1983, p. 62), which led to a focus on what theories and experiments 'tell us about the world' (i.e., a focus on theories and experimental facts as re-presentations of the world). Such emphasis on science as re-presentation worked to the detriment of science as material interaction with, and transformation of, the world—a practical dimension paradigmatically instantiated by *experimental* practices. In Hacking's terms, traditional philosophy of science showed "a single-minded obsession with representation and thinking and theory, at the expense of intervention and action and experiment" (Hacking, 1983, p. 131).

New experimentalists performed detailed, empirically informed studies of experimental practices. They examined processes involved in the invention of new instrumental devices; the manipulation and transformation of scientific apparatuses; the acquisition of skills for using instruments and interpreting their outputs; the discrimination between real phenomena and experimental artifacts; and the production of new phenomena by means of previously stabilized instrumental techniques. The conclusion was that, according to Hacking's phrase, "experimentation has a life of its own" (Hacking, 1983, p. x). This slogan sums up several emblematic claims of the new experimentalists, which Mayo (1994) specifies. First, the aims of experimentation need not involve the testing of theories, as "actual experimental inquiries focus on manifold local tasks: checking instruments, ruling out extraneous factors, getting accuracy estimates, distinguishing real effect from artifact, and estimating the effects of background factors" (Mayo, 1994, p. 271). Second, experimental results can be justified independently of theory and need not be theory-laden in an invalidating way. Third, experimental outcomes are sometimes more robust than theories, in the sense that they are retained despite theory change.

Overall, New Experimentalism gave strong impetus to the practice turn by directing attention to a *certain kind* of scientific practice, namely *experimental* activity taken as emblematic of a pragmatic activity, and by providing detailed, empirically informed characterizations of actual, past and contemporary diversified experimental activities. From a *methodological* point of view, this effort to provide empirically informed characterizations of actual experimental practices took several forms, which were largely congruent with the ones put forward in relation to sociology of science.[20] The new experimentalists' historical studies were finely contextualized and attempted to recover actors' aims, doings, and ways of thinking as they operated in the historical situations, by relying not only on scientific publications, but also on various other more informal sources, such as laboratory notebooks, memoranda, drafts, slides used in conference talks, and so on.[21] Whenever possible, they favored direct interactions with the scientific actors involved in the episodes under study (through interviews, discussions, etc.).[22] New experimentalists also sometimes attempted to acquire some kind of direct experience with particular scientific actions, such as when Hacking (1983) was introduced by scientists to the utilization of microscopes, thereby gaining a 'first-hand' appreciation of the nature and difficulties involved in the complex processes through which a novice learns to 'see' through a microscope (that is, learns to distinguish between artifacts and manifestations of natural phenomena).

New Experimentalism involved a turn to practice, not only because it studied experimental practices instead of reducing experimentation to experimental outputs (used as data in philosophical characterizations of abstract scientific methods for verifying or falsifying theories), but also because it drew attention to fundamentally practical aspects *of science in general*—in fact exhibited in and through the particular case of experimentation, but in principle not restricted to experimental sciences. In this respect, the *general* message carried by New Experimentalism can be characterized as follows: analysts of science not only should consider the contemplative-representational-ontological dimension of science, but should also examine the transformative-technical-pragmatic dimension of science, with its material, somatic, skilful and utilitarian aspects. In principle, this general message applies to any scientific activity, including nonexperimental, 'purely theoretical' activities. However, in fact, new experimentalists initially generally focused on experimentation as the most relevant candidate for advancing their aim to reintroduce practical aspects into the picture,[23] and this emphasis on experiment as practice in reaction to theory-centrism initially produced a paradoxical effect in philosophy of science:[24] it initially contributed to reinforcing a classic opposition between theory and practice and to precluding a full attention to *theories themselves from a practical point of view*. As Woody notes in her chapter: "Hacking's popular book *Representing and Intervening* (1983) laid out the agenda for a turn to practice, . . . but in doing so it further entrenched a dualism between theory and experiment—the realm of the conceptual/linguistic versus the realm of practical action".[25] Accordingly, in the early stage of the practice turn in philosophy of science, globally, "[w]hereas experiment was conceptualized as practice, theory essentially was not"[26] (Woody, this volume, p. 124). It was only later that theories and theorizing themselves became widely considered and analyzed *as practice*. As Rouse (2002, p. 163) notes: "an important complement to the rediscovery of experimental practices has been to recognize" that scientific theory itself "is better understood in terms of theoretical practices of modeling particular situations or domains", "rather than in the classical sense of *theoria*".

We can note that this initial neglect of theoretical activities in historically inspired philosophy of science at the beginning of the practice turn did not hold to the same extent *in sociological and ethnographical studies of science*. Admittedly, a majority of works remained focused on practices related to experimentation, as Pickering and Stephanides stressed in a 1992 paper which, as the title indicates, precisely intended to draw more attention to the ". . . Analysis of Conceptual Practice": "Perhaps in compensation for the traditional emphasis on theory, the analysis of practice has so far focused on experimentation and on the construction of the sociotechnical networks that link the laboratory to the outside world" (1992, p. 139).[27] However, theories and theorization were considered from the standpoint of practical actions by *some* influential scholars in social studies of science, *especially in relation to mathematics*. In SSK and ethnomethodology, mathematics was studied in a practical perspective from the start. Especially Bloor, in the early 1970s, and Livingston, in the mid-1980s, considered mathematics as practices, using a Wittgensteinian approach to mathematics as conventional (Bloor, 1973, 1976/1991) or as sequences of practical action (Livingston, 1986). The target of Pickering and Stephanides, in the 1992 paper just quoted, was broader: It encompassed "conceptual practice in both mathematics *and science more generally*" (p. 140, italics added). The authors, however, also relied on a case study from the history *of mathematics* to develop their more general interpretative model of conceptual practice. They motivated the choice of mathematics by the fact that "mathematics offers a particularly clean instance of conceptual practice, free from the material complications of experimental practice in science and from the esoteric subtleties of, say, recent theoretical physics" (pp. 142–143). Their aim was to show that conceptual practice "is amenable to the same kind of analysis as that already developed for experimental and sociotechnical practice" (p. 140). In particular, they wanted to show that an analogue to the "resistance" found in experimental practice is also found in 'purely' conceptual practice. Compared with sociological and ethnographical studies of science, in philosophical studies, the interest in mathematical practices did not develop until much later[28]—and is still a relatively marginal trend in the philosophy of mathematics (see the analysis of Van Bendegem in Chapter 7 of this volume).[29]

   Another important global difference between the sociological and philosophical orientations to the practice turn lies in their relation to constructivism and relativism. Although, as stressed above, social constructivism was a dominant trend in the sociological branch of the practice turn, this was not the case for the philosophical branch. Allan Franklin, in particular, waged a spirited debate with sociologists of science over their constructivism and relativism. In a number of books and articles, he attempted to refute Collins' thesis that the "experimenters' regress" comes to an end not through evidence alone, but through social factors (Collins 1985/1992); Franklin (1986) argued instead that the job is done by "epistemological strategies".[30] In the same vein, Franklin (1998/2012) criticized Pickering's constructionist account of high-energy physics and accused Pickering of leaving out scientific practices that are crucial for the scientific process, viz. the application of "epistemological strategies . . . to argue for the correctness of an experimental result", and the "performative act" of writing scientific papers. Galison and Pickering also had extended debates about the closure of experimental controversies and the role of evidence in constraining experimental judgment.[31]

IV. AN INITIAL GLOBAL CHARACTERIZATION OF THE PRACTICE TURN

IV.1. Uniformity in What Is Rejected

We have stressed above that the major works commonly associated with the practice turn are far from homogeneous. However, at least at the level of general slogans, such works advise us to eliminate some possible stances as partial, insufficient, impoverished, distorting, untenable, and sometimes even dangerous. Rather than being unified as a positive program, the practice turn is more united by statements about what must be avoided, and by a view of the sorts of inquiry that are devalued. This uniformity concerning negative claims is visible in diverse contributions to this book, and is a good starting point for grasping the specific identity of the practice turn.

The most recurrent critical themes are directed against (often a caricature of) the logical or empiricist positivist approach and against other classic philosophies of science, such as those of Popper, or even Lakatos. These studies are often called the "mainstream philosophy of science", or "traditional philosophy of science", or "the received view", or again "the standard positivist view" (but since authors like Popper and Lakatos developed their views in reactions against positivism, although it is commonly done, it actually is not appropriate to lump them in with "the standard *positivist* view"). Most of the time, the main target of criticism is a certain way of practicing *philosophy* of science. But sometimes, certain ways of doing *history* of science also are the target (the positivist history of science, or "Whig historiography", see Section V.4). Furthermore, the traditional *sociology* of science has also sometimes been criticized (the main target in sociology was Merton's structural-functionalism).

The content of the shared negative claims that characterize the practice turn will be specified in more detail in Section V, and is also exemplified by several chapters of this book (see especially Chapters 2 (Chang), 3 (Lynch), 4 (Woody), 6 (Bucciarelli and Kroes), and 9 (Rouse)). Briefly summarized, the main message is that traditional perspectives on science have developed a priori, overly idealized and truncated accounts of science, which prove untenable when treated as descriptions of actual science (of actual scientific *theories*, of actual scientific *methods*, of actual scientific *development*, etc.). The limitations ascribed to such perspectives include their focus on "science-as-knowledge" (Pickering, 1992) and more generally on established scientific products (see Section V.6), associated with their tendency to provide retrospective reconstructions of scientific development through ex-post facto accounts of successful science (rational reconstructions and Whig historiography—see Section V.4).

IV.2.   An Attempt to Formulate a Positive Identity for the Practice Turn

In addition to the specification of a shared negative idea of what must be avoided, it is possible to extricate some 'positive' shared positions and some common themes addressed in much of the scholarship associated with the practice turn, that can be considered as characteristic of the practice turn compared with other ways of approaching science. These positive features may not be universal, but more in line with Wittgensteinian family resemblances. Section V provides an analysis of conceptual and methodological 'shifts' that the practice turn introduced into science studies, but before we elaborate upon these shifts, let us provide a brief overall characterization of the practice turn.

In contrast to traditional perspectives on science, one common aim of supporters of the practice turn is to provide *descriptively adequate* accounts of scientific knowledge and scientific activities (see Section V.3). This descriptive ideal goes with the requirement of *empirically based* studies of science—in opposition to a priori, *excessively idealized* reconstructions of what scientists do or should have done (see Section V.2). The shared ideal to produce empirically based and descriptively adequate accounts of science goes hand in hand with certain methodological approaches. Two such approaches are intimately associated with the practice turn, to the point that they sometimes are *sufficient* to identify a particular study with *the practice turn:* ethnographic studies of contemporary science-in-the-making, which typically require spending extended periods of time in one or more scientific laboratory; and historical studies that attempt to reconstitute the actual circumstances that held for past scientific actors. The former methodology has been specifically introduced by pioneers of the practice turn (see Section III.1), and is often viewed as an essential characteristic of this turn. History of science, of course, was established long before the practice turn, but there are many ways to do history of science. Supporters of the practice turn rejected some of them (Whig history, rational reconstruction, etc.) and valued others as appropriate ways to achieve descriptively adequate accounts of science (see Section V.4).

Beyond any divergent understandings of "practice", for both ethnographic and historical studies, the maxim "pay attention to scientific practice" conveys a crucial methodological ideal: to recover detailed actions and reasoning—including uncertainties, conflicting interpretations, and so on—*as they operate in the situation,* in contrast to retrospective reconstructions of actions and results provided by scientists and traditional philosophy of science. In other words, the ideal is to recover important aspects of actual scientific activity that are left out of scientific publications (see Section V.6) and, more generally, that tend to be 'forgotten' after controversies are settled and facts are established. This is what advocates of the practice turn mean when they stress the need to start from "scientific practice" or "science in action" (Latour, 1987). When the target is a *present-day* scientific development, *contemporaneous with the analysis of the relevant science,* the ideal to recover actions and reasoning as they operate in the situation requires one to follow the unfolding of this development in real time, paying attention to the details and multiple dimensions of the situation, without presupposing that scientists act in accordance with 'purely rational algorithms' or are uniquely motivated by the selfless pursuit of truth (or some substitute such as empirical adequacy, degrees of falsifiability, or the like). When the target is *past* science, the same ideal implies an effort to reconstruct the situation 'from inside', as it held at the time, without presupposing that the subsequent history of science was inevitable and that presently taken-for-granted results should have been recognized at the time as correct (see Section V.4).

The core contention associated with the practice turn is that these methodological approaches reveal a distinct image of science from the received view. Globally, this novel image showed scientific development as more chaotic, multi-faceted, variable in space and time, contingent, and open than was the case for the received view. Moreover, the multiple dimensions involved in actual scientific activities (however categorized—as material, tacit, explicit, linguistic, propositional, exploratory, justificatory, methodological, normative, social, psychological, and so on) appeared to be deeply intertwined, interactive, co-evolving, co-stabilized, and thus uneasily separable in practice. Accordingly, many traditional dichotomies were criticized as inoperative or pointless. These include the internal/external, cognitive/social, and context of discovery/context of justification dichotomies; but also those between theory and experiment, science and technology, or representation and intervention. More holistic, 'network-based' (as in the "actor network" approach of Callon and Latour), "symbiotic" or "mangled" (Pickering's vocabulary (1995a)) interpretative schemes were elaborated. Scientific products (i.e., taken-as-reliable scientific items used in a given stage of scientific development) appeared strongly related to, and not easily detachable from, the details of multi-faceted, multi-dimensional, "mangled" scientific processes. Scientific processes themselves, in turn, proved to be variable from one historical period and one community to another at all levels, including beliefs, presuppositions, language, norms of validation, general ideas about science, instruments, techniques, and so forth. Consequently, relativist, constructivist stances acquired more and more importance—social constructivism, in particular, had often been seen as essential to the identity of the practice turn.[32]

V.   SHIFTS INTRODUCED BY THE PRACTICE TURN

V.1.   Analyzing the Practice Turn in Terms of Shifts

To analyze the practice turn further, we now will characterize a series of shifts with respect to previously dominant perspectives on science. For the most part, the shifts introduced by the practice turn present two sides. On one side, they appear to be shifts *of interest* that are not incompatible with previous studies of science. Viewed in this perspective, they involve a criticism of traditional perspectives on science as excessively focused on certain aspects of science at the expense of others, coupled with an incentive to enlarge the domain of objects of interest for science studies. On the basis of such a reading, nothing would prevent us from maintaining the analyses and conclusions of traditional perspectives on science, while supplementing them with the analyses and results that issued from the practice turn. Accordingly, the movement from traditional perspectives on science to the practice turn could be seen as following a cumulative pattern. But on other side, the shifts almost always convey more aggressive criticisms. Very often, the message is that science is not what traditional perspectives on science suggested it was. In that case, the shifts are *substantial*, and *destructive* with respect to traditional conceptions of science.

V.2.   From A Priori and "Too" Idealized to Empirically Based and Empirically Adequate Accounts of Science

A first shift that has accompanied the practice turn consists in moving away from accounts of science that are based in a priori conceptions of science and are too idealized. The criticized accounts are too idealized both in the sense of being too disconnected from empirical reality (i.e., they just do not seem to resist analyses grounded in detailed case studies), and in the sense of being 'embellished' (i.e., the a priori conception of science in which they are based is largely illusory).

The 'too idealized' criticism has been directed toward multiple targets within traditional philosophy of science. Traditional characterizations *of scientific method for theory-choice*, such as verificationism, falsificationism, methodology of research programs, and so on, have been criticized as ignoring local and historical contexts, material and somatic aspects, and various tacit elements that are in play in actual theory appraisal, and as giving a too narrow and deterministic conception of scientific method when compared with the actual plurality of positions and divergences among practitioners. Traditional views *of scientific development and scientific progress* (typically through "rational reconstructions"—see Section V.4) have been criticized as too linear for suggesting a continuous and straightforward historical path to the science that is accepted today. Traditional treatments *of scientists* have been criticized for giving a disembodied and unambiguously altruistic picture of scientists: It was as though scientists embodied a 'rational algorithm' that delivered a unique solution to a given problem, with nothing about motivations such as career advancement, quests for prestige, or favoritism for some scientists over others. In such a picture, scientists became as though omniscient (possessing all there is to scientific knowledge at a given time), without taking into account limitations in background knowledge, inferential capacities, and so on.[33]

At best, such traditional accounts are simply too crude and impoverished to address actual science in particular circumstances of practice. Many critics have also pointed out that their poor empirical adequacy rendered them descriptively valueless and uninformative. At worst, such overly idealized accounts could even be harmful when they convey a distorted and deceitful image of certain aspects of science.[34] The practice turn was a reaction to such drawbacks in its proposal to develop empirically based accounts of science, scientific agents, and scientific activities.

This shift—to avoid over-idealization from a priori conceptions of science—is perhaps the most general formulation of the criticism directed by the practice turn against traditional philosophy of science. For this reason, each of the other shifts analyzed below conveys, in addition to specific contents, a particular form of the more general shift from a priori and too idealized, to empirically based and empirically adequate accounts of science.

V.3. From Normative to Descriptive Perspectives on Science

A first variation on the movement "from too idealized to empirically adequate accounts" is a shift from normative to descriptive perspectives on science. This shift aims to move from characterizations of science *as it should be* to characterizations of science *as it actually is*. It comes as a reaction to a strong tendency, within traditional philosophy of science, to take a *certain kind* of normative stance.

This normative stance typically took the form of demarcation principles and criteria for drawing a boundary between science on the one hand, and non-science or pseudo-science on the other. For instance, the logical empiricist characterization of science proposed the verification principle in order to demarcate science from metaphysics. Popper criticized this solution and, instead, developed his falsificationist criterion, according to which only claims refutable by empirical tests are of the scientific type. By doing so, Popper claimed to demarcate genuine science (e.g., physics) from pseudo-science (e.g., psychoanalytical theory) and non-science (e.g., religion). Demarcation criteria in traditional philosophy of science also came with general prescriptive rules supposed to define scientific method for theory-choice—for example, comparison of degrees of confirmation, of degrees of falsifiability, and so on. Often, in such contexts, greater value was assigned to what falls on the side of science compared to non- or pseudo-science.

Prior to the practice turn, general rules for theory-choice were already criticized along different lines, but following the turn, one of the most widespread criticisms was that these rules were too far from what actual scientists, *including those who cannot be suspected to be bad scientists and who commonly work as highly valued models* (Galileo, Newton, etc.), actually do. Certainly, a normative characterization is not intended to apply to all historical cases, since the standard defines an ideal case. However, norms should at least be approximately instantiated in some paradigmatic cases, or should have some specified relevance for actual situations, and this is what has been challenged.[35]

A related criticism of normative methodological prescriptions is that they adopt the privileged viewpoint of a 'science of science': an overarching position from which the analyst presumes to tell practicing scientists how to do proper science. In this vein, Rouse (1996, 2002) criticizes such a position as an attempt to achieve a standpoint that he calls "epistemic sovereignty" (an expression inspired by Foucault): "a theoretical position 'outside' and 'above' scientific practices from which to establish or undermine their legitimacy and authority once and for all"; the will to ascertain "from the 'outside' what science 'really is', its indispensable nature or goal", and to provide "ground for assessing its successes and failures". Rouse refers to such a normative position in traditional philosophy of science as "the legitimation project" (Rouse, 2002, p. 180).

In contrast to such a normative perspective, proponents of the practice turn advocated a descriptive perspective. This was especially the case in sociology of scientific knowledge during an early phase of the turn to practice. Rouse (2002), for instance, insists that the Strong programme involved "a rejection of a priori normative principles" (p. 135; also see Lynch, this volume, Chapter 3, p. 96, who pictures one important aspect of this programme as a "descriptive turn"). Such a descriptive turn permitted analysts of science to adopt a much more modest attitude towards science, avoiding the privileged position of claiming to know better than the scientists they studied.

After this descriptive turn, however, many analysts of scientific practice came to realize that a 'purely descriptive' ideal had damaging effects, and that *some kind* of normativity had to be reintroduced in the science studies. As a result, and as several of the chapters in this book elaborate (see Section VI for a summary), normativity remains important for current science studies, or more precisely, just *how* to be normative in light of the practice turn currently is a lively and contested issue.

V.4.   From Present-Centered Reconstructions of Past Science, to Historically Adequate Reconstitutions of Past Science 'From Inside'

Another important sub-genre of the "from too idealized to empirically adequate" shift targets the way present-day science often is used for the interpretation of past science, such as in so-called "rational reconstructions" in traditional philosophy of science, as well as in what has been labelled "Whig" historiography or "present-centered" history of science.[36] The criticized narratives retrospectively reconstruct past science from a present vantage point that presumes what is true, rational, and successful.[37] Consequently, anything that cannot be considered as a step toward presently accepted scientific knowledge is ignored or denigrated. In contrast, proponents of the practice turn attempt to understand past science 'from inside'—that is, to reconstitute what was known and valued, and possible to know and to do, for actors at the time in question. The shift from present-centered reconstructions to historically adequate reconstitutions of past science was encouraged before the practice turn, in different fields and from different perspectives, and some scholars have categorized the corresponding tendencies in philosophy of science as an "historical turn".[38] But at the very least, the practice turn can be said to have reinforced and systematically developed the prior criticisms.

From a *methodological* point of view, critics of present-centered perspectives on science oppose the dismissal of participants in a given historical episode who endorsed positions that are not nowadays considered as scientifically valid. Instead of considering that such agents were misled by psychological and social factors that deflected their judgments from a rational path, critics of present-centered perspectives call for a reconstitution of the scientific situation *as it appeared to the actors of the time*. Accordingly, an adequate characterization of the historical situation requires that analysts of science reconstitute the relevant context—scientific, technical, social, cultural, and so on—at the time, and that they examine past episodes, especially concerning the "closure of controversies", as if they did not know about the subsequent history.[39]

From a *substantial* point of view, present-centered accounts of scientific development were criticized for creating the *illusion* that science progresses in a cumulative way toward 'what we now know' and that science has always been the same kind of enterprise as present-day science—involving the same kinds of questions, methods, idea of science, institutional organization, and so on. In contrast to this idealized picture, the practice turn resulted in more variable, discontinuous, chaotic, and open-textured accounts of the history of science.

V.5. From Decontextualized, Intellectual, Explicit, Individual, and 'Purely Cognitive' to Contextualized, Material, Tacit, Collective, and Psycho-Social Characterizations of Science

The three shifts associated with the practice turn that we have discussed thus far take different points of departure: They move away, respectively, from a priori and overly idealized, normative, and present-centered accounts of science. They nevertheless converge, however, on roughly the same aim of giving *faithful* accounts of science: that is, the aim to provide empirically based, historically adequate (for past science), descriptively adequate (for present and past science), and 'more realistic' (contrasting with 'too idealized') characterizations of science.

This aim has been intimately associated with adopting a *local scale* of analysis and paying close attention to the *specific contexts* in which scientific results are produced, used, and disseminated. This has encouraged a move from types to tokens and from decontextualized-global-panoramic to contextualized-local-microscopic accounts of science. As a result, the practice turn has produced numerous detailed and finely contextualized micro-studies of science.

The same aim also has led philosophers to pay attention to dimensions of science that had largely been neglected, because they were considered as irrelevant, in previous philosophies of science. Below, we highlight four of these dimensions of science: the material, tacit, collective, and psycho-social. The introduction of these dimensions into the picture can be seen as four ways to instantiate the request of 'more realistic' accounts of science. Supporters of the practice turn have claimed that such dimensions, far from being anecdotal, play essential roles in science—although different authors conceive these roles and their philosophical implications somewhat differently.

One first widespread way to provide a 'more realistic' characterization of science has been to take into account the diverse *material aspects* of scientific practice. The corresponding characterizations are 'less idealized' in the sense that they move from ethereal, 'purely' intellectual, conceptual, abstract, disembodied perspectives on science to more 'down to earth', practical, concrete, and embodied pictures. Typically, studies associated with the practice turn have focused on instrumental devices, experimental setups, protocols, and manipulations (see Section III). Other material aspects that have been included in the picture are the particular "formats" of scientific theories, models, and representations, such as images, diagrams, tables, and so forth (see Section V.7).

A second way to treat science more realistically has been to focus on *"tacit"* or *"unarticulated"* aspects of scientific practice. Among these are the various sorts of "know-how", whether they concern manual skills, such as technical gestures (knowing how to take a blood sample from a mouse, for example), or intellectual ones, such as knowing how to perform certain complex computations in an efficient way, or seeing a new problem as similar to another already mastered one. Other partially unarticulated features of science concern norms and values that often remain in the background, yet condition scientists' goals, preferences, options, and beliefs. These are among the constituents of so-called "tacit knowledge".[40] Characterizations of science that take into account tacit aspects are less idealized in the sense that they move from the explicit, well-articulated, visible, and easily accessible dimension of science (for example, what appears in scientific publications, or what scientists say about scientific method), to

V.6. From Scientific Products to Scientific Processes

Traditional philosophy of science has been criticized for focusing too much on scientific products to the neglect of scientific processes.[43] In this context, scientific products refer to stabilized outcomes at a given time in the history of science (theories that are taken for granted, experimental facts, mathematical theorems, material artifacts, published experimental and technical protocols, published mathematical proofs, and so on). By contrast, scientific processes refer to the dynamic unfolding of actual scientific activities and, in particular, to the actual historical process through which particular achievements come to acquire the status of knowledge.

The criticized neglect of scientific processes can be interpreted in two different ways. The first interpretation targets a global *disinterest* in processes. In this criticism, traditional approaches fail to consider the relevance or importance of scientific processes and thus fail to study them *at all*. This criticism is featured in the New Experimentalism (see Section III.2). According to that perspective, experimental processes have not been an object of philosophical study; experiment has been reduced to experimental products, typically observational statements. This is the sense of the title of Franklin's famous book, *The Neglect of Experiment,* or of Hacking's line that experiments "have been neglected for too long by philosophers of science" (1983, p. xiv). New Experimentalism, as well as various laboratory studies discussed above, rejected the traditional treatment of experiment as a 'black box' delivering unproblematic experimental facts. They proposed a move from accepted experimental products (facts and black-boxed material artifacts) to examining how experimental processes constitute such products in laboratory life. The first interpretation of the neglect of scientific processes is also involved in early criticisms of the disinterest in the so-called "context of discovery" within traditional philosophy of science. Reichenbach, Popper, and other philosophers who consigned the context of discovery to psychology, while focusing on the context of justification, were accused of deliberately ignoring crucial aspects of scientific processes that are responsible for the validation and stabilization of the scientific products that those philosophers attempted to logically reconstruct.

The second interpretation of the neglect of scientific processes acknowledges that traditional philosophy of science did *attempt* to study *some* processes, but considers that the resulting characterizations are so overly idealized that they cannot stand for *actual* scientific processes; accordingly, it is *as if* scientific processes had not been studied at all. A typical example of this criticized tendency is the treatment of the problem of theory-choice in traditional philosophy of science. The very effort to focus on the problem of how scientists come to choose one particular theory over its competitors does involve an interest for scientific processes. However, the way the problem is set up was criticized as too idealized to capture what is at stake in the *actual* processes through which theories come to be established or undermined. Expressed in a more radical form, what traditional philosophy of science says about processes of theory-choice is not informative about the scientific processes through which theories are actually instituted in the history of science.

From a methodological point of view, the shift from products to processes was accompanied by a request to enlarge the sources of science studies and to move from retrospective accounts provided by scientists in publications to more diverse materials. Scientific publications such as textbooks or papers are, by nature, reconstructed versions of actual historical processes. Scientific publications do not aim to indicate the details of the path that led to the successful results they report, and they definitely are not the place to describe the doubts and dead-ends that scientists faced during the research. The overwhelming majority of such publications only present the stabilized outcomes. Thus, by their very nature, they are selective and provide reconstructed, idealized, and linear versions of the actual courses of action that led to the results (for example, the final version of an experimental protocol after it was stabilized at the end of an experimental sequence). Medawar, in his 1964 paper, "Is the scientific paper fraudulent?" went so far as to suggest that the scientific paper is a "lie". His provocative language suggested that retrospective accounts of methods and scientific episodes are misleading if taken as descriptions of actual sequences involved in scientific work. The lesson for analysts of science is that if they aim to reconstitute actual scientific processes, they should not just rely on scientific products as they appear in publications, but should also value other sources.

V.7.   From Science as Contemplation of the World to Science as Transformation of the World

Hacking's book title *Representing and Intervening* captures the spirit of this shift—but, because the term "representation" is used in multiple, heterogeneous senses, we prefer a formulation of the shift that avoids that term. The emphasis, in this shift, is on the insufficiency of a "spectator theory of knowledge"[44] that ignores the aim to manipulate, probe, and transform the worldly materials being investigated. Hacking rejects "the false dichotomy between acting and thinking" (Hacking, 1983, p. 130), stressing that "we represent in order to intervene, and we intervene in the light of representations" (p. 31).

Hacking and other new experimentalists paid special attention to the non-contemplative dimension of science. They studied experimental activities such as those that occur when more or less new instruments and techniques are first produced and used (see Section III.2.). Hacking insisted on intervention as a means of investigating the world through probing and manipulating, rather than simply perceiving and contemplating what exists. For example, using a microscope is not just a matter of looking through the instrument and reporting on what one sees. It also involves efforts to understand how the instrument represents the world, and scientists do so by manipulating both the instrument and the specimen that is observed through it. For instance, microscopists inject chemical stains into a cell to 'label' organelles of interest, and they embed the cell in wax or plastic so that it can be sliced into thin sections that are transparent to the instrument's light source or electron beam. If such interventions accord with expectations, they reinforce scientists' belief that the microscope shows the cell and shows it in a reliable way (see also Lynch, 1985).

New Experimentalism does not exhaust the ways in which the practice turn has linked representation to practical action. Another striking example concerns a change in the status of *theories*—from abstract propositions supposed to describe the world to tools for action with respect to specific cognitive tasks. Ways of thinking commonly applied to material-technical objects and to technological activities have been transposed to theories and theorization themselves. In this perspective, theories not only are representations that must be assessed for their empirical adequacy, but they also are, perhaps primarily, material means to achieve cognitive aims. Consequently, they are significant as "theoretical artifacts" (cf. Woody's paper in Chapter 4 of this volume), "theoretical technology" (Andrew Warwick's term; see Warwick, 1992, 1993, 2003), "paper-tools" (Ursula Klein's term; see Klein, 2001a, 2003), and "calculational techniques" (see David Kaiser, 2005), which are designed, manipulated, and transformed in order to progress in specific problems. Theorization and theory evaluation thus are governed by practical values such as efficiency, commodity, and computability, and the comparative merits of several theoretical artifacts are assessed relative to particular problems. Such a framework shifts the meaning and centrality traditionally assigned to the requirement of theory-choice, understood as a choice of *the* best theory *tout court,* and encourages more pluralist stances.

Beyond its relevance to the notable case of *theories*, the contemplation-to-transformation shift has had a similar impact on understandings of *all kinds of representations* involved in the natural and formal sciences. Such representations include "graphs, diagrams, equations, models, photographs, instrumental inscriptions, written reports, computer programs, laboratory conversations, and hybrid forms of these" (Lynch & Woolgar, 1988, p. 99). The move is from representations as descriptions of the world to representations as means for doing things, tools for intervening, and material artifacts for transforming the world. This move went hand in hand with a growing interest in models and modeling.[45] The conception of representations as material artifacts led to an interest in the various "formats" of representation in science, where "format" refers to the specific material form of a representation (a given type of diagram, of table, of graph, etc.). Where traditional philosophy of science often tacitly assumed that formats were indifferent to the epistemic contents they conveyed, most analysts associated with the practice turn argue that "the differences between drawings, instrumental inscriptions, autoradiographs, etc. are not incidental" (Lynch &

Woolgar, 1990, p. 105), and that the visual and verbal formats in which 'scientific contents' are embodied have a major impact on the specificity, significance, pragmatic import, and success of these 'contents'.[46]

VI.   A SELECTION OF CURRENT ISSUES AS THEY APPEAR THROUGHOUT THIS BOOK

The less idealized image of science that emerged from the practice turn generated new issues and trends, sometimes described as new turns (see Lynch's contribution in Chapter 3 of this book for an overview of a series of claimed turns and trends in STS). The different chapters of this book offer a selection of current issues associated with the practice turn in science studies. The first two chapters (by Salanskis and Chang) analyze the notions of scientific practice and human action. They provide a general framework, and introduce tools and useful distinctions for those of us who are interested in the study of science from the perspective of practice. Chapters 3 (Lynch), 4 (Woody), and 5 (Andersen) discuss important conceptual and methodological changes that the practice turn brought about in science studies. They identify dominant trends, and present lessons and challenges, associated with the practice turn. They also consider how some central philosophical issues and debates—such as those associated with descriptive versus normative perspectives or with different versions of the nature and roles of representation and explication in science—have been transformed by the practice turn. Whereas Chapters 3, 4, and 5 are primarily concerned with the *natural* sciences, Chapter 6 (Bucciarelli and Kroes) is dedicated to *engineering*, and Chapter 7 (Van Bendegem) and 8 (Chemla) to *mathematics*. These chapters address the form the practice turn has taken in specific domains, and give elements about what form it could take in the future. Finally, Chapter 9 (Rouse), based on some lessons drawn from the practice turn elaborates a very integrative conception of scientific practice and an original, naturalistic interpretation of the new "scientific image" following the practice turn.

VI.1.   Chapter 1. Some Notions of Action, by Jean-Michel Salanskis

Salanskis' chapter addresses the question of what counts or does not count as *practice*. This is a fundamental issue for practice-oriented philosophy, yet a relatively neglected one. Salanskis tackled this issue from a perspective fed by multiple sources of inspiration. He was first trained to be a mathematician, and was then initiated to philosophy by Jean-François Lyotard. His early work was accomplished in the field of philosophy of mathematics, partly in the line of French historical epistemological school. Subsequently, he has been importantly committed with phenomenology, writing about Husserl, Heidegger, and Levinas, but has at the same time always been open to analytic philosophy. In the last recent years, he has become interested in the practice turn in philosophy of science, as a member of the PratiScienS project (see note 2 of this introduction).

Salanskis stresses that there has been a widespread and damaging tendency, within contemporary work inspired by the practice turn, "to regard anything as practice". He adds that this tendency is far from unique to the practice turn, and is, rather, "a typical danger threatening radical pragmatisms in general", to which Marxism already succumbed in its day, and which today threatens to render the practice turn "meaningless". To avoid this risk, the practice turn must rely on a "non-universally encompassing notion of practice". In line with this suggestion, Salanskis proposes to tie practice to action, stating that "something deserves to be named practice only if we are able to understand some *action* in it". The whole problem then requires a specification of what counts or does not count as an action in science.

Salanskis grounds his analysis in his previous investigations on issues related to action (see especially his 2000 book, *Modèles et pensées de l'action*). The general definition he elaborates distinguishes two aspects of an action: its "morphological" and its "behavioral" aspects. On the morphological side, an action is defined as a "resultative impulse". This means that an action necessarily involves a result (the end-point of the process), an impulse (the source of the process), *and* "some continuity between impulse and result, allowing us to see each in the other". This aspect of the definition can be regarded as a specification of the shift from scientific products to scientific processes (see Section V.6), with an emphasis on the reciprocal determination of product and process. On the behavioral side, an action must involve some form of instigation, commonly called the "agent", but that Salanskis prefers to call a "substrate". An action is not an action without a substrate for which the action means something. This leads to the following general definition of action: "a resultative impulse, in the course of which some substrate gets involved and draws itself together". This general definition is later instantiated with three models of action: the *physical* model of an action (involved for example when an agent punches someone else); the *linguistic* model (involved in speech acts); and the *mathematical* model (understood as the "fabrication of some object on the basis of primitive given objects", using some fixed conventional building rules). This permits Salanskis to offer an operational notion of action that can be used to determine what counts as practice. Relying on his frame of action, Salanskis discusses some ways in which contemporary analysts of science inspired by the practice turn, in particular Hacking and Pickering, have identified and treated scientific practices.

VI.2. Chapter 2. Epistemic Activities and Systems of Practice: Units of Analysis in Philosophy of Science After the Practice Turn, by Hasok Chang

In Chapter 2, Hasok Chang, a leading figure among historians and philosophers of science in practice, and a co-founder of the Society for Philosophy of Science in Practice (SPSP), shares the same general concern as Salanskis, which is to conceptualize scientific practice in a precise and productive way. Although critics of "standard Anglophone philosophical analyses of science" agree that philosophers of science should move away from an exclusive focus on "*theories* as organized bodies of propositions" to examine scientific practices, Chang observes that "no clear and widely adopted alternative philosophical framework has emerged". Accordingly, Chang aims to fill this gap. In this purpose, he exploits critically and creatively the proposals of some notable predecessors (Fleck, Polanyi, Kuhn, Lakatos, Dewey, Hacking, Pickering, Gooding, Rheinberger, and others). His goal, Wittgensteinian in spirit, is to offer a "philosophical grammar of scientific practice", which, hopefully, will have both a sufficiently broad scope and operational power to be widely adopted for diverse practice-based studies of science, whatever their domain, target of interest, and perspective (philosophical, historical, sociological, etc.).

Elsewhere, Chang has developed his framework for activity-based analysis, and tested its fruitfulness in relation to detailed analyses of episodes in the history of chemistry, including the Chemical Revolution (Chang, 2012). Here, Chang offers a more systematic characterization of the framework. The framework consists of several hierarchical, interrelated levels of what Soler and Catinaud call, in their commentary on Chang's chapter, "action-type units", each unit of action being associated with an "inherent aim". Chang introduces three distinct levels of action-type units. Starting from the most encompassing and proceeding to the most particular unit, we find "system of (scientific) practice", "epistemic activity", and "operations". A system of practice is a *coherent* set of epistemic *activities* performed in order to achieve certain aims, and, in turn, an epistemic activity is a *coherent* set of *operations* performed in order to achieve certain aims.

One major challenge, in order to make this multi-level framework complete, is to elaborate on the latter notion of "coherence". Such coherence must depart from the traditional notion, since it holds between units of the type "action" rather than between propositions. As Chang analyses it, coherence is a complex, multi-dimensional notion which, when roughly characterized, is related to the more or less good "synergy" of multiple action-type units (for example, the more or less smooth "coordination" of different activities constitutive of a system of practice), and is measured in terms of the resulting efficiency with which the aims of the corresponding system of practice are achieved. Chang's chapter thereby can be viewed as an attempt to take forward a difficult issue, arising from the practice turn, of how to re-conceptualize the crucial value of scientific coherence now that the content of science is no longer equated primarily with propositions.

Chang's work instantiates most of the shifts discussed above. In particular, Chang's program exemplifies the shift from idealized to empirically adequate accounts of science, through a requirement to develop a less impoverished treatment of scientific actors, including their desires, expectations, genuine freedoms, unarticulated embodied skills, and the like, and through a methodological injunction to "concretize" "overall evaluative notions like truth and success", "talk of the aims of science" and "general epistemic values/virtues", by showing the specific forms they take for actors "in the context of each activity and system" under study.

VI.3. Chapter 3. From Normative to Descriptive and Back: Science and Technology Studies and the Practice Turn, by Michael Lynch

In Chapter 3, the contribution of Michael Lynch has two parts. First, Lynch offers an appreciation of the practice turn and its aftermath in Science and Technology Studies (STS), following its culmination in the 1990s. This is particularly welcome with respect to the aim of the present volume, since it shows how STS researchers have reacted to several shifts introduced by the practice turn, and it thereby helps us to assess the significance of this turn. Second, Lynch addresses an issue that has recently become particularly salient in STS following the practice turn—namely, the alleged necessity to assume a normative stance, which, as a corollary, raises the question of the appropriate way to be normative in light of the practice turn.

Taking stock of his recent experience as an editor of the journal *Social Studies of Science* during ten years (2002–2012), as well as of his active involvement in the practice turn from the beginning as a student of Garfinkel in the field of ethnomethodology, Lynch reflects upon the origins and main lessons of the practice turn, and outlines a set of ten current trends in STS that followed from, succeeded, or displaced the practice turn. Among these trends, we find, for instance: "a widespread move towards political engagement"; a "growing interest in public engagement in science"; or "an increase in enthusiasm for normative engagement in controversies about science, technology, and medicine" that goes hand in hand with a pervasive interest for the topic of "expertise".

Lynch offers an analysis and assessment of the diverse "normative turns": trends in favor of normative, often politicized, treatments of science and expertise, which have been advocated by many STS scholars, focusing specifically on the normative program developed by Collins and Evans on expertise. Such recent normative turns in STS appear to reverse what has been described above in Section V.3. as the "normative-to-descriptive shift". Indeed, recent normative aspirations in STS have often been explicitly directed against "the (apparently) disinterested" or neutral stance of the sociology of scientific knowledge. A major problem with non-evaluative descriptions is that, not only do they turn away from developing criteria for demarcating 'sound' science from 'pseudoscience', but moreover, they offer no basis for defending accepted science against ill-intentioned lobbies, religious groups, or other "merchants of doubt" (Oreskes & Conway, 2010)—for example, in arguments over climate change or evolution. The problem is increased by the fact that following the practice turn, sociologists of science often have underlined the lack of ultimate scientific certainty in established 'mainstream' sciences, and have stressed the possibility of dissent even for widely accepted scientific 'truths'. Many scholars in STS have promoted normative versions of science, but often they reduce all sciences indiscriminately to politics—once again begging the question of how to distinguish between contending positions on issues such as climate change, when both sides claim that the other is 'playing politics'. The challenge then became: How to be normative after the practice turn? How to avoid a "re-turn to classic epistemology", which had been so severely criticized by actors of the practice turn?

With respect to this challenge, Lynch contrasts two ways to be normative. The first one is exemplified by the program on expertise of Collins and Evans, and is criticized by Lynch for adopting a position of experts on expertise—that is, a position of the type mentioned above in Section V.3, which had already been widely criticized by supporters of the practice turn, and which Rouse characterizes as an "outside" and "above" normative position aspiring to "epistemic sovereignty". Such a position is, according to Lynch, highly problematic, because it fails to recognize that "expert", "possession of tacit knowledge", "scientific", and so forth, are not neutral terms and attributes (i.e., "descriptive categories") that simply depict an objective realm of competent persons and activities; rather, they are vernacular terms that commonly are used to promote professional authority or defend the autonomy of a profession (including a scientific profession). As such, they are used by actors in conflicting situations as "rhetorical resources". Furthermore, Lynch insists, STS scholars should recognize that they are not necessarily "on the side of the angels". When claiming to possess criteria for demarcating real and pseudo-expertise, or claiming to be able to distinguish the genuine experts from those who are not in a particular technoscientific controversy, analysts of science, whether they like it or not, participate in the situation (more or less directly according to the case), and thus take part in the "political actions and discourses" with which expertise, like any other normative category, is bound up.

Lynch favors an alternative way to take normative positions and to deal with expertise and tacit knowledge. Instead of a "science of expertise", he proposes "a form of casuistry", "more akin to expertise about *something* in particular" because the normative claims only hold for particular situations. More concretely, his version of casuistry can be analyzed in two steps. The first one is descriptive: "Initially, and predominately . . . it is a descriptive approach that focuses on actions and discourses that often are explicitly normative". Then, in a second step, it is sometimes possible to use the previous thorough understanding of the particular case under scrutiny as a basis for making "particularistic judgments about the credibility of particular participants and the plausibility of their arguments", and to "motivate normative critique and advice of a rather down-to-earth sort". This form of modest normativity seems to do justice both to one major recommendation of the practice turn—namely, to pay attention to practice in detail—while attempting to overcome the problems raised by exclusively applying this recommendation.

VI.4. Chapter 4. Chemistry's Periodic Law: Rethinking Representation and Explanation After the Turn to Practice, by Andrea Woody

Chapter 4 shows how the practice turn has modified the classical questions of representation and explanation in science. Taking stock of her multiple, widely recognized, historically grounded philosophical works on representation and explanation in chemistry,[47] Andrea Woody puts two elements in dialogue: on the one hand, a systematic characterization of some important transformations and gains induced by the practice turn, framed in terms of four changes and two challenges; on the other hand, a detailed discussion of a historical case study. Her list of four changes brought by the turn to practice is essentially in agreement with the shifts presented above in Section V, but does not map the situation in exactly the same way. One of the challenges she puts forward is also at the heart of Lynch's chapter: How to return to normative claims about science after the practice turn? Woody recognizes that "it is not clear how this may be achieved".

The first change, "from conception to representation", is perhaps the most relevant to Woody's chapter, and is the only one we shall discuss in this brief summary. In relation to this change, Woody describes how theories, traditionally understood as "abstract conceptual objects", are now best appreciated as artifacts, rendered into particular representational formats. This change corresponds to one particular aspect of our shift "from contemplation to intervention" (Section V.7), or, more precisely, to the application of this shift to a particular target in which Woody is especially interested: *theories*. Regarding theories, Woody situates her work on formats in continuity with Hacking's seminal *Representing and Intervening* (1983), with the aim of developing "a parallel conception of theoretical practice [parallel to Hacking's conception of *experimental* practices], one engaged with the active construction and manipulation of representational artifacts and shaped by practical concerns and contingent, contextually determined goals", and one in which it becomes clear that "representing *is* intervening of a conceptual sort".

Woody's case study, related to the establishment and entrenchment of the periodic law in the 19th century, shows how attention to representational practices, and more specifically to representational *formats* in science, leads to consequential transformations of traditional philosophical issues, such as "theory-choice" and the role of predictions in theory-choice, the nature of representation, or the nature and functions of explanation. The representational formats in which, historically, the periodic law has been embodied, are diverse: tables in Mendeleev work; graphs in Meyer's research; but also spiral and cylindrical representations, algebraic formulations, and so on. Why did Mendeleev's version quickly become "the dominant expression of periodicity"? Woody's answer is, in substance: because the tabular format of Mendeleev's version provided *specific virtues and powers that fit well with the collective practices of the chemical community* at the time ("the skills, interests, aims, and background knowledge" of this community), and that fit *better* than the resources provided by competing theoretical artifacts.

According to Woody, one other crucial point concerning representational practices in general, and representational *formats* in particular, is their role in what appears explanatory to practitioners. She discusses this point both in the case of Mendeleev's table and in general, advocating for a *functional perspective on* scientific explanation. Such a perspective illustrates another important change brought about through the practice turn, namely the product-to-process shift, since "in comparison to traditional accounts of scientific explanation, there is a shift in focus away from explanations, as *achievements,* toward explaining, as a coordinated *activity* of communities". Beyond the plurality of kinds of explanations and senses in which a scientific 'contents' might be viewed as explanatory by practitioners, the main function that explanatory discourses serve in science is, according to Woody, a "methodological role". By this she means that the role is to achieve and maintain "coherence within and across communal scientific activities", through the inculcation of "patterns of reasoning" which "sculpt and subsequently perpetuate communal norms of intelligibility". Once again we meet, as in Chang's chapter, the important issue of the coherence of scientific practices.

VI.5. Chapter 5. Epistemic Dependence in Contemporary Science: Practices and Malpractices, by Hanne Andersen

Hanne Andersen is head of a research project in Denmark entitled "Philosophy of Contemporary Science in Practice". Like Lynch in Chapter 3, but with a different perspective, in this chapter, Andersen is concerned with normative perspectives on science. Indeed, her contribution is devoted to "malpractice", understood in a broad sense that includes both intentional fraud and "the so-called 'grey zone' behavior such as sloppiness and incompetence". One of Andersen's aims is to "point to a new area for philosophy of science in practice to address". Existing studies have, according to Andersen, focused mainly on intentional deceit, and treated this restricted class of malpractice primarily as "a topic for ethical analyses". Andersen suggests including also "reckless as well as negligent actions" in the discussion, and argues that the corresponding discussion has an important *epistemological* dimension, beside the ethical one, that should not be ignored in philosophy of science following the practice turn. Moreover, after having stressed that scientific research involves pervasive epistemic interdependence, both synchronously (peers relying on contemporary peers) and diachronically (scientists relying on past results), Andersen focuses on malpractice *in situations of epistemic dependence*. But, whereas most other accounts of epistemic dependence have been primarily concerned with conditions under which a given scientist concludes that another scientist is *trustworthy* (i.e., with "calibration of trust"), Andersen concentrates on the issue of how a scientist can establish that another scientist is *not* trustworthy (i.e., on calibration "of *distrust*").

In relation to these purposes, Andersen begins with a critical discussion of existing philosophical accounts related to "calibration of trust" in situations of epistemic dependence. Following Hardwig's analysis, she claims that epistemic dependence includes both a moral *and an epistemic* component: Scientist A must trust that scientist B is both honest and competent. Then, she emphasizes that prominent analyses (particularly those given by Kitcher and Goldman) of how scientists "calibrate" trust in particular relationships do not distinguish between moral and epistemic components. She argues that, while such analyses are sufficient for the analysis *of trustworthiness* (since trustworthy scientists are *both* competent and truthful), they are too coarse to analyze *untrustworthiness*. This is because "an untrustworthy scientist may either be untruthful or unknowledgeable (or both)". Therefore, the moral and epistemic character of scientists may be subject to distinct moral and epistemic "calibrations" in situations in which scientists have to decide when they do *not* want to become epistemically dependent. Andersen concludes that the analysis of malpractice is not merely problems for research ethics, but also is important for epistemology. Andersen does agree with one important aspect of Kitcher's and Goldman's account, however: Calibration of trust is "relational", which means that it depends, among other things, on the degree of expertise possessed by the person doing the calibration, so that different standards should be expected—and demanded—from more or less expert (or experienced) scientists.

One interesting aspect of Andersen's paper, which illustrates the fruitfulness of philosophy of science in practice, is that she connects epistemological analyses with well-documented cases about how calibrations of malpractices are dealt with by practitioners themselves. Relying both on testimonies of individual scientists and on reports of diverse committees in charge of investigating cases of suspected malpractices, Andersen appeals to a rich set of what seem at first sight to be heterogeneous situations, ranging from highly publicized "spectacular cases"—such as the famous fraud case of William Summerlin, who claimed to graft skin from a black to a white mouse but was found to have painted the black patches on the white mouse—to more confidential, and more delicate to analyze, cases of misconduct, in which negligence, ignorance, or incompetence appear to have been involved. The cases she analyzes confirm a "relational account of calibration"; in particular, they show that higher degree of vigilance in calibration, together with the application of stringent standards, is expected from senior researchers than from junior researchers. The cases also reveal that more vigilance is expected from co-authors in the case of collaborations, since co-authors tend to have access to more information than other parties. Andersen concludes the chapter by discussing the extent to which misconduct implicates a research group in which only one collaborator is found guilty of fraud. Her answer is that it is *epistemically* expected from collaborators to be responsive and vigilant towards the inputs of their co-authors. Failing to identify the fraud of a collaborator is therefore an *epistemic* fault, not a moral one.

VI.6.   Chapter 6. Values in Engineering: From Object Worlds to Socio-Technical Systems, by Louis Bucciarelli and Peter Kroes

In Chapter 6, Louis Bucciarelli and Peter Kroes contribute to the ongoing discussion about engineering education on the basis of the pioneering ethnographic studies of engineering practices Bucciarelli carried out in the second half of the 1980s (Bucciarelli, 1988, 1994). Bucciarelli, who at the time was a professor at MIT in the school of engineering, has been one of the first researchers to apply the ethnographic approach to the engineering sciences through participatory observations in engineering firms. Peter Kroes is an analytical philosopher of technology who, in his book *The Empirical Turn in the Philosophy of Technology*, co-published in 2001 with Anthonie Meijers, called for a more empirically based philosophy of technology.

In their contribution, Bucciarelli and Kroes observe a mismatch between the engineering curricula and the daily practices of professional engineers. They diagnose a neglect of "social features and social values" in engineering education. First, engineering curricula neglect the crucial fact that professional engineering design requires teamwork. Engineers with radically different backgrounds and technical competencies have to negotiate their diverging methods, responsibilities, and interests if the design process is to result into one successful object of design—one object of design, different "object worlds". To achieve their common design, they must be able to communicate their ideas clearly and convincingly, be open to alternative solutions, and be constructively critical. Second, engineering curricula neglect the fact that the technical creations of engineers end up in society and, because of their social impact, are always embedded in perspectives which involve conflicting and evolving societal norms, values, and stakes.

The neglect of the previous aspects renders engineering education, according to Bucciarelli and Kroes, "deficient". One way to repair this deficiency would be to broaden the kinds of exercises assigned. Engineering curricula offer "well-posed, single-answer problems" and teach students to solve them using "instrumental rationality" (i.e., choosing the most efficient means to achieve a certain end without discussing these ends). Instead, students should be taught to frame problems, to prioritize them, and to discuss the goals and societal values involved—which would require the development of, and respect for, qualitative reasoning that is often neglected in engineering curricula.

Another way to repair the deficiency would be to study socio-technical systems (e.g. energy grids, traffic infrastructures, or international air traffic control systems). Such systems comprise both human actors and artifacts as indispensable constituents. Due to their composition, their development explicitly requires elaborating juridical, ethical, and technological norms. Moreover, the assessment of these systems is ordinarily accomplished "on the fly"—that is, during the development of these systems. It is an intractable complicated process in which many stakeholders with diverging "perspectives, priorities, methods and competencies" are implicated in large societal networks, which display multi-faceted feedback loops. The focus on socio-technical systems would force students to take a societal perspective on technology. At the end of their paper, Bucciarelli and Kroes reflexively propose to consider engineering education itself to be a socio-technical system. They do not claim to have a finished design brief for a new curriculum, suggesting only that engineering educators should be open to the possibility of the design of such "on the fly".

The work of Bucciarelli and Kroes instantiates some shifts analyzed in Section V. Bucciarelli's urge to participatory observation and Kroes' call for empirically informed philosophy of technology exemplify the shift towards more empirically adequate local studies of science and technology. Moreover, Bucciarelli explicitly claims to consider technology and design as processes and not just as products. His slogan, "designing is a social process", explicitly epitomizes, simultaneously, both the product-to-process shift (see Section V.6), and the shift from individualist to collective accounts of scientific and technological developments (see Section V.5). Bucciarelli and Kroes' proposal for how to deal with the deficiency of engineering curricula lies at the intersection of two recent research lines in science and technology studies. One is the tradition of extrapolating lessons drawn from ethnographic studies in engineering *professional practices* toward engineering *education* (see also Downey & Beddoes, 2011, and Trevelyan, 2009); the other is the current trend in social study of science that attempts to return from the descriptive stance to renewed, empirically informed ways to be normative (see above the presentation of Chapter 3). At this intersection, Bucciarelli and Kroes advocate a turn towards a new engineering curricula paradigm in which the social dimensions are taken at least as seriously as the technical ones. Putting socio-technical systems on the forefront of such curricula, they set the research agenda for the renewed interest in the definition, identification, and roles of technological and societal values in today's science and technology. Moreover, their analysis of socio-technical systems as objects of study epitomizes the extraordinarily complex ways in which, in practice, descriptions and normativity go hand in hand and are almost impossible to disentangle.

VI.7.   Chapter 7. The Impact of the Philosophy of Mathematical Practice on the Philosophy of Mathematics, by Jean Paul Van Bendegem

Chapter 7 is devoted to the "philosophy of mathematical practice". Jean Paul Van Bendegem offers a panorama of the different trends that currently constitute this domain, including historical elements about landmark developments. The result is a fairly complex picture, involving eight, not all obviously homogeneous, or easily connectible, approaches. For instance, among the various empirically based approaches to mathematical practices listed in the article, those informed by sociology of science, ethnomathematics, evolutionary biology, or cognitive psychology seem remarkably different perspectives. Van Bendegem's picture raises the issue of the relations among the different approaches, and more generally the issue of the unity of practice-based studies of mathematics. Chapter 7 also provides insights into the relations between the philosophy *of mathematical practices* and the "*traditional* philosophy of mathematics". Although the two perspectives are often viewed as opposed, if not "incommensurable"—focused on different questions, favoring different methods, and speaking different languages— Van Bendegem is convinced that interactions between the two not only are possible, but could also be fruitful for both. He sketches some proposals about possible bridges.

Van Bendegem's view of philosophy of mathematics after the practice turn is particularly interesting, given his curriculum and the actual position he occupies in the field of the philosophy of mathematics. He is a leading figure in the practice-based philosophy of mathematics,[48] and has played an active role in establishing this new trend within the philosophy of mathematics— in particular, he is one of the founding members of the Association of Philosophy of Mathematical Practice, created in 2009. His training as a mathematician allows him to draw upon his own experience with mathematical practices. Moreover, when he turned to the philosophy of mathematics, his research initially developed within the *traditional* approach, so he is familiar with the questions, methods, and language of both the traditional and the practice-based approaches, and is capable of switching from one to the other.

Although Van Bendegem does not systematically discuss this problem, his contribution provides clues with which to consider whether mathematics enjoys special epistemic status as a practice. He notes that practice-based approaches have barely penetrated the philosophy of mathematics, that the philosophy of mathematical practice currently is an "emergent" and still marginal domain of investigation, and that neither its development nor its survival are assured. In contrast, the interest in scientific practice is pervasive and well-entrenched in the philosophy of *natural* science. Moreover, in the philosophy of natural sciences, *sociological* studies are taken into account and discussed; though contentious, these studies participate in the shaping of various philosophical debates associated with the practice turn. In contrast, thus far, philosophers of mathematical practices have largely ignored studies of mathematics in practice conducted by sociologists and social historians such as Bloor, MacKenzie, Livingston, and others. Van Bendegem discusses Bloor, stressing that the sociological approach does not "merge easily" with the other approaches of mathematical practices and suggesting that "one of the corner elements is the internal-external debate"; he mentions some rare "brave attempts" to establish connections. All these elements suggest that to the eyes of many philosophers of mathematics, even those who are sympathetic to the practice turn, mathematics remains a special case, and remains a special case for reasons that have to do with the traditional idea of the universality and necessity of mathematics.

VI.8. Chapter 8. Observing Mathematical Practices as a Key to Mining Our Sources and Conducting Conceptual History, by Karine Chemla

Karine Chemla's chapter jointly addresses two issues. The first one relates to the specific domain of inquiry of Chemla, namely the history of mathematics in Ancient China: What can we say about mathematics in Ancient China from the little source material that we have? Chemla's answer, as surprising as it may appear, given the nature of her sources—a *very few* number of *texts*—confers a pivotal role to mathematical practices, and shows how the desiderata of historically adequate reconstitutions 'from inside' of past science (see Section V.4) can be applied even in very poorly documented situations. The second issue is more general, although it is also addressed on the basis of the special case of mathematics from ancient China. It relates to the product-to-process shift (see Section V.6), and concerns the relationship between mathematical practices and mathematical results: How can we think this relation?

Chemla addresses these issues by focusing on division and other arithmetic operations. For this she relies on two sets of sources. In one of them, a classical Chinese book, *The Nine Chapters on Mathematical Procedures,* probably composed in the 1st century CE, plays a key part. In the other set, a manuscript recently excavated from a tomb sealed at the beginning of the 2nd century BCE, the *Book of Mathematical Procedures,* is the main piece of evidence. In contrast to mathematical texts composed in other cultural contexts, the *Book of Mathematical Procedures* and *The Nine Chapters* proceed through a sequence of problems, apparently practical[49] (246 in total, divided into nine chapters), followed by the answers. Moreover, a procedure—or algorithm— is systematically attached to each problem and solves it. In this context, commentators provide the means to understand the algorithm used in general.[50]

Throughout her case study, Chemla attempts to substantiate, and argue for, three claims. The first is that relying on close scrutiny of mathematical texts, the historian can, to some extent at least, "restore" some of the mathematical *practices* involved at the time. The second is that the restoring of these practices enables the analyst to reconstitute aspects of the mathematical *knowledge*—in the case study, knowledge about arithmetic operations—the authors of the book possessed but did not explicitly state in the text. Chemla formulates these two points as follows: "Reconstructing practices in relation to which sources were produced enables us to uncover and interpret clues that give indirect insight into the results known to actors (e.g., how to divide) and more generally, into the nature of actors' knowledge (e.g., a way of structuring a set of operations), when these bodies of knowledge only leave traces in the documents". The third claim is that through the whole iterative activity of restoring practices and disclosing new hidden knowledge—an activity which could be re-described as a version of the hermeneutic circle, although Chemla does not use the term—the historian experiences, and is in a position to show, the "intimate relationship" that exists between mathematical practices and mathematical knowledge. As Chemla puts it, "practices and bodies of knowledge" "are intertwined", "shaped conjointly", jointly produced and jointly transformed. Chemla's contribution therefore emphasizes the possible circularity (a fruitful, and not a vicious, circularity) between practices and products. Through an iterative process, the analyst can successively learn about practices from a book, which in turn can be reinterpreted in light of practices. By offering a careful reconstitution of mathematical practices and knowledge in Ancient China, Chapter 8 also broadens the (still largely Eurocentric) scope of science studies.

VI.9.   Chapter 9. Scientific Practice and the Scientific Image, by Joseph Rouse

The aim of Rouse's chapter is to determine how the "scientific image" should be modified, in a naturalistic perspective, following the practice turn. The term "scientific image" is taken from Sellars and refers to "how the sciences understand the natural world", in the sense of what the sciences tell us about the world. Rouse's naturalist project, developed in his *How Scientific Practices Matter: Reclaiming Philosophical Naturalism* (2002) and *Articulating the World (forthcoming)*, intends, like Sellars' project, "to reconcile our scientific understanding of ourselves as natural beings with our philosophical sense of ourselves as answerable to conceptual norms". But Rouse criticizes Sellars' solution—in particular, Sellars' "conception of scientific understanding as representing the world" and Sellars' scientific image as "a conception of a body of scientific knowledge". Taking into account the lessons of the practice turn, what remains, according to Rouse, is to elaborate a revised "conception of scientific understanding" that goes hand in hand with a different idea of how the sciences understand the natural world. Accordingly, Rouse's chapter is dedicated to developing his answer to the following question: "How would an emphasis upon science as practice revise the scientific image, in its dual sense of a scientific understanding of the world and a conception of scientific understanding"?

When assessing the outcomes of the practice turn, it may seem as though the main lesson regarding this question—the lesson drawn by what Rouse calls the "disunifiers" such as Cartwright, Dupré, Giere, Hacking, Lange, Teller, and so on—is that there is no such thing as one all-encompassing scientific image of the world. According to Rouse, however, the most fundamental lesson of the practice turn—congruent with the shift "from contemplation to transformation of the world" introduced in Section V.7—is that scientific understanding *is not primarily representational*. Rouse then develops his alternative naturalistic account of scientific understanding, in contrast to traditional, pre-practice turn perspectives on science. According to this account, the practice turn teaches us that "sciences are first and foremost research enterprises", and the corresponding kind of research enterprise is conceptualized by Rouse as "a distinctive form of [biological] niche construction". In such framework, scientific understanding is conceived as an open-ended evolutionary process of extension and reconfiguration of a complex space, a process through which humans transform their environment, and are themselves transformed, in unpredictable ways. The complex space in question involves multiple heterogeneous kinds of items. It involves material, technical, and practical elements and practices ("equipment", "practical skills", "perceptual discriminations", "creation of new material arrangements in specially prepared work sites", "material transformations of the world" inside of scientific laboratories, but also outside—for example, through industrial practices and products). The space also involves diverse social practices, structures, and actors, which extend far beyond the strictly speaking scientific ones (they, for instance, include "the universities, hospitals, institutes, government agencies, or corporations in which research activities are situated", the "sources of financial and other material support" which "shape the priorities and direction of research itself", "disciplinary and other professional associations, journals and other publishers", and "scientific education"). Scientific understanding as a form of niche construction leads to a holistic picture, but one that cannot be reduced to a *linguistic* holism of the type defended by Quine when he introduced his web of beliefs (see Trizio's commentary on Rouse's paper). For Rouse, scientific understanding does not primarily identify with empirically justified beliefs supposed to represent the world, or with some "intralinguistic" "set of positions within the space of reasons". Instead of such "idealized and disembodied philosophical conception of scientific understanding", Rouse's holistic entanglement is much wider than previously foreseen by traditional epistemology. In full conformity with the shifts introduced by the practice turn—in particular, the shift from too idealized to empirically adequate accounts of science (Section V.2) and the shift from decontextualized, intellectual, explicit, individual, and 'purely cognitive' to contextualized, material, tacit, collective, and psycho-social characterizations of science (Section V.5)—the dynamic of scientific understanding is viewed by Rouse as a process involving not only the space of reasons, but a complex space including the material, social, practical, technical, industrial, institutional, educational, cultural, and other dimensions too. Rouse moreover insists—this is another important point that the notion of scientific research as niche construction is intended to capture—that the world itself, and not just our representation of it, is at stake in the holistic process of scientific understanding, so that there is more sense to say that the world itself changes in this process, rather than to think of the targeted object of scientific inquiry as fixed and "already articulated into entities and properties, which may or may not be discernible to us". This is so because as a research enterprise, science creates "new ways of interacting with the world that mutually reconstitute us as organisms and the world around us as our biological environment".

Rouse also insists on another important lesson of the practice turn, which is the fundamentally *prospective* nature of scientific research. Contrary to what philosophers of science have traditionally assumed, scientists do not primarily struggle to reach a stabilized consensual representation of what the world is, which would be deposited in scientific papers and textbooks, so that scientific understanding could be captured by philosophers of science through a retrospective compilation of scientific publications considered in isolation from their site of production. Instead, actual researchers have first and foremost a *prospective understanding* of their field. This means, first, that practitioners do not primarily look for a consensual system of knowledge about the world, but, rather, for achievements that are seen as promising tools for advancing in current scientific problems or that open new interesting opportunities of inquiry. This also means that the content and significance of scientific "topics, claims, tools, and issues"—including scientific propositions about the world—are not fixed once for all, or even univocally determined by their relation to a currently established system of knowledge, but, rather, are largely determined by possibly divergent judgments of practitioners about the role they could play in future research. As a corollary, this implies that one "cannot take the primary journal literature as a distributed repository of the scientific image". This literature is better seen as a set of future-oriented, selective, and partial, possibly conflicting accounts, elaborated by scientists for specific purposes and specific audiences, on the basis of possibly divergent aims and assessments. Consequently, the project of traditional philosophers of science of an "all-purpose or no-specific-purpose compilation" that would enable access to the scientific image, identified with a supposedly consensual representation of the world in a given stage of scientific development, can only lead to an "idealized retrospective reconstruction" that is a creation of philosophers and does not exist as such for actual researchers.

Rouse's naturalistic framework and its pivotal notion of niche construction also goes with an unpredictable, open-ended, and contingent conception of scientific development, scientific achievements, and the very existence of science. Against the traditional idea of science as "a 'god's-eye view' of ourselves and the world, and the correlative 'sciences' supposed transcendence of the local and the human", Rouse claims that "[s]ciences are . . . precarious and risky possibilities that only emerged in specific circumstances, and could disappear", and argues for "[t]he contingency of conceptual understanding generally and scientific understanding specifically".

NOTES

1. See especially the title of (Schatzki, Knorr-Cetina, & Von Savigny, 2001), *The Practice Turn in Contemporary Theory* (this book includes studies that target science, but its scope is much broader); "turn to practice" is employed as well (cf., for instance, Rouse, 2002, p. 163); "practical turn" is also sometimes found (see, for example, Pickering, 1995b, p. 43, or David Stern's paper, "The Practical Turn" (2003), in which "practical turn" and "practice turn" are used interchangeably).
2. The aim of the PratiScienS project is to develop a systematic analysis of the consequences of the practice turn in science studies and to investigate philosophically significant issues about scientific practices. Central issues are how something acquires the status of a robust result in the empirical and the formal sciences (Soler, Trizio, Nickles, & Wimsatt, 2012), the role of tacit aspects in the constitution of scientific achievements (Soler, Zwart, & Catinaud, 2013), how we should assess the contingency or inevitability of robust scientific achievements (Soler 2008a, 2008b; Soler, Trizio, & Pickering, in progress), and the nature and specificity of mathematical and logical practices (Giardino, Moktefi, Mols, & Van Bendegem, 2012). The project started in 2007 and is pursued by a small interdisciplinary group of France-based historians, philosophers, and sociologists, of natural sciences, mathematics and logic, and engineering sciences. For more on the PratiScienS project, http://poincare.univ-lorraine.fr/fr/operations/pratisciens/accueil-pratisciens
3. They include Hasok Chang, Karine Chemla, Harry Collins, Peter Galison, Thomas Nickles, Andrew Pickering, Claude Rosental, Jean Paul Van Bendegem, and William C. Wimsatt.
4. On the issue of what is new in the recent interest in scientific practices, and of different conceptions of practice, see the remarks of Brian Baigrie (1995), especially in section 2, entitled "The Babel of Practice". On the multiple understandings of practice, see Stern (2003).
5. On Wittgenstein's and Heidegger's (through Dreyfus' interpretation) notions of practice, and more generally for landmarks concerning the "practical turn" in the philosophy, history, and social sciences in general, see Stern (2003).
6. According to Foucault, these "practices are characterized by the demarcation of a field of objects, by the definition of a legitimate perspective for a subject of knowledge [and] by the setting of norms for elaborating concepts and themes" (1997, p. 11).
7. According to which, the "basic domain . . . is neither the experience of the individual actor, nor the existence of any form of societal totality, but social practices ordered across space and time" (1984, p. 2). Giddens' structuration theory has notably been applied to the study of technology; see Desanctis and Poole (1994), Orlikowski (2000), or Workman, Ford, and Allen (2008).

8. Bloor (1973, 1976/1991, 1981); Barnes (1974, 1977); Barnes and Shapin (1979); Shapin (1979, 1986). Donald MacKenzie (1981) carried out research into the shaping of statistics in Great Britain around the turn of the 20th century. Andrew Pickering worked on high-energy physics in the 1960s and 1970s (Pickering 1981a, 1981b, 1984a, 1984b).

9. Transversely Excited Atmospheric Pressure CO_2 lasers.

10. Collins (1974, 1975, 1981a, 1981b, 1981c, 1984, 1985/1992); Pinch (1977, 1986); Travis (1980, 1981). Collins (2004) gives an overview on the author's work on the gravitational waves episode.

11. Garfinkel, Lynch, and Livingston (1981); Lynch (1982, 1985); Lynch, Livingston, and Garfinkel (1983); Livingston (1986). On ethnomethodology, see also Heritage (1984); Button (1991).

12. For a global analysis of the situation of SSK in the early 1980s, including many aspects relevant to the practice turn, see Knorr-Cetina and Mulkay (1983a).

13. Hacking (1999, p. 48) recommends the term "constructionism" to avoid confusion with other versions of "constructivism" (such as constructivist programs in mathematics), but most STS writers use "constructivism"; in any case, the two terms are largely interchangeable in STS, and we will use them interchangeably here.

14. As Latour and Woolgar put it: "reality (ideas in Plato's terms) is the shadow of scientific practice" (1979, p. 186, note 19). For the first constructivist volume on technology, see Bijker, Hughes, and Pinch (1987).

15. For a discussion between Lynch and Bloor about this difference of opinion, see Pickering (1992). See also Button and Sharrock (1993).

16. In relation to this point, it is worth mentioning that some authors (e.g., Gilbert & Mulkay, 1984/2003) warn against the goal of wanting to arrive at *the* ultimate and best account of some scientific episode because of the variability of actor accounts. They propose to apply the methodology of "discourse analysis" to understand how the discourse of scientists is socially produced and to avoid "the authorial voice of the sociologist" (Gilbert & Mulkay, 1984/2003, pp. 2–3). All accounts should be dealt with much more equally. Note that discourse analysis was importantly put forward by Harris (1952) and was given a boast by Foucault (1972). In many papers, Mulkay and Gilbert applied its method to sociology of science, which culminated in Gilbert and Mulkay (1984/2003). An interesting description and critique of the discourse analysis tenets can be found in Shapin (1984).

17. The term has been coined by Robert Ackermann (1989), in a review, entitled "The New Experimentalism", of Allan Franklin's 1986 book, *The Neglect of Experiment.*

18. Early exemplary works include: Cartwright (1983); Hacking (1983, 1992a); Ackermann (1985); Franklin (1979, 1981, 1986); Galison (1983, 1987); Giere (1988). See also Achinstein and Hannaway (1985) for a collection of papers.

19. Deborah Mayo explicitly borrows Ackermann's term "New Experimentalism" and uses it in the title of her 1994 paper, "The New Experimentalism, Topical Hypotheses, and Learning from Error".

20. Although from a substantial point of view, Franklin and Galison, for example, have critically engaged in debates with sociologists of science, such as Pickering—see the end of this section.

21. Galison (1983), for example, describes his work on the discovery of weak neutral currents in the 1970s as a "contemporary history of physics" that is "based on an archival record with a different character from that of even

thirty years ago" (in addition to being different from the scientific correspondence "that formed the backbone of the classical works of the history of physics from Descartes to Einstein and Bohr"). He pursues: "Some of the older types of sources remain, such as draft manuscripts, transcripts of conference meetings, and occasionally notebooks. In addition new sources material can be exploited: preprints, computer simulations, computer calculations, internal memoranda, minutes of collaboration meetings, log books, experimental proposals, and grant applications. I have used those sources in addition to the more usual ones, along with the interviews" (p. 481).

22. See, for instance, the end of Galison's quotation in the previous note, or Hacking (1983, p. vii).

23. This claim captures a global tendency about what was the central focus, but of course, it does not mean that theories were completely excluded from the corresponding studies. Hacking (1983), for instance, dedicated a whole chapter to theories (Chapter 12, entitled "Speculation, calculation, models, approximations", which stresses that "There is not one activity, theorizing. There are many kinds and levels of theory, which bear different relationships to experiment" (p. xiii)). Galison's notion of "trading zone", which he began to formulate in the 1980s, includes theorists in dialogue with experimentalists and instrumentalists. Cartwright (1983) could also be mentioned. However, the activity of theorizing was, initially, rarely scrutinized as the central target of the analysis and considered in its specifically practical dimensions. Such perspective fully developed only later, during the 1990s and after. See, for instance, Galison and Warwick (1998); Warwick (1992, 1993, 2003); Kaiser (2005), who provides historical landmarks about the interest of historians and philosophers of science for practices of theorizing, as well as multiple references; and Giere (2006b, Chapter 4). See also the references mentioned in Section V.7 in relation to practices of representation and theoretical "formats".

24. The effect was paradoxical in the sense that it was in tension with the spirit of New Experimentalism.

25. However, Hacking himself did not subscribe to this dualism; see for example (Hacking 1983, p. 62 and p. 130), where Hacking follows Dewey despising all dualisms, including the one between theory and practice (or thinking and acting).

26. In the same vein, Rouse, after having stressed that "one crucial mistake to avoid is reading the concept of scientific practices in terms of a distinction between theory and practice", notes that this mistake is "especially tempting" because of New Experimentalism, which "emphasizes the autonomy of experimental and instrumental practices from theoretical determination" (2002, p. 162).

27. The authors mention Latour (1987) and Livingston (1986) as exceptions.

28. Although some rare works existed, which explicitly considered mathematics as practices; the most famous one is certainly Kitcher (1983).

29. The issue of the early forms taken by the practice turn in philosophy, history, and social studies *of engineering sciences* is systematically addressed in Zwart's commentary on Bucciarelli and Kroes' paper (Chapter 6 of this volume). For this reason, we do not develop this aspect of the practice turn in the present introduction.

30. For the dialogue between Collins and Franklin about experimentation, see also Franklin and Howson (1984), Collins (1984, 1994), Franklin (1994, 1997, 1998/2012).

31. See Pickering (1984b, 1988, 1989a, 1995b), Galison (1987, 1995), and Baigrie (1995) for an analysis of disagreements between Pickering and Galison.
32. As Lynch writes in Chapter 3 of this volume, "the practice turn [in STS] was marked as a development from social constructionism".
33. See, for instance, William Wimsatt, who criticizes the identification of the subjects of science to "Laplace's demons", "omniscient and computationally omnipotent" or "the vision of an ideal scientist as a computationally omnipotent algorithmizer" (Wimsatt, 1981, p. 152). See also the quotation of Kuhn, note 35.
34. In this vein, Rouse (2002) stresses that, according both to sociologists of scientific knowledge and many feminist studies of science, "important aspects of traditional epistemologies and philosophies of science are not merely false, *but ideological*" (p. 139, italics added), and that according to SSK, traditional perspectives on science "misconstrue the actual development of science as revealed by empirical sociological studies, *and thereby unjustifiably legitimate the cultural and political authority of the sciences*" (p. 139, italics added).
35. See already Kuhn's criticisms of Popper's normative characterization of theory-choice in Kuhn (1970) (p. 146). See also Kuhn (1977, p. 326), in which Kuhn stresses, against the traditional philosophical ideal of scientific justification as "an algorithm able to dictate rational, unanimous choice", that "even an ideal . . . if it is to remain credible, requires some demonstrated relevance to the situations in which it is supposed to apply". For a recent example, see Rouse (2002, p. 14), where the author criticizes "the formal theories of confirmation and explanation proposed by logical empiricists", pointing that "these inferential norms were disconnected from the practices of theory acceptance and explanation . . . that they supposedly governed".
36. Herbert Butterfield (1931) first introduced the expression "Whig" history, where "Whig" refers to an English political party. On "Whig" and "present-centered" history of science, see Wilson and Ashplant (1988) and Ashplant and Wilson (1988).
37. Most of the time, this is done relying on present scientific publications. As Lynch writes (this volume, Chapter 3): the practice turn criticized "present-centered histories of science that treated currently accepted textbook accounts as a basis for reconstructing past developments".
38. Kuhn is often credited with having initiated an "historical turn" in the Anglo-American philosophy of science (see, for example, Bird, 2008a). Hacking (1983) gives a striking formulation of the historicizing stage of the philosophy of science (a stage he also associates to Kuhn's work): "Philosophers long made a mummy of science. When they finally unwrapped the cadaver and saw the remnants of an historical process of becoming and discovering, they created for themselves a crisis of rationality. That happened around 1960" (Hacking, 1983, p. 1). Sometimes the historical turn is understood as a first step toward, and subsequently a component of, the practice turn. In any case, some supporters of the practice turn were inspired by the historical turn (or by historical trends in philosophy of science), and sympathies for historically based and practice-based studies of science often go hand in hand.
39. Methodological principles of this kind have been formulated in different contexts under different names. The famous Strong programme in the sociology of scientific knowledge proposed the "symmetry", "impartiality", and "causal" principles (see Section III.1). The French "historical epistemology" also developed a historiography 'from inside'; see, for instance, Koyré (1961/1973), Canguilhem (1968), Canguilhem (1970, 1971), Fichant & Pêcheux (1969).

40. For pioneering works, see Polanyi (1958/1962, 1967). For a selection of "classical" writings, see Collins (1974, 1985/1992, 2001a, 2001b); Dreyfus and Dreyfus (1986); MacKenzie and Spinardi (1995); Pinch, Collins, and Carbone (1996); Turner (1994). For recent works with furnished bibliographies on the topic, see Collins (2010); Soler (2011); Soler, Zwart, and Catinaud (2013).
41. Fleck (1935/1979), with his notion of "thought collective", is often viewed as a pioneering work. See Goldman (2010) and Fuller (1988/2002).
42. See, in particular, the last paragraph of his paper.
43. Many examples of claims of this kind could be provided. Just to give a few, Pickering, in the introduction of his edited collection (1992, p. 3), writes: "Their primary concern [of "most of the twentieth century Anglo-American philosophy of science", including "philosophers who have opposed mainstream thought" such as Hanson and Feyerabend] has always been with the products of science, especially with its conceptual product, knowledge"; Batens and Van Bendegem (1988, p. vii), after having stressed that the "scholars of the Kuhnian period" "instigated a shift in attention that changed history and philosophy of science in a irreversible way", note that a "clearcut example" "concerns the study of science as a process, and not only as a result". See Section VI of this introduction for further examples.
44. "Perhaps all the philosophies of science that I have described [they include Carnap, Popper, Lakatos, and Putnam], are part of a larger spectator theory of knowledge" (Hacking, 1983, p. 130).
45. See, for example, Cartwright, Shomar, and Suarez (1996); Morgan and Morrison (1999); De Chadarevian and Hopwood (2004).
46. For references before 1990, see Lynch and Woolgar (1988, note 3). For more recent works and references see, for instance, Vorms (2010).
47. In close relation to the themes of Chapter 4, see, in particular, Woody (2000, 2004a, 2012).
48. For works more specifically related to the topic of Chapter 6, see, in particular, Van Bendegem (1993, 2004) and Van Bendegem and Van Kerkhove (2008).
49. Chemla (2009) describes the practice of mathematical problems as it can be restored on the basis of historical evidence.
50. On the practice of mathematics in the context of commentaries, see Chemla (2008). Chemla (2012) describes the practice of mathematical proof to which commentaries bear witness.

1   Some Notions of Action

Jean-Michel Salanskis

This paper is written by someone who remains external to the so called *"practice turn"* in epistemology[1] (taking for granted that something like it did indeed happen, if not necessarily for all researchers, nor necessarily to the same degree for everyone). From such a position, I am tempted by both positive and negative reactions towards that which I do not feel part of.

A positive one, because practice turn encourages researchers to think of science in a holistic way and not only in terms of its contribution to truth: to take into account the meaningful embodiment of science in culture. Therefore, it leads (or it could lead) scholars to re-discover continental philosophy of science, which never limited itself to what I might call the "veritist"[2] or the theoretical horizon.

A negative one, because I perceive in contemporary work inspired by practice turn a tendency to regard anything as practice, in such a way that it renders practice turn meaningless. And this is nothing new; I recognize here a typical danger threatening radical pragmatisms in general. Marxism, in the time of its splendour, was a victim of it. And one may say that, under the banner of practice turn, we often witness something like a return of Marxist fallacy.

If the very program of practice turn is to be significant, it must recommend looking at science in a specific way, which can then be contrasted with other ways. This absolutely requires that we have at our disposal a non universally encompassing notion of practice. But this in turn probably calls us to overcome the opacity brought about by the common use of the word "practice". When we mention practices, our discourse seems to refer to some dimension of reality, to already available data recognizable as such. I would rather argue that something deserves to be named practice only if we are able to understand some *action* in it. Action is the general and simple category that is at stake. If everything is not to be drowned in some kind of practical soup, we must know what we call action and why, and we also need to distinguish between kinds of action.

This is what I aim to further in this paper, taking up analyses and considerations I included in my (2000) book *Modèles et pensées de l'action*. I shall first expose my general definition of action and the three typical models illustrating it. Then I will add some thoughts about other definitions or distinctions pertaining to action in philosophical tradition. And in the last part, I will make an attempt at evaluating whether science studies, when they highlight practices inside the global process or event of science, do so in conformity with one of my models of action, while also, on the other hand, examining what they call practice in line with classical philosophical discussions of action. This last part will be strictly tentative and exploratory, as is inevitable in the context of such a short paper and also given that my knowledge of the field is clearly limited.

I.   A GENERAL DEFINITION AND THREE MODELS

First, let us circumscribe a little better our ambition. What I try to characterize is action not in its broadest possible meaning, but more or less "human action": I do not wish to take into account, for example, the physical energetic concept of action. The actions I am thinking of should not be impersonal, in the sense that they should involve something like an agent, even if I don't wish to strongly identify this agent as some consciousness, some "subject" according to traditional philosophy, or some body in the context of a more naturalistic approach. Actions are meant to be human actions, but we wish to avoid prejudicial decisions about which figure of humanity is involved.

With respect to such human action, I distinguish between two facets or dimensions: the *morphological* one, and the *behavioural* one.

As for what concerns morphology, I begin with a distinction that words built with the *–ation* termination indicate rather well: the distinction between the action as available result and the action as initiating impulse. I think any word built with *–ation* has the ability to suggest both: I may say "translation" and mean the process of translating, or I may mean the result (an actual text, in that case). When I have to signal some translation in my footnotes, this second meaning prevails. Taking another example: If I speak of observable *deteriorations* making the roads of some country less secure, I refer to the result. But if I say that my body goes through a slow and constant *deterioration*, I refer to the process. Certainly words ending by *–ation* cannot mean, or do not usually mean, the initiating impulse in which the process originates. But such an origin is implicit, as a kind of symmetry of the result in the global history of the action-process. Just as by "the action" we mean "the process", so also is such an impulse implicitly evoked. More importantly, if there is no such impulse, then there is nothing I may grasp as a local and particular course of events. Therefore, to name something an action means considering an impulse and a result at the same time. But they have to be in some way connected. I express this requirement by saying that an action is a "resultative impulse", by which I mean an impulse, which is already considered at the level of the result it triggers. Certainly, I could formulate the same idea by taking an action to be a result, regarded in the light of some impulse leading to it. I keep "resultative impulse" only for its being the simplest and the strongest formulation. That an action has to be a resultative impulse tells us everything about its morphology: All action must display some continuity between impulse and result, allowing us to see each in the other.

But this is not enough, because what we have so far could still be an impersonal process. An action also needs to have some behavioural feature, as I call it. This means that an action should appear as the action of some agent. What does this mean? First, it means that some instance should be involved in the action. Let us call this instance the substrate. For an action to be an action, its issuing must mean something for some instance, which places its bets in the course of the action, so to speak. This instance dedicates its ontological weight to the action. In the course of the action, such a substrate may certainly gain something or win some additional substance—in any case, it is likely to get some identity: In the course of the action, the diversity of the involved substrate is united under the banner of what the action shows. Going through the action path, the substrate draws itself together and displays some official unified figure. I may sum up what precedes by saying that an action is "a resultative impulse, in the course of which some substrate gets involved and draws itself together". The part "in the course of which some substrate gets involved and draws itself together" expresses the behavioural aspect of action, so that it is not impersonal. The part "resultative impulse" expresses the morphological aspect of action, saying what kind of morphology actions enjoy, what distinguishes them among processes at the morphological level.

Now, this general definition has some conceptual interest, but, in my view, it benefits from being connected to three canonical illustrations. This leads us to three models of action, which I will now describe.

I.1.   Physical Model

This would be the common-sense notion, or the concrete one. An action would be typically what happens when I hit someone through some gesture: My hand moves unto my neighbour's skin. Here we have some impulse originating in my body, maybe in my neurophysiology; it can be represented as a speed vector, affecting my hand at some instant t, arising in my hand at time t. My hand follows some path, beginning in such a way that the preceding vector is actually a tangent to the resulting curve. This path ends when it transversally reaches the surface goal: my neighbour's skin, in the example. It is the transversal property, here, which allows such a trajectory to wound or pierce. So this very elementary scheme would be the one indicated by Figure 1.1.

It is clear that such a model is appropriate for many cases of concrete or material action. Even a large collective action, such as when Russian

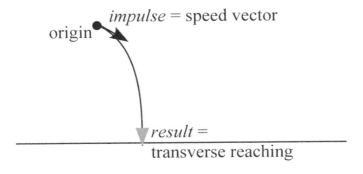

Figure 1.1   Elementary scheme of action

people storm the winter palace, seems open to deconstruction into a lot of actions of this kind: fragments of running to the end of some street or corridor, individuals' hitting gestures against resisting soldiers, breaking of doors, and so on. It is not only the morphological side of action that is illustrated in such cases, but the behavioural side also, insofar as the agent is involved each time, as a body or a body part, even drawing herself together through the gesture, identifying with the hitting hand or with the running body.

This physical version of action may appear to be suited only to accounting for concrete and effective actions, precisely those actions which common sense privileges as being the only real ones. But it may be adapted to more sophisticated cases: cases where action means the appearance of some quality.

What I have in mind here is attractor modelling, which has been made popular by connectionist approaches in cognitive science, and which was generally defined in terms of dynamical systems by René Thom at the end of the 1960s, drawing conclusions from his mathematical "catastrophe theory" (Thom, 1972/1976, 1974). The basic idea is quite simple. Any quality is seen as emerging through something like a competition between available qualities. The theory of dynamical systems then offers some modelling for such competitions: Given a vector field on some differentiable manifold, let us suppose for simplicity that the flow is defined everywhere. At each point of the manifold, we have a trajectory determining an infinite future and an infinite past of positions for the point. An attractor of the field is defined as a region A such that there is some open set U (called the basin of the attractor) such that for every point of U, the point ultimately arrives in A. When this is the case, it stays in A; because A must also be stable under the action of the flow, every future of any point of A remains in A (cf. Figure 1.2). We may imagine that the manifold is divided into the basins of various attractors, each of which incarnates some possible quality. We only need to know which point of the manifold stands for the thing or the system under consideration: The quality this thing or system receives will be the one corresponding to the attractor capturing the future of our point. Which quality shows is then clear, at least when our point does not lie on the frontier between basins.

In this context we can easily re-introduce our action-morphology: The selection of some point in the manifold determines some impulse (which we could represent using the couple formed by the point and the value of the vector field at that point). The attractor to which such an impulse leads will be the result. And the trajectory, as in the previous elementary case, substantiates the continuity between impulse and result. The only difference is that the goal-reaching condition no longer corresponds to transversal intersecting, but rather to entering the attractor's area.

What then of the behavioural aspect? Can we really conceive such a dynamical plot-line as involving some substrate in the correct way? This

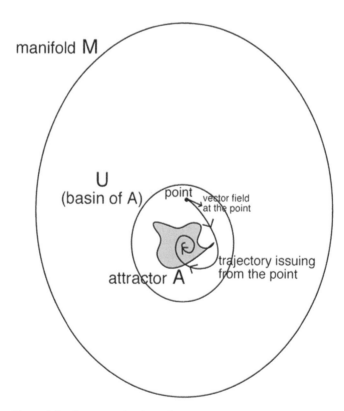

Figure 1.2   Attractor, Basin and Trajectory

is precisely the problem of using attractor modelling in cognitive science: What scientists want, in such a case, is to use their model to illustrate something like a psychological, or at least cognitive, event, which could or maybe should be considered as the organism's performance. But that requires us to connect our dynamical systems, our differential manifolds, to human psychology or neurophysiology. It must be shown that the dynamic is either the psychological dynamic of idea forming, or the neurophysiological dynamic of colour perception, for example. In the famous "Thom-Zeeman" model, states of mind are equated with points in some hypercube $[0, 1]^N$, upon which some vector field is defined, and "having an idea" is then identified with arriving, from some point within the space, to some attractor (Thom, 1974, pp. 153–154). Is such an explanation sufficient for asserting that something of us gets involved, and draws itself together along such a trajectory or performance? We have difficulties in answering positively, and this shows how and why applying this physical model, in its qualitative and sophisticated version, is far from straightforward. The modelling induces a kind of third person reduction, which seems to make it quite difficult to envision behaviour in a second phase.

I.2. Linguistic Model

Maybe it is more promising to illustrate the general definition of action in the context of language and language use. It so happens that we have nothing to look for or to invent here: a standard theory of action as arising through language, and as belonging intrinsically to language in that case, has been transmitted to us by Austin (1962). I refer here, of course, to the theory of *speech acts*. Therefore we only have to examine these "acts" in the light of our definition.

Something, to begin with, has to count as an impulse: A natural candidate is what we may call *enunciation*. The word has the ability to mean the very act of "throwing away" some phrase. In that specific case, the *–ation* word is normally understood as naming the impulse. At least, it is so in French—partly because we have two words, *énoncé* and *énonciation*, the former one meaning clearly the result, the uttered statement, and the latter the throwing away of the former. My guess is that enunciation and statement may be opposed in English in a similar way. Statement means the available piece of language, likely to be quoted. Enunciation is the gesture which delivers it: Through enunciation, we grasp a statement at the level of its original arising. But we remember that a requirement, in the typical morphology of action, is for impulse and result to be continuous: Some tear-proof bond should connect them. Here, such a bond is given by a kind of logical and metaphysical relation: We would not say "enunciation" for something which fails to deliver some statement, and we would not say "statement" for something which is not borne by some enunciation (even if it be only the writing down of it, in the case of a textual statement). Enunciation and statement are but two sides of the same event, or at least they are originally conceived as being in such a relation.

What now about the behavioural facet? Here, we use Austin's conception of illocutionary force. Any uttered sentence achieves some specific action, exactly insofar as it evokes commitment, triggers asymmetrical shared commitments. If I ask a question, I commit myself to waiting for the answer, and I commit the other to producing an answer. If I utter some promise, I commit myself to actually doing in the future what I have promised, and I commit the other to relying on the fulfilment of my promise. If I relate a description, I commit myself to the enfolded truth claim, and I commit the other to choosing among approval, contestation, requiring evidence, and so forth. The involvement dimension in speech acts is therefore quite clear.

But can we also recover the "drawing together" part? By making some speech act, I posit my individuality under the banner of some role that I endorse insofar as the performer of that particular speech act: I display myself as promiser, questioner, and so on. Illocutionary drama thus appears as a way of gaining some unified figure in the eyes (or ears) of others: The diversity of our being (possibly incoherent) will be pushed aside; only our role will count in the context of our speech act.

So, indeed, *speech act* does fit with our general definition, illustrating it in all of its specificity. Perhaps because of what has been called the linguistic turn, we could even say that speech act has become, at least in philosophical discussions, paradigmatic for action. If "practice turn" is to be taken seriously, one consequence for speech act would be the loss of this paradigm status or, at the least, its becoming questionable. My general definition with its three basic theoretical realizations provides tools for relativizing the speech act model.

I.3.   Mathematical Model

What I have in mind here is mathematical construction. The concept of construction was given to us by Brouwer, although it is now understood minus its reference to mind and mind space, and in quite an unfaithful way too, in that respect. I am not sure contemporary conceptions were right in parting from Brouwer on that point, but let us not go through that debate, at least not now. A construction is the fabrication of some object on the basis of primitive given objects, using some fixed, given building rules. So constructions always happen in the context of some specified construction convention, a *recursive clause*, as we may call it. Such a clause gives meaning to some constructive class, the class of all constructions obeying the clause. As it is well known, natural numbers may be seen as constructions in this sense: For example, I take as my primitive object the empty inscription or the "|" inscription, and I give myself only one rule, allowing me to build the inscription "S|" if S is any already obtained string, any already constructed natural number. As it is also well known, any constituent of formal languages is also a construction in that sense: What we call *terms*, in first order languages, what we call *formulae* in first order language or in propositional calculus, what we call *theorems* or *proofs* within an inference system, have to be defined through some recursive clause which shows that each of them is nothing other than some construction in some constructive class.

What I have to show, now, is that constructions possess both the morphological and the behavioural feature: that they are indeed actions, in our sense.

As far as the morphological part is concerned, all we have to do is to reflect seriously on what the construction-tree of some construction displays. For example, the formula $P \rightarrow (Q \wedge (\neg R))$ of propositional calculus admits of the construction-tree shown at Figure 1.3.

Each node of that tree records some choice: either of some primitive formula (a propositional variable) or of some building rule (corresponding to some binary or unary connector). The construction as a whole appears as the result of some finite set of elementary choices, associated with the nodes of the tree. These elementary choices are pieces of impulsion, bringing together the final formula as a result. If we think that way, we have a double reading of the completed tree: On the one hand, it is the result, the intended formula; it displays the formula and its structure (which determines also how the formula has to be read, understood, used). On the other hand, it summarizes the impulse-history standing behind such a result. The tree unifies and displays in one and the same symbolic setting both impulse and result, making them "continuous" as required. So construction deserves to be read as a "resultative impulse".

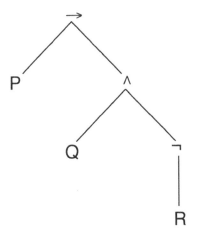

Figure 1.3   Construction tree of formula P → (Q ∧ (¬R))

But are constructions also behaviours in our sense? Here, I think we should recall their foundational role. What we agree upon with the highest degree of certainty, inside the rational circle, is that something (some formula) has been correctly derived in the context of some inference system: This amounts to acknowledging that something is a regular construction according to some recursive clause. Our use of formal reasoning rests upon our ability to share these recognitions. And this itself points to our ability to mentally accompany the constructions concerned, verifying that the rules have been respected step by step. We indeed experience constructions as being one and the same, as legitimate in the same sense, whether they are accomplished in our minds, in our speech, or on a sheet of paper. What we feel about constructions is that they are the same when internally experienced, temporally disclosed by some discourse, or laid down with symbols on some inscription surface. Constructions are the objective witnesses of "formal agreement". And such agreement implicitly involves some arbitrary consciousness, likely to "experience" the corresponding act of construction. In some cases this "experiencing" is understood potentially; it can even be delegated to some computer. But this means that we are satisfied with mentally envisioning what could happen, or in the case of the computer, with controlling what actually happens in the machine, through some genuinely "accompanied by consciousness" construction—for example, the construction of the program.

But indeed, accompanying consciousness is also relevant for common intersubjective use of constructions: Understanding some sentence of ordinary language enfolds the apprehension of the construction-tree shaping it at the syntactic level (as Chomsky taught us), reading formulas of some formal language and building proofs out of them requires our seeing them as organized by their structure, which again amounts to having their construction-tree in mind. So we indeed have the involvement property: Consciousness is implicitly involved in constructions, rightly considered in the light of their value and their function.

We could say that construction is the defining behaviour of the formal agent: She offers her construction up to the collective work, be it formal or mathematical. This is particularly clear when this construction is a routine written in some programming language. In that case, the agent adopts the figure of the contributor providing the input/output capacity of the routine. In that sense, as in the case of speech acts, the agent of construction unifies herself or himself under the banner of some role, even if it be of a different nature.

II.   MORE CLASSICAL VIEWS

From Aristotle, we inherit, at least, the distinction between *poièsis* and *praxis,* and the idea of "practical syllogism".

Considering the former, let us recall that *praxis* is an action which has its end in itself. It contrasts with actions not having their end in themselves, of which the typical example is the fabrication of objects: These objects ultimately survive the action, detach themselves from it, and thus appear to be its goal. *Poièsis* is the generic name of such actions.

Considering the latter, let us remind ourselves that practical syllogism is that figure of reasoning which, given our aim of realizing the state of affairs expressed by proposition *Q*, leads us to look for the relevant *P* such that *P→Q*, to the effect that we substitute to our original goal, as an intermediate goal, the state of affairs expressed by *P*. Aristotle himself had already observed that in a way this amounts to reasoning "backwards", and compared it to what geometers do and call analysis.

More recently, we have the threefold distinction formulated by Hannah Arendt among *labour, work,* and *action,* in the elevated sense.

—*Labour* names what slaves perform. They exhaust their vital strength the whole day long, and the product of their work is immediately consumed (the typical case is slaves working for agriculture). This way of "acting" entails inserting oneself in natural cycles (sleep/wake, eating/excreting, sexual drive and its satisfaction). By going through labour, we experience ourselves as living, even if we are working like a beast of burden.

—*Work.* This overlaps, apparently, with Aristotle's *poièsis (fabrication),* but Arendt's notion is more radical. Through work we build objects that may last a long time, even longer than us. They are likely to be shared; they constitute the furniture of a common world. For Arendt, knowledge and science are made possible by the interposition of this stable world between us and the Heraclitean flow of nature: Such interposition provides for the necessary distance between the latter and us.

—*Action* in the proper sense, action in its authentic and noble figure, cannot be anything else, for Arendt, than the political behaviour of the individual. I disclose myself by initiating something in front of others, showing myself as distinct from any of them and at the same time as equal to any of them. Speech is the canonical vehicle of this public self-disclosure. It opens the way to the unpredictable relaunching of my action by others, through a complicated reaction-network. Such higher-level action constantly renews the world, while taking responsibility for it at the same time.

Beside Arendt in contemporary philosophy, I could also evoke the phenomenological and analytical conceptions of action. Yet I have the feeling, perhaps in a very unfair way, that they do not help us in seriously contrasting action with non-action.

Phenomenology has been committed, since Heidegger, to interpreting any "intentional" relation to the world, any "aiming at", as practical; such a stance more or less derives from describing the original status of all things as the status of "tool". In the perspective of phenomenology, as intentionality is always practical, practice, correlatively, is always intentional. I would say that Merleau-Ponty and Sartre very much confirm this view. But for phenomenology, everything is intentional as related to intentionality. It seems that practice, therefore, becomes universally encompassing.

From analytical philosophy on the same subject, I basically retain David-son's work (2001). He locates action in the realm of purported behaviour, in respect to which Aristotle's practical syllogism is relevant. We also find in his writings at least the wish to make the theory of action more naturalist, either by taking our reasoning premises for action as *cause,* or by positing at the origin of any action some bodily movement, related with a description of the action as a "primitive" one. But nothing of that strategy helps very much in discriminating action from non-action: We only have at our disposal the criterion of purported behaviour, which is obviously too selective for a "practice turn" project.

And neither phenomenology nor analytical philosophy show great interest, or so it seems from what information I have, in defining and distinguishing great types of action.

III.   SOME QUICK REMARKS ABOUT HOW EPISTEMOLOGISTS APPROACH SCIENCE AS PRACTICE

Some practices, identified as such by epistemologists, are clearly speech acts: to take an example, in Hacking's scheme of "laboratory science" (1992a, pp. 29–64), *questioning* and *formulating* theories. In the second case, we have a sequence of "declarations". But insofar as the corresponding text has sometimes to be written in mathematical language, we perhaps also have *construction* behind it (far behind: mathematics in physics do not usually come in formal vestment).

One may ask whether epistemologists consider the fabrication of artefacts as scientific practice. Of course, we are tempted to answer in a positive way, because a lot of authors advocating for practice turn have insisted on the importance of *instruments,* claiming that the history of technique is fundamental to history and philosophy of science in general. It is enough here, for example, to evoke Pickering when he highlights the part played by instruments and their readjustment in fundamental physics.[3] Still, in Hacking's scheme of laboratory science again, the first mention of devices in the laboratory refers to formulating the theory of the device. More generally, Hacking describes the filling of the laboratory with devices as complicating some kind of network. We first have targets, then devices able to send energy into targets, then tools that are added to the devices already given in order to make everything operational, then finally data generators. What counts as practice is more dealing with these internal references inside the laboratory than the fabrication of some particular tool or device. So Hacking's paper seems to say, at the same time, that the laboratory is the place for science as practice par excellence, and that the fabrication of instruments is not, as such, a part of it. On the other hand, he talks about fabrication of instruments quite extensively, and gives to it much importance, in his (1983) *Representing and Intervening.* I think specialists should address, at least, two questions:

1) When epistemologists describe the fabrication of instruments, do they locate the corresponding action at the heart of scientific practice? 2) When they do so, do they implicitly refer to Arendt's notion of work,[4] or to some other notion of action (to be made precise)? I think the whole theme would be made richer if researchers interested themselves in facing such issues.

Sometimes, what is called practice is the introduction of a representational or notational medium, something we could call "symbolic tools". This also is not work in the basic Arendtian sense. More, it is some kind of official proposition inside some community: a speech act, and maybe an action in Arendt's third noble sense? And behind it, construction (think of the proposition of diagrammatic devices or of formal languages).

There is something that is very easily taken as practice: what I would call *reintentionalization*. Aiming at something in a different way. Uncovering something under some new aspect. In Hacking's description of laboratory science, again, the work on data is described in this way: We display them differently. Now, mathematics gets more or less constantly used in that manner. But when Alexander Bird (2008b) comments on Kuhn's paradigm shifts by offering the analogy to the *Gestalt* switch in perception, we are on the same track. It is interesting that such a notion of action be used in our context, because it is implicitly phenomenological, although epistemologists today, for the most part, are not phenomenologists.

But we have another example of underlying phenomenological influence. I am thinking of the tendency to consider any description, unveiling the situation of scientific work at some moment, as a description in terms of practice. How this may be relevant, I think, is best explained in Sartrian terms: Human reality is always situated, but what the situation is may only be grasped and understood in terms of the project which is about to lead beyond it. The rock in the mountain is my situation as far as it lies on the slope I am actually attempting to climb (Sartre, 1943, pp. 526–534). If we take such Sartrian considerations as true to our human condition, then the situation of science is already indicating how scientists are going to or may exceed it, thus it disengages their "practice" in that sense (which, it may be added, converges with the *reintentionalization* sense previously considered).

Now comes our last remark. Epistemologists, it seems, do not spontaneously recognize formal proof as a case of action through the model of construction. They rather consider as examples of scientific practice precisely everything in mathematical activity that stands beyond formal proof. Still, on the other hand, the new games that mathematicians play with computers are categorized as "practical aspects". When construction connects with technical material devices it recovers its dignity of action, or so it seems. Here we could think, again, of Arendt: Maybe epistemologists consider formal proof as a kind of slave labour, as a way for mathematicians to spend their energy functioning like machines (Arendt, 1958, p. 171–172). Proofs would be comparable to bureaucratic documents absurdly produced by civil servants of the ancient Egyptian empire:[5] Papers only meant to be gathered and forgotten in obscure rooms, or destroyed to leave room for new ones. So true practice would only arise when mathematicians make use of constructive processes while staying outside of them, typically dealing with computers and assigning them tasks according to some strategy. Practice, here, begins with a kind of back and forth play with constructive machines. But such perspective does not do justice to mathematical activity. Formal proofs are already, from the onset, strategic experiences. Following the path of some formal proof involves freedom at each step, and enfolds existential aspects (like when one takes stock of what has been done along some sequence and introduces → in natural deduction, or like when one picks up a very old result in order to apply it, freely reactivating some chosen part of the past). So, not only do formal proofs, understood as constructions, remind us that both the mathematical world and mathematical discourse result entirely (and in a related way) from some act of *making*, but, furthermore, construction, understood as the very mode of this *making*, emerges as being an existential process.

NOTES

1. I use this word, the most neutral as I see it (probably under the influence of French tradition), in order to simultaneously evoke philosophy of science, history of science, and science studies of various type (sociological, anthropological, political, etc.). Admittedly, the word may have been used and understood in a quite different and restricted way in some contexts and discussions; I just ask from my readers that they do not take it that way.
2. Borrowing the adjective from Alvin Goldman (1999).
3. The following quote has been very kindly transmitted to me by Léna Soler:

> I have focused upon the interpretative aspect of experimental practice because that was the most conspicuously arguable facet of the experiments in which the weak neutral current was discovered. I do not want to suggest, however, that other "instrumental" aspects of experimental practice are unproblematic. By "instrumental", I mean to refer to manipulative operations in the laboratory like setting up the apparatus, tuning it, making sure it "works" and so on. Such instrumental practices are surely bound up with both interpretative practices and the phenomena to which they give rise. As an example of such an instrumental practice, one can think of the episode in which the HPW group "improved" their apparatus, replacing the 4 foot iron shield with 13 inches of steel. At the time, this replacement was thought to be irrelevant to the phenomena at issue. It did, of course, serve to obliterate the neutral current signal, but one can well imagine that in a different context (e.g., if the Gargamelle group had not already published their discovery) this would simply have been taken to indicate that a genuine improvement had been made. As it was, the disappearance of the signal led the experimenters to reconsider their interpretative practice with regard to "punch through" and this eventually culminated in the reappearance of the signal. Thus, in this instance one can argue that instrumental practice, interpretative practice and natural phenomena had to be assessed together. (Pickering, 1984b, p. 113)

4. For example, in the case of the preceding quote of Pickering, the difficulty is that the produced *artefact* does not seem to be considered as a stable part of a common world to be shared. It is taken into account in the context of its replacement. Scientific *poièsis* seems devoted to consumption, like labour in Arendt, even if this be for quite different reasons.
5. I take up Arendt's (1958) narrative on pages 92–93.

Action in Science

Commentary on "Some Notions of Action", by Jean-Michel Salanskis

Jean-Luc Gangloff and Catherine Allamel-Raffin

According to Jean-Michel Salanskis, action is an essential conceptual key to give a relevant account of practices, be they scientific or not. He writes: "... something deserves to be named practice only if we are able to understand some *action* in it. Action is the general and simple category that is at stake" (p. 44). However, a brief survey of works in science studies for the past forty years shows that action is probably not an *explicit* central feature, neither in their developments, nor in their conclusions. This observation should not be considered as an objection against the perspective Salanskis suggests we should adopt. Salanskis' aim is to clarify the word "practice" by finding some definitional foundations.[1] According to him, this is a necessary step in order to better grasp the meaning of "practice" and the sense of the "practice turn".

   In the three sections of our commentary, we will address the following questions: (1) Which place has been given to the concept of action by "practice turn" epistemologists in their works?[2] In which particular type of writings did the concept explicitly appear and for which purpose? (2) What are the main features of a definition of action, if we take into consideration some major works in recent philosophy of action? (3) What does Salanskis tell us about action and how is his proposal related to the literature on action?

I. THE PRACTICE TURN AND THE CONCEPT OF ACTION

Let us start with some historical insight. Authors participating to the "practice turn" originally defined their way of studying science by following those who, in a sense, are their immediate predecessors, namely authors of the "historical turn" (T.S. Kuhn, I. Lakatos, P. Feyerabend, etc.).[3] Unquestionably, both groups of authors denounced the overemphasis on the *product* and the relative neglect of the *process* of scientific investigation.[4] Their common aim was "to counter the then dominant conception that scientific knowledge was simply the end product of the application of logic, observation and experimentation, the results of which were somehow an adequate representation of the real world existing outside and independently of the scientist" (Gingras, 1995, p. 123). Beyond this shared opposition to the "received view", the specificity of the practice turn can be located both in the diversification of the methods employed and in the wide variety of objects under interest. More specifically, while authors of the historical turn had already understood that "(s)cientific concepts do not pop out of heads, but are constructed in response to specific problems by utilizing methods appropriate to the solution of the problem" (Nersessian, 1988, p. 41), their discussions remained focused on processes *of theory change* and related issues. In contrast, supporters of the practice turn combined the rejection of a static and ahistorical account of scientific investigation *with the multiplication of targets,* often neglecting theories in favor of *other topics,* such as experimentation, models, instrumentation, and so forth. They shed light on these components of empirical sciences, advocating for much more contextualized and dynamic models of scientific investigation than those of the historical turn.

Thirty years ago, in their book *Science Observed* (1983b), Karin D. Knorr-Cetina and Michael Mulkay listed some of the main currents involved in the practice turn at the time: "the ethnomethodological study of scientific practice . . .; the discourse analysis of scientists' talk, writing and pictures . . .; the ethnography of scientific work . . .; the relativist programme in science studies . . .; and the sociology of knowledge perspective in 'strong' and 'weak' versions" (Knorr-Cetina & Mulkay, 1983b, p. 1). Today, we may add to this list many other trends: philosophy of experiment, actor-network-theory, third wave of science studies, and so on. Knorr-Cetina and Mulkay moreover noticed: "It is impossible to identify a single set of characteristics shared by all of these analytical positions. Nevertheless, there are certain broad themes which link them together through a series of family resemblances" (1983b, p. 1).[5] Thus, a few decades ago, Knorr-Cetina and Mulkay tried to identify family resemblances by gathering works realized by epistemologists who had initially different aims and programmatic agendas, and who sometimes even ignored each other's work. At least, all these works shared one characteristic: They focused on practices approached from a bottom-up perspective. Up until the 1990s, however, the notion of practice itself was rarely a subject of analysis. One significant example of this absence could be Buchwald's 1995 edited volume, entitled *Scientific Practice.* It included papers on many different topics related to the "interactions between experimenters, instruments, apparatus, and different layers of theorizing" (Hacking, 1995, p. 2). Surprisingly, however, the book contained no detailed analysis of the concept of scientific practice, not even in Baigrie's (1995) paper, despite its title and the title of the entire book—namely, "scientific practice". This exemplifies the fact that during a programmatic stage, from the 1970s to the 1990s, the word "practice" was commonly used as a *banner against* the philosophical tradition that favored the study of scientific theories, theory-change, and theory-choice, but was *not itself an object of conceptual and definitional investigation.*

This started to change in the early 1990s, when a trend of more reflexive or "meta" accounts developed with the so-called "practice theorists" such as Theodore R. Schatzki (2001), Joseph Rouse (2007), and Stephen Turner (1994). These authors did not restrict their scope to science studies; they considered also human and social sciences. In their works, the concept of practice started to be systematically explored, as was also the relevance of the label "practice turn", and definitions of "practice" were confronted to the contents of bottom-up analyses (i.e., of case studies taken as particular instances of scientific practice) previously conducted within the practice turn. Salanskis' approach may be included in this trend of "meta-epistemology".

The more thorough attempts of meta-epistemologists to characterize practice often involve the notion of action, as in Schatzki's definition: "a practice is, first, a *set of actions*" (Schatzki, 2001, p. 48), though such definitions are usually completed by other features (among which we find rules, norms, skills, forms of life, and so on).[6] Salanskis' suggestion is that it is worth strongly focusing on the notion of action to define practice. Because of this advocated association of practice and action in both Salanskis' and other meta-epistemologists' contributions to the proper definition of practice, we propose to focus, in the next section, on some crucial recurrent issues raised by the notion of action in the literature.

II.   WHAT IS ACTION?

In the huge literature on action in contemporary philosophy, we find multiple, sometimes conflicting, proposals in which we can identify several recurrent problematic points. Offering a (non-exhaustive) list of these points will permit us to better evaluate where Salanskis stands with respect to already existing conceptions of action.

(1) *A change in the world.* "The snow is falling", "A mosquito is biting me", "The biologist is cleaning the Petri dishes": In each case, something happens or occurs. These three examples are examples of *events*. An event consists of a change in the world, a change in some object or other (Lombard, 1995). Generally, the third sentence only is considered as an action. Events like the falling of the snow or the biting of a mosquito are commonly not viewed as actions. Conversely, all actions are usually seen as events. In this perspective, actions are a subclass of events. Note that this common view is sometimes rejected by those who think that actions are ontologically distinct from impersonal events. Let us, however, accept the common view, namely an event-based account of action. Then the issue is: Under what condition can an event be considered as an action?

(2) *Agency.* This condition is the capacity of some agent X to do or to perform an action A. Which basis can we find to distinguish minimal

agency from non-active consequences within extended causal chains? One candidate is "a certain kind of direct (motor) control over the goal-seeking behavior of [X's] own body" (Wilson & Shpall, 2012). So the subclass of events constituted by actions does not include all things that Xs are doing: Some bodily movements (blinking, coughing, and sleeping) are considered as mere happenings, and not as brought about by something people genuinely do. Some actions, however, do not consist of bodily movements. For example: "I'm waiting for my uncle Peter". In relation to this point, David Rayfield notices: "Subscribing to the view that action involves bodily movement can only lead to an overly restrictive analysis of action in that it must exclude forbearances" (1968, p. 134). Forbearing consists in omitting a doing (Van Dijk, 1974, p. 295), letting the situation be as it is. For example: "I'm not eating while I'm hungry", which typically is the case in a hunger strike. Does the negative character of such cases allow us to nonetheless talk about an action?

(3) *Initiative.* A controversial feature of the definition of action concerns psychological factors, which invariably turn out to be "acts of will": volitions, intentions, choices, decisions to act, and the like. These mental items, which are supposed to explain actions, are problematic. Consider, for example, idiosyncratic habits like fiddling with one's glasses in seminars. "It seems plausible that there are instances of such actions when one will have nothing much to say except that 'I did it out of habit', and that need not entail that it was attractive to one in any way. It could be utterly indifferent to it" (Pollard, 2005, p. 5). In such instances, we cannot identify any explicit or conscious intention. But in *some* instances, an agent X *could have decided* to fiddle with his glass. A widely accepted minimal position is that an item of behavior deserves to be considered as an action of X, if and only if X could have had the intention to do or choose to do or decide to do A (which is, for example, not the case for "X's pupils are dilating"). To be described as actually acting, an agent X must have the *initiative* to do or not to do A.

(4) *Free will.* Is the agents' feeling that they have the initiative just an illusion that they act freely, while they are in fact influenced by factors of which they remain completely unaware (Nahmias, 2007)? A negative answer to this question entails that when X makes a choice, X *really* has the capacity to choose (i.e., is not pre-determined by any internal or external factor). We are confronted here with the classical debate about the existence of free will (i.e., self-determination or spontaneity). Issues arising from this debate are strongly related to the question of moral responsibility.

(5) *Individuation.* We can distinguish types and tokens (Collins & Kusch, 1998; Goldman, 1970), but also ask the question of whether there are some "basic" actions (Danto, 1973). Is it possible to identify the most minimal unit of action?

(6) *Temporality.* Related to the problem of individuation is the problem of temporality. We may think that an action has a beginning and an end. But what are the relevant criteria to determine the temporal limits of an action?

(7) *Causes/reasons.* It is generally accepted that an agent X is performing an action because he has a reason to do so. One famous definition is from Donald Davidson: Whenever X does something for a reason, she has (a) a pro attitude (desires, urges, moral or aesthetic views, economic or social prejudices or conventions) towards actions of a certain kind, and (b) a belief that her action is of that kind (1963, p. 685). There is a great controversy about the nature of reasons: Can they be taken as the causes of our actions?

(8) *Results/consequences.* Considering the temporal dimension of action, one additional distinction might be useful: The distinction between result and consequence. According to Collins and Kusch, ". . . the *result* of an action is the state of affairs that has to obtain for the action to have been carried out, whereas the *consequence* of an action is a further state of affairs that has been brought about by the attempt to carry out the action" (1998, p. 9). If Mary stands up in the bus and an old lady takes her seat, the result of the action is that Mary now is standing up and the consequence is that the old lady is seating in the bus—unless Mary intended to give her seat to the old lady by standing up from it—in which case, the fact that the lady is sitting now is also a result.

(9) *Interpretation.* Like discourses, actions are meaningful. They are "open worlds", inherently subject to interpretation (Ricoeur, 1986). This characteristic has to do with their intrinsic historicity: Actions are historically situated, and their meaning is not fully determined by their performers and their immediate audiences (Dauenhauer, 2011).

Some subfields of philosophy—philosophy of action and moral psychology—are strongly concerned with all the key issues listed above. Some of these issues (e.g., point 4, free will) imply metaphysical questions that are beyond the scope of the practice turn studies of science. Some other issues (e.g., point 6, temporality, or point 8, results/consequences), are crucially at stake in the practice turn in studies of science, though not always presented with an explicit reference to the concept of action. In the following section, we will analyze Salanskis' own definition of action and evaluate how it stands in relation to the key issues listed above.

III. SALANSKIS' DEFINITION OF ACTION

In *Modèles et pensées de l'action* (2000), Salanskis advocates for a transcendental definition of action. The definition developed in this book is also presented in the paper commented upon here: "action is a resultative impulse, in the course of which some substrate gets involved and draws itself together" (p. 46). On the one hand, action is defined as a *process* (the "morphological dimension"): It connects a beginning (the impulse), which "throws" it to a final state where the process ends (the result). On the other hand, action is defined as a *behavior* (the "behavioral dimension"): The behavior of an entity that Salanskis designates as a "substrate".

What grounds the motivation for this transcendental definition of action? First, Salanskis underlines the necessity of such a definition in order to do justice to the very notion of action. He believes that we can begin our reflection without any reference to the available contemporary theories of action. Indeed, these theories will predetermine the content of the definition and direct the analytical work into some dead ends. Salanskis puts the emphasis on the a priori conditions for the recognition of action as such: What do we anticipate and prescribe as the topic of action? Which types of events deserve to be called "action"?

Without any pretense of being exhaustive, let us specify briefly how Salanskis' definition relates to the key issues listed in Section II, which are frequently involved in definitions of action. Some of these issues are not explicitly treated by Salanskis. For example, classical issues of free will, intentionality, causality, or interpretation are not *directly* addressed in his paper. This does not mean that they are absolutely irrelevant, but they are left in the background.

Concerning agency (point 2), Salanskis chooses to adopt a liberal position. He just stresses the opposition between two types of changes in the world (point 1), impersonal processes on the one hand, and events categorized as "more or less 'human action(s)'" (p. 45) on the other. The lack of precision concerning the notion of "substrate" (the word he uses instead of "agent") is deliberate. Salanskis wants to avoid strong preliminary metaphysical commitments. As he states: ". . . I don't wish to strongly identify this agent as some consciousness, some 'subject' according to traditional philosophy, or some body in the context of a more naturalistic approach" (p. 45). What a substrate may be must be determined all along the different steps of the inquiry. This is the path Salanskis follows in his paper, when he develops three paradigmatic models illustrating his definition (the physical, linguistic, and mathematical models). Agency—or, to use Salanskis' terminology, the behavioral dimension of action—is a plastic notion. Relying on Salanskis' paper, one important criterion for the recognition of some agent's figure is the function of unification: "in the course of the action, the diversity of the involved substrate is united under the banner of what the action shows" (p. 46). This unification is also presented as a "drawing together". What do these definitional terms mean? From an extensional point of view, the class of things the concept of "substrate" applies to might be very broad. If we limit our study to the linguistic model put forward by Salanskis, the illocutionary dimension of any speech act "appears as a way of gaining some unified figure in the eyes (or ears) of others . . .: only our role will count in the context of speech act" (p. 50). The notion of "role" implies no strong metaphysical thesis, but only a speaker's commitment shared with his interlocutor: "If I ask a question, I commit myself to waiting for the answer, and I commit the other to producing an answer" (p. 50). As Salanskis emphasizes it, linguistic conventions play here an essential part. Many other figures may constitute relevant types of "substrate" if they satisfy one requisite: to include a human part. Salanskis thus leaves open the class of candidates to the status of substrate and engages us to recognize the relative conceptual flexibility of the notion of agency.

A major concern, in Salanskis' view of action, is related to our point 3 (initiative) and point 6 (temporality). In a previous paper, he wrote: "An action is a process in which a result is readable, this result itself revealing the impulse from which it is tear-proof . . ." (Salanskis, 2010). Salanskis emphasizes the necessity of thinking about action from the point of view of a strong *continuity* between the two terms of its definition: "all action must display some continuity between impulse and result, allowing us to see each in the other" (p. 46). Against all the presuppositions of what he calls (2010, p. 163) a "deliberationist" model of action (one which establishes a clear demarcation between three steps: deliberation, decision, and action), Salanskis thinks that the dissociation of the impulse from the result is an artificial operation, which fails to capture the nature of action. Some philosophers try to identify and to isolate one point in the course of the process that constitutes the "decision" (point 3). But according to Salanskis, this is an illegitimate reification of an alleged element of the process, and an element that may correspond to nothing. Consequently, as regards our point 6 (temporality), he argues that it is not relevant to disconnect the beginning and the end of the action. Instead, it is the whole process that is constitutive of an action.

Another definitional point that relates to Salanskis' assumption of an essential continuity between impulse and result is our point 8 (consequence/ result). Salanskis does not introduce this distinction as such in his conceptual analysis, but he is very concerned with the idea that the result is not to be considered apart from the impulse. In particular, his criticism of some common philosophical views about action leads him to reject any account of action that gives an essential importance to the *material result* associated with an action. In *Modèles et pensées de l'action,* he develops an example of action, to dig a hole, which illustrates the substance of his criticism. The action shows its effectiveness in the hole itself, the materiality of which will interfere with a causal history including natural elements (water falls in the hole) and individual-cultural-historical elements (a tribe of nomads takes the habit to live near the hole, poets write about the landscape in which the hole now takes place) (Salanskis, 2001, p. 224). But to focus on the materiality of the hole moves us away from the consideration of the digging *as an action*. What counts here is not the material result itself. This reductionist way of thinking is precisely what Salanskis tries to avoid. What is essential, according to him, has to do with the narrative that relates the material result to the resultative impulse, where a substrate gets involved and draws itself together.

One can wonder how Salanskis' general definition of action and his three models of action are reflected in the practice turn literature. Salanskis does not provide us with many examples, as his subtitle underlines it ("Some quick remarks about how epistemologists approach science as practice", p. 54). However, his categories might be used in order to classify and reorganize the whole set of writings put under the banner of practice turn. It would help us to determine which of these writings really include the dimension of practice in Salanskis' sense, by tracking the descriptions of action they contain (or not).

For example, Salanskis rightly observes that in Hacking (1992a), "what counts as practice is more dealing with these internal references [targets, devices, data generators, etc.] than the fabrication of some particular tool or device" (p. 54). In some of Hacking's other writings, however, there are descriptions of physical actions, especially in the chapter devoted to microscopes in *Representing and Intervening* (1983). The place occupied by action-based practices depends on the specific aim of each text. To propose a taxonomy of the elements involved in experimentation in order to characterize stability in laboratory sciences, as in Hacking (1992a), is not the same goal as directing the attention to the dimension of manipulation and skills in scientific experimentation in order to fight against theory-centrism and a one-sided intellectualist approach of science, as in Hacking (1983). Thus, even in the work of a great figure of the philosophy of experiment such as Hacking, action-based practices can be more or less in the foreground, depending on whether the description of practices is what the paper is about, or just an element of an argumentation that has a different target.

Moreover, a lot of contributions pertaining to the practice turn are probably not, strictly speaking, concerned with practices understood in Salanskis' sense. One prominent case is constituted by works pointing out the embeddedness of scientific activities in a determined cultural context, where the emphasis is on the institutional dimension of science (and particularly on the relationships between laboratories, editors, political instances, etc.). Considerations relative to social conditions of research do not involve a notion of practice directly analyzable in terms of action, though they are valuable contributions that are undoubtedly focused on practice. Thus, to confront contributions of the latter sort with Salanskis' framework is probably irrelevant.

IV. CONCLUSION

In his paper, Salanskis advocates an attention to the concept of action in order to assess the significance of the practice turn. He argues that every practice necessarily involves a dimension of action. Consequently, practice turn epistemologists cannot fully grasp scientific practices if they do not pay attention to action. While we support Salanskis' plea for the study of action and believe that he provides an interesting account of it, Salanskis' proposal only applies to a restricted class of practice turn studies, namely those which involve an action-based notion of practice, and not to various other contributions that aim to analyze the epistemological consequences of certain political, economical, or cultural factors. Accordingly, Salanskis' approach could be fruitfully complemented by other approaches.

NOTES

1. "First, I think that philosophy has always a foundational function, because I believe that the foundational enterprise is essentially a question of sense . . ." (Salanskis, 2003, p. 30).
2. We use the word "epistemologist" in the same sense as Salanskis (see note 1 of his paper): "Epistemology" includes philosophy of science, history of science, and science studies.
3. See the introduction of this book, Section V.4.
4. See the introduction of this book, Section V.6.
5. We leave aside two facts which would deserve to be fully analyzed in a more extended survey: (1) other "micro-turns" have characterized different steps of the development of the practice turn itself. An example of micro-turn is the one evoked by Knorr-Cetina and Mulkay, the "linguistic turn" in social studies of science, which includes the essays of Latour and Woolgar (scientific work is a form of writing), of Lynch, Knorr-Cetina, and Woolgar again (scientific writing is embedded in scientist's informal practical reasoning), and of Mulkay, Yearley, and Potter (". . . generalizations about scientific practice derived from scientists' accounts are only as dependable, as precise and as valid as the accounting practices on which they are based", in Knorr-Cetina & Mulkay, 1983a, p. 10). (2) Some other "turns" are intertwined with the practice turn: the "naturalistic turn" (cf. Callebaut, 1993); the "cognitive turn" (cf. Fuller, De Mey, & Shinn, 1989; Thagard, 1993, etc.). Significantly, the reader who looks at the contributors list of many books that include these labels in their titles often finds the same authors' names as those who appear in practice turn studies of science.
6. The three concepts of practice, action, and activity are generally defined on the basis of the following relationships: Compared to an action, an activity is characterized by its *temporal development*. Indeed, it is generally not reducible to a single instantaneous action and requires a set of actions. We do not use the word "practice" to qualify an activity performed just once in his life by an agent. The concept of practice implies an *idea of regularity* in the performing of the activity.

2 Epistemic Activities and Systems of Practice

Units of Analysis in Philosophy of Science After the Practice Turn

Hasok Chang

I. INTRODUCTION: MAIN AIM OF THE PAPER

My objective in this paper is to take a step toward creating a *structured* and *precise* philosophical framework for thinking and talking about scientific practices. For that purpose I articulate the concepts of "epistemic activity"[1] and "system of practice" as units of analysis for framing discussions of science. Standard Anglophone philosophical analyses of science have been unduly limited by the common habit of viewing science as a body of propositions, focusing on the truth-value of those propositions and the logical relationships between them. The premier subject of discussion in such philosophy of science has been *theories* as organized bodies of propositions. This has led to the neglect of experimentation and other non-verbal and non-propositional dimensions of science in philosophical analyses. A number of historians, sociologists, and philosophers, including those who made the "practice turn" that forms the starting point of the discussions of this volume, have pointed out this problem. However, so far no clear and widely adopted alternative philosophical framework has emerged. What I aim to spell out is a philosophical grammar of scientific practice—"grammar" roughly as meant by the later Wittgenstein.[2]

For those who have been influenced by the "practice turn", I assume it is not a subject for too much dispute that a serious study of science must be concerned with what it is that we actually *do* in scientific work. This requires a change of focus from propositions to actions. I begin with the recognition that all scientific work, including pure theorizing, consists of actions—physical, mental, and "paper-and-pencil" operations, to put it in Percy Bridgman's terms (1959, p. 3). Of course, all verbal descriptions we make of scientific work must be expressed in propositions, but we must avoid the mistake of only paying attention to the propositional aspects of the scientific actions.[3] That is a sure path to disconnection from practice, and it is precisely the path that Anglophone philosophers on the whole have taken. What I am complaining about is our habit of focusing on descriptive statements that are either products or presuppositions of scientific work, and our commitment to solving problems by investigating the logical relationships between these statements. I take heed of Bridgman's conviction (1954, p. 76) that "it is better, because it takes us further, to analyze into doings or happenings rather than into objects or entities"—or propositions that simply describe the properties and relations of objects and entities. When we do pay attention to words, it would be better to remember to think of "how to *do things* with words", to recall J. L. Austin's (1962) famous phrase.

II.   COLLECTING PRE-EXISTING CLUES

My first task is to stand on the shoulders of giants: to make a collection of clues from past and present scholars who have made important moves in the direction I am suggesting. I regret that this part of the paper is very brief and sketchy, neglecting many relevant authors.

(a)   Various Types of Elements

In moving away from propositions, the most obvious clue may be that there are different *types* of elements that constitute scientific work. Thomas Kuhn (1970, pp. 182–187) gives an indication of this heterogeneity in his notion of paradigm as "disciplinary matrix", "composed of ordered elements of various sorts", including symbolic generalizations, metaphysical beliefs, values, and exemplary problem-solutions. Also instructive is Hans-Jörg Rheinberger's (1997, p. 238) notion of an "experimental system", defined as "a basic unit of experimental activity combining local, technical, instrumental, social and epistemic aspects". Ian Hacking (1992a, pp. 44–50) lists as many as 15 different kinds of elements that enter experimental practice (see Table 2.1), whose interplay produces the stability of laboratory sciences that he is trying to explain. Léna Soler (2012a, p. 24) gives a similarly heterogeneous list of the ingredients of scientific practice, and she identifies the "thesis of heterogeneity" as a leading characteristic of practice-oriented science studies (Soler 2009, ch. 9, sec. 2.1). A similar sense of heterogeneity is given in Mike Fortun and Herbert Bernstein's (1998) discussion of how scientists "muddle through".

Table 2.1   Hacking's Elements of Laboratory Practice

Ideas
1. Questions
2. Background knowledge
3. Systematic theory
4. Topical hypotheses
5. Modeling of the apparatus

(*Continue*)

Table 2.1 (Continued)

Things
6. Target
7. Source of modification
8. Detectors
9. Tools
10. Data generators

Marks and the Manipulation of Marks
11. Data
12. Data assessment
13. Data reduction
14. Data analysis
15. Interpretation

(b) Systematicity

One thing that Hacking's view lacks is *structure*. It is as if he gave us the vocabulary of scientific practice without a grammar to go with it. But clearly these elements combine and interact with each other in a systematic way, and various suggestions have been made on how to describe that systematicity. Imre Lakatos (1970b), even though he only deals with propositions, makes an important distinction between the "hard core" and the "protective belt" of a research program, and describes the heuristics that guide the development of the program. David Gooding (1990, 1992) provides a diagrammatic representation of how elements of scientific practice fit together in a flow of experimental decisions.

(c) Styles

It has also been commonly recognized that there are *different* ways in which the elements of scientific practice are brought together. This has often been termed a matter of "style"—for instance, by Ludwig Fleck (1935/1979), Ian Hacking (1992b), and Alistair Crombie (1994). Similar ideas are advanced by John Pickstone (2000) in terms of "ways of knowing".[4]

(d) Purpose

What is *implicit* in all the ideas presented so far is the fact that scientific work consists in actions carried out by agents. Before we can truly understand epistemic activities, it is crucial that we make a full and coherent account of the epistemic agent. Gooding (1992) makes a strenuous and effective advocacy of "putting agency back" into descriptions of scientific activity, as do Andrew Pickering and Hans-Jörg Rheinberger. Quite memorably, Pickering (1995a, p. 23) says that we ought to describe how "the contours of material and social agency are mangled in practice, meaning emergently transformed and delineated in the dialectic of resistance and accommodation".[5]

In thinking about the basic attributes of scientific agents, first we need to recognize teleological or purposive behavior in the agent, at least in terms of instrumental rationality. The most basic thing about an agent is that she carries out her intentions, which are formulated on the basis of her desires and beliefs: The agent takes the kind of actions that she believes will contribute to the satisfaction of her desires. That seems to be the "standard story of action" in philosophical accounts, as Jennifer Hornsby (2004, p. 2) calls it. Hornsby criticizes this account as not giving a truly active role to the agent, but within the kind of philosophy of science dominant in the Anglophone world, even recognizing properly that the epistemic agent has *desires,* rather than just beliefs, would already be a distinct advance! What kind of desires do epistemic agents have? Many types of pleasures motivate scientists, and human beings in general, including physical comfort and sensual well-being, abstract and concrete understanding, love and conviviality, self-esteem, security, legacy, and a sense of beauty, order and coherence. It is important to keep in mind such plurality of desires. And for a detailed understanding of scientific practice, we need to think in terms of desires that are sufficiently concrete, and see more clearly how actual scientific activities are shaped by various proximate aims that help agents achieve their ultimate desiderata.[6] Harry Collins and Martin Kusch (1998, esp. ch. 3) give a very helpful discussion of "the shape of actions", analyzing particular forms of action defined via specific intentions.

(e)   Freedom

We also need to find better ways of describing agents with genuine *freedom,* giving real meaning to words like "choice" in the phrase "theory-choice" and "decision" in "decision theory". In the typical analytic philosopher's picture, the scientist only enters as a ghostly being that either believes or doesn't believe certain descriptive statements, fixing his beliefs following some rules of rational thinking that remove any need for real judgment. All the things that do not fit easily into this bizarre and impoverished picture are denigrated as pieces of "mere" psychology or sociology. We need a more serious understanding of the scientist as a real agent, not as a passive receiver of facts or an algorithmic processor of propositions. Valuable clues in this regard might be found in the works of various authors, ranging from Marjorie Grene (1974) on the knower as an agent to Stuart Hampshire's (1959/1982) reflections on the relation between thought and action.

(f)   Capabilities and Limitations

We do not exercise our freedom *freely,* as it were, but only under the constraints of our limited capabilities. Hornsby (2004, p. 21) complains that in the standard story of action, "not only are agents removed, but also their capacities cannot be recorded". This gets at something similar to Michael Polanyi's emphasis on the role of skills in scientific work (1958/1962, ch. 4). The consideration of human capabilities affects many philosophical issues, such as observability, testability, simplicity, and incommensurability—and therefore also realism, demarcation, confirmation, theory-choice, scientific revolutions, and so on. Capabilities have much to do with scientific rationality in general, because rational decisions should be based on an accurate sense of the agent's own capacities and skills. For Polanyi, skills were very much in the tacit dimension, and a key aspect of tacitness is the *embodiment* of knowledge. To be sure, there are mental skills, but many of the important skills for scientific work are firmly located in the body, and therefore not reducible to beliefs or knowledge in the form of propositions.[7] On the other side of the coin from our capabilities are our limitations.

Part of this picture of capabilities/limitations is the expectations with which and within which we act. Expectations concern the future, which is all-important in the context of action—as John Dewey (1917, p. 12) put it, "we live forward". But expectations are not simply beliefs about the future. They are often not beliefs at all, if by "belief" we mean a conscious assent to an articulated proposition. Expectations often exist on the "horizon", as Edmund Husserl would have it, or in the tacit dimension, according to Polanyi. They can even consist in an absence of any particular belief— I may be just walking along as normal, and my expectations involved in that activity will not be exposed or even formulated until they are met with something incoherent with them, such as the tremors of an earthquake, or the left arm grabbed by an excited old friend, or a gaping hole in the pavement. Scientific practice is also full of that sort of expectations, sometimes guiding our activities smoothly, sometimes preventing certain activities, sometimes making us attempt something repeatedly without a clear sense of why.

III.   FRAMEWORK FOR ACTIVITY-BASED ANALYSIS

III.1.   Activities and Systems

As seen in the overview just given, the demand to talk about practice has been bursting through the seams of analytic philosophy. The task I have set myself is to create a more commodious structure for housing these bursting insights, and I have made a few preliminary attempts in that direction. For the sake of consistency, I will begin by quoting from my latest attempt in my 2012 book (pp. 15–16): "I propose to frame my analyses in terms of 'systems of (scientific) practice' that are made up of 'epistemic activities'. . . . An epistemic activity is a more-or-less coherent set[8] of mental or physical operations that are intended to contribute to the production or improvement of knowledge in a particular way, in accordance with some discernible rules (though the rules may be unarticulated). An important part of my proposal is to keep in mind the aims that scientists are trying to achieve in each situation.[9] The presence of an identifiable aim (even if not articulated explicitly by the actors themselves) is what distinguishes activities from mere physical happenings involving human bodies, and the coherence of an activity is defined by how well the activity succeeds in achieving its aim[10] . . . . Epistemic activities normally do not, and should not, occur in isolation. Rather, each one tends to be practiced in relation to others, constituting a whole system. A system of practice is formed by a coherent set of epistemic activities performed with a view to achieve certain aims. . . . Similarly as with the coherence of each activity, it is the overall aims of a system of practice that define what it means for the system to be coherent. The coherence of a system goes beyond mere consistency between the propositions involved in its activities; rather, coherence consists in various activities coming together in an effective way toward the achievement of the aims of the system. Coherence comes in degrees and different shapes, and it is necessarily a less precise concept than consistency, which comes well-defined through logical axioms".

III.2.   Aims and Coherence

The interrelated notions of aims and coherence are central to the framework I am proposing, and they need further elaboration. To consider more carefully what coherence means, and to distinguish it sharply from logical consistency, it is helpful to think about the simplest kinds of activities that one finds in scientific practice. So consider, say, the act of lighting a match—the root of so much that took place in the history of chemistry![11] Most people can probably bring up the memory of learning how to do this simple act, which actually takes some skill and coordination (and it is surprising to observe how many people actually can't do it very well). One hand holds the matchbox with the rough strip facing the other hand; the other hand holds the match tightly, just so, and pulls the head of the matchstick across the rough strip on the box (no, no, the correct move is to push it), at an appropriate angle and at the right speed, and hold it still once the flame comes on. My two hands need to be highly coordinated in order to accomplish this task; in fact a large number of my muscles (I don't know anatomy well enough to tell you how many) need to be carefully orchestrated. Much else needs to be coordinated at the same time: While doing this hand-motion, I should not be running or blowing on the match, or have my face too close to it, or have anything flammable around other than what I'm trying to light; the speed and strength of my hand-motion need to be adjusted according to the dryness of the match-head, the roughness of the strip, the humidity of the air, and so on. All this is a good illustration of what *coherence* means in the realm of action. It is not at all reducible to the consistency between any propositions that might be involved in the act. What defines the coherence of match-lighting is whether all the things that I do (or don't do) lead to the match being lit.

Now, while such coherence pertains to each *act* of match-lighting, it is more interesting to consider the coherence of the *activity* of match-lighting. While an individual act may be performed in a haphazard way and even succeed in its aim somehow, to call something an "activity" implies a routinized and repeated performance of the act, which the agents involved carry out according to a reasonably fixed set of rules governing their attempts to achieve the aim of the activity.[12] When we consider activities (rather than acts) there are, then, two further aspects to coherence: the various rules (spoken or unspoken) governing the activity should exhibit good synergy, at least not counteract each other; and there should be a harmonious relation of learning and adjustment between earlier and later instances of the activity.[13] So coherence is going to be multi-dimensional, even in relation to a single epistemic activity; the picture is sure to get further complicated when we consider whole systems of practice.

Before discussing systems, however, let's consider more carefully the aim of an activity that defines its coherence. If we asked someone who is trying to light a match, "What are you trying to achieve?", we may receive two different kinds of answers. The answer may be "I'm trying to light a match", or "I'm trying to get a combustion-analysis of an organic compound going". Both are cogent answers, but they get at two different kinds of aim. The first answer addresses what I will call the *inherent purpose* of the activity:[14] Getting the match to light is the whole point of the activity itself, regardless of why one is engaged in that activity—that may be to light a Bunsen burner with it, or to burn down a house, or just to watch and admire the marvelous process that combustion is. These latter reasons might be called the *external functions* of the activity—namely, what various agents might want to achieve by means of the successful execution of that activity. An activity, as I conceive it, is partly *defined* in terms of a specific aim, which is what I have just called its inherent purpose; match-lighting is not match-lighting if one does not at least intend to light a match. The inherent purpose of an activity exists regardless of any external functions that the activity may or may not serve. Both inherent purpose and external function fall under the rubric of aims, so the talk of aims needs to be disambiguated accordingly.

The consideration of external functions brings us to a consideration of how different activities can come together into a system of practice, where they can serve larger or more complex aims. A system is something *put together* by agents, and maintained by specific effort; it only exists because someone upholds it by means of some mechanisms for propagation and maintenance. A system of practice is not guaranteed to be completely coherent. It will be *more-or-less* coherent; because hardly anyone would persist in trying to maintain a system that has little coherence to it (any such attempt would be unsuccessful, by definition). But what exactly does the coherence of a system of practice consist in? The key here is what I propose to call *functional coherence,* which is achieved when the external functions of the various activities that constitute the system are coordinated effectively. But effectively for what? This is where the aims of the system come in: systems of practice are crafted in order to achieve certain aims which go beyond the inherent purposes of the activities that are pulled together to constitute the system.

These aims may be easier to discern in systems of applied science; for example, in industrial pharmaceutical research or soda manufacture, there are clear practical aims such as better yield, economy, or safety. But such overall aims are present in systems of pure science, too. For instance, Lavoisier worked to create a system of chemistry aimed at understanding the composition of various substances, and explaining chemical reactions in terms of composition. The main activities of this system included collecting gases produced by chemical reactions, measuring the weights of the ingredients and products of reactions, tracking chemical substances through those weight-measurements, combusting organic substances for analytical purposes, and classifying compounds according to their compositions. Each of these activities had their inherent purpose, but they also had external functions that came together well to serve the aims of what I have elsewhere called "composition-ist chemistry".[15]

The coherence of a system of practice has several layers to it: (i) Is each constituent activity of the system coherent within itself, in serving its inherent purpose? (ii) Do the inherent purposes of different activities constituting the system not interfere with each other? (iii) Are the presuppositions or implications of different activities consistent with each other? (iv) Are the external functions of the activities coordinated so as to achieve the overall aims of the system?[16]

III.3.   Levels and the Activity–System Relation

In the remainder of this paper I would like to illustrate how the analytical framework I have just laid out can be applied. Before that, however, a couple of clarifications are in order. First, it may seem difficult to make a sharp distinction between an epistemic activity and a system of practice. The activity–system distinction is indeed relative and context-dependent; a system of practice, if it has a clear inherent purpose, can be taken as a single activity that may form part of a larger system.[17] For example, organic combustion-analysis is a system of practice consisting of various other activities: burning the target substance; absorbing the combustion-products using other chemicals; weighing the resulting compounds with a balance; making percentage-calculations; and so on. But it can also be regarded as a unified epistemic activity with the inherent purpose of determining the composition of organic substances in terms of the relative abundance of oxygen, hydrogen, and carbon; as such, it is one of the main activities constituting the Lavoisierian system of chemistry. In contrast, the Lavoisierian system as a whole did not have anything that can be identified as its inherent purpose, though it (and its practitioners) did have various aims. To take a different kind of example: The game of soccer does not have an inherent purpose, so it is not an activity in the strict sense as I define it here. This is not to deny that some particular aspects of soccer have inherent purposes: For example, the inherent purpose of goal-keeping is to prevent the other team from scoring. But isn't winning the inherent purpose of the game? That depends on one's philosophy (some might say the real point is to participate, and to build teamwork or character, or to get exercise, rather than to win). And winning each game may be the main purpose for each team involved, but it is not clear that it is the main purpose for an individual player, or for the collective of everyone who is playing each game, or for the whole institution of soccer, or for the whole society that supports this institution.

It is also useful to stress the non-reductive nature of the analysis to be made in this framework. The structure of actions and processes is not atomistic in a reductive way, unlike the structure of things and statements. Each epistemic activity can be analyzed into components, but the so-called "component" activities are not necessarily simpler than the "whole" activity in an absolute sense, and the analysis can go on indefinitely, as those component activities themselves consist of other activities. For example, the activity of weighing-with-a-balance has component activities such as placing samples and weights on balance-pans, and reading the number off the scale. If we stop there, it may seem that we are getting to simpler and simpler activities as we continue in our analysis of actions, on our way to reaching a rock-bottom of atomic operations. But there are less convenient things in the picture. For instance, consider the activity of certifying the standard weights, which is an essential component of the activity of weighing-with-a-balance. This certification-activity may consist in ordering the weights from a reliable supplier, or comparing them to a more trusted set of weights, or checking them against certain natural phenomena (e.g., the weight of a certain volume of water at a certain temperature). Whichever option we go with, it is clear that this component activity is not simpler, in any clear sense, than the main activity of weighing with a balance. The relation between various epistemic activities is ultimately non-reductive and reticular, although in many situations we can gain useful insights from analyzing an activity into its *apparent* components. There is no lowest level of description, and no clear end to the process of activity-analysis. Most of all, I want to avoid implying a reductionist metaphysics in which a system is made by a simple addition of various activities that do not really have any connection with each other; even a brick wall is not built like that! Rather, the analysis should be carried out wherever, and as far as, it is productive.

IV. HOW TO APPLY THE FRAMEWORK

How can the framework of activities and systems be employed in our work in history, philosophy, and sociology of science, and what difference can it make?

(1) The most straightforward use of the framework is its use as a descriptive tool in historiography. I hope it will provide a welcome relief to those who have felt that the theory–observation dichotomy (even with theory-ladenness), Kuhnian paradigms, Lakatosian research programs, and so forth, were limited or ill-fitting categories for describing the bits of past or present science that they were interested in. I also hope that the template of activities and systems will draw attention to various aspects of scientific practice that historians have often passed over, and highlight various connections between different strands of scientific work that may not have been obvious so far. I have certainly begun to benefit from all of those effects in employing the activity–system framework in my recent historical work (e.g., Chang, 2012, ch. 1–3).

(2) One step away from concrete historiography, the activity–system framework will also help inform our philosophical views on the nature of scientific knowledge. The habit of analyzing science in terms of actions will make us recognize (to the extent that it is actually the case) that even the most abstract aspects of science are rooted in doings. For example, take something that is usually considered by philosophers perhaps as far removed from actions as possible: the definition of a concept. Instead of thinking about the abstract nature of a definition, we can consider what one has to *do* in order to define a scientific term: formulate formal conditions for its correct verbal and mathematical use; construct physical instruments and procedures for measurement, standard tests, and other manipulations; round people up on a committee to monitor the agreed uses of the concept; and devise methods to punish people who don't adhere to the agreed uses. In one stroke, we have brought into consideration all kinds of unexpected things, ranging from operationalism to the sociology of scientific institutions. Examples can be multiplied easily. Confirmation can be understood better through via various *processes* of hypothesis-testing. The nature of models is best discussed by considering the actions that take place in modeling. And going beyond thinking about scientific explanation in terms of logical relations between *explanandum* and *explanans*, we can consider how the demand to explain something arises, and how it is best met. Michael Scriven (1962) has already given such an analysis of explanation, though it seems mostly forgotten nowadays.

(3) Because of the central place given to the consideration of aims (including the definition of coherence in terms of aims), my framework also provides a natural basis for normative evaluations. It helps us understand the evaluations that scientists make of their own and each other's practices, and also helps us make our own independent judgments on scientists' work. Evaluation is complex business, and I think the structure of activities and systems that I have outlined helps us make sense of the complexity involved. Overall evaluative notions like truth and success need to be dissected significantly in order to become applicable to systems of practice, and the talk of the aims of science and general epistemic values/virtues need to be concretized in the context of each activity and system we are considering. Thinking about evaluation in science in this framework will give us an entirely different view on a whole range of issues in the philosophy of science, including theory-choice, scientific change and progress, realism, and pluralism.[18]

(4) Activity-based analysis has an application reaching well beyond philosophy of science narrowly conceived. In that sense, my ambition is comparable to that of Bridgman or the Vienna Circle, reaching beyond philosophy into the rest of life. This is a very brief glimpse at the kind of shape that philosophy can take after the practice turn. Most standard philosophical topics can receive a new lease of life by being re-conceptualized fully in terms of activities. When you have a philosophical issue, ask what kinds of activities are relevant to it. For each activity, query its purpose, and ask who is carrying it out, and so on. This rapidly produces a checklist (Table 2.2), a set of questions to apply to any given topic. I can almost guarantee that applying this recipe to any philosophical topic that seems irrelevant or dull will generate a take on it that is quite a lot more interesting.[19] There is nothing surprising about this broad applicability of activity-based analysis: When the "practice turn" in science studies recognized science as practice, we came to recognize more clearly the continuity between science and the rest of life. Along with that, we can also recognize how mainstream analytic philosophy had come to see life along the lines of its own impoverished image of knowledge and science.

Table 2.2 Checklist for Activity-Based Analysis (Recipe for the Transformation of Boring Philosophical Issues)

- Activity: What is being done in the practice in question?
- Aims: What is the inherent purpose of this activity, and what external functions does it serve?
- Systematic context: Does the activity constitute part of broader systems of practice?

(Continue)

Table 2.2 (*Continued*)

- Agent: Who is doing the activity?
- The second person: To/with whom?
- Capabilities: What must the agent be capable of, in order to carry out this activity?
- Resources: Which tools are necessary for this activity to be successful?
- Freedom: What kind of choices does the agent make?
- Metaphysical principles: What must we presume the world to be like, in order for this activity to be coherent?
- Evaluation: Who is judging the results, and by what criteria (in addition to coherence)?

NOTES

1. This is a term explicitly used by Hans-Jörg Rheinberger (2005, p. 409).
2. For my preliminary attempt on this, see Chang (2011a). On Wittgenstein on grammar, see Baker and Hacker (1985): Wittgenstein's *Philosophical Grammar* itself is not explicit on what is meant by "grammar".
3. In the course of resisting the "singular science model", John Pickstone (2007, pp. 490–491) also aligns himself with the "practice turn" in the history of science, and explicitly opposes the assumption that "formal knowledge is prior to technical action".
4. For my attempt to express along Pickstonian lines the ideas presented in this paper, see Chang (2011b).
5. Joseph Rouse's notion of practice (1996) instructively goes *beyond* an agent-focused view, but I am not able to reach that stage of sophistication in this paper.
6. For instance, if someone is driven by the desire for the Truth, what will she do? There isn't an activity of truth-seeking per se, so she will need to engage in some concrete activities that she deems as truth-conducive. This reminds me of Richard Rorty's pronouncement that there is no such thing as the "love of truth" (1998, p. 28): "what has been called by that name", Rorty says, "is a mixture of the love of reaching intersubjective agreement, the love of gaining mastery over a recalcitrant set of data, the love of winning arguments, and the love of synthesizing little theories into big theories".
7. For recent literature on this topic see, for example, Noë (2005), and references therein.
8. In a different publication (Chang & Fisher, 2011, p. 361), I used the word "system" here instead of "set"; that usage is avoided here, as it is not coherent with the explication of "system" below.
9. Note here the similarity with Léna Soler's notion of an "argumentative module", which is "individuated and defined as a unit on the basis of its aim: on the basis of the question it is intended to answer, or the problem it tries to solve" (Soler 2012b, p. 235).
10. The dependence of success means that coherence is not a matter entirely within our control. We might distinguish *intended* vs. *achieved* coherence.
11. In life, the most primitive acts come in the satisfaction of physiological urges (drinking a sip of water, etc.); these acts also require the same kind of coherence as I discuss here.

12. The term "operation" (used in the passage quoted from my earlier work) is ambiguous between act and activity.
13. The latter point is helpfully dissected according to different types of action in Collins and Kusch (1998, ch. 3).
14. Cf. the morphological condition of "resultative impulse" in Jean-Michel Salanskis' theory of action in Chapter 1 of this volume.
15. See Chang (2012, sec. 1.2.3) and Chang (2011b).
16. It is interesting to note that the cluster of Lavoisierian activities mentioned above were not so coherent with the Lavoisierian activity of caloric-based analysis or oxygen-based acid-theorizing.
17. In my book (Chang, 2012, pp. 16–17) I made the following statement on that issue, which is correct but exaggerated: "The distinction is only relative and context-dependent. In each situation in which we study a body of scientific practice, I am proposing to call the overall object a *system;* when we want to study more closely more specific aspects of that system, we can analyze it into different subordinate *activities.* What we take as a whole system in a given situation may be seen as a constituent activity of a larger system, and what we see as a constituent activity in a given situation may in a different situation be analyzed as a whole system made up from other activities. . . . In this way, my framework is applicable at all levels, and can be zoomed in and out to suit any level that we want to focus on. At each focus-point of analysis, we can call the overall practice 'system' and its constituents 'activities' . . . without intending to stick those categorical labels to anything on a permanent basis".
18. For my own initial steps in this direction, see Chang (2012, ch. 4, 5).
19. For instance, see Chang and Fisher (2011) on the ravens paradox.

Toward a Framework for the Analysis of Scientific Practices

Commentary on "Epistemic Activities and Systems of Practice: Units of Analysis in Philosophy of Science After the Practice Turn", by Hasok Chang

Léna Soler and Régis Catinaud

In his contribution to the present book, Hasok Chang extends a project initiated in several of his recent publications (Chang, 2011a, 2012): to build a broad-ranging and operative framework for the analysis of scientific practices. The framework has a broad scope: It is intended to apply to any scientific practice, whatever the scale and whatever the field. The framework is operative, in the sense that it helps to analyze and reveal previously unnoticed important features of these diverse practices.[1] In the present commentary, we first briefly reconstruct Chang's reading of the practice turn—what lessons can be drawn from this turn, and what remains to do now, according to Chang as we understood him. Next, a brief overview of the main pieces of Chang's framework is offered. Subsequently, some general difficulties of any project of activity-based analysis of science are emphasized. Finally, some aspects of Chang's particular proposals are discussed. The discussion is more especially focused on the nature and place of aims in Chang's account.

Let us start with a reconstruction of Chang's answer to the central question of this book.

I. WHAT DID THE PRACTICE TURN TEACH ANALYSTS OF SCIENCE ABOUT SCIENCE AND WAYS TO STUDY SCIENCE?

The negative side of Chang's answer is completely congruent with the dominant discourse of advocates of the practice turn. Negatively, analysts of science in practice learned that both the methods and the picture of science issuing from the "mainstream analytic philosophy" (p. 77) of science must be questioned. They came to realize that this philosophy only pays attention to the propositional dimension of science. They also came to realize that traditional philosophy of science treated the epistemic agent as a "ghostly being" (p. 70)—that is, treated scientists, or more generally rational beings, as sort of algorithms which, once fed with experimental facts and background established knowledge, should deliver one single universal solution regarding theoretical propositions to believe.[2] So globally, Chang—as many others before him—criticizes the traditional approach to science as "unduly limited" (p. 67), too narrow, too idealized, inattentive to pluralism and to context-dependency. The result—expressed in Chang's diplomatic way—is an "impoverished image of knowledge and science" (p. 77).

Positively, what did scholars interested in science learn? The focal point of Chang's answer is captured by the claim: "all scientific work, including pure theorizing, consists of actions" (p. 67). When analysts of science look at science as an activity, they realize the *crucial importance* of aspects previously neglected or completely ignored by traditional analytic studies of science. In this vein, the practice turn taught them how important it is—at least if the aim is to give a not-too-distorted picture of science, scientific outcomes included—to take into account aspects such as the details of experimental processes, including material, non-propositional aspects; tacit skills and tacit expectations; the limited capacities of real epistemic agents; the bounded and contextual character of actors' judgments; the purposive behavior of these actors, as well as the rich plurality of their aims, desires, evaluative criteria, and expectations; and, of course, human freedom. So globally, people interested in science have already learned that science, scientific actors, and scientific work are much more diversified and complex than the "typical analytic philosopher's picture" (p. 70) suggests. They furthermore learned that there is much more "continuity between science and the rest of life" (p. 77).

Given this appreciation of the situation, what remains to do now? According to Chang, it remains to articulate "a *structured* and *precise* philosophical framework for thinking and talking about scientific practices" (p. 67). This is the "kind of shape" (p. 77) Chang would like to give to philosophy of science after the practice turn. Let us have a closer look at Chang's framework for activity-based analysis.

II. GENERAL STRUCTURE, MAIN ANALYTICAL ELEMENTS, AND EPISTEMOLOGICAL STATUS OF CHANG'S FRAMEWORK FOR ACTIVITY-BASED ANALYSIS

Figure 2.1 gives a graphic representation of the overall structure of Chang's framework. At the upper level, there is the "system of practice", represented by the external bold circle. At an intermediate level, a system of practice can be decomposed into different "activities", represented by the dotted circles. At the lowest level, each activity can be further analyzed into several sub-activities (the small circles on Figure 2.1) that Chang sometimes calls "operations" (Chang, 2011a, 2012).

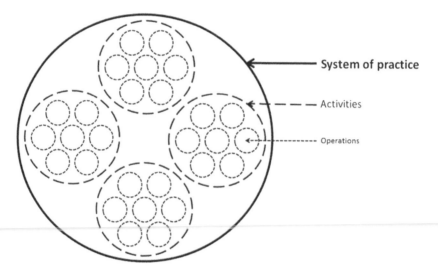

Figure 2.1   The structure of Chang's framework for activity-based analysis

The circles of all sizes depicted in Figure 2.1 fall under the rubric of "action". We call them "action-type units". Three ingredients are central to Chang's characterization of any action-type unit: (a) aims (in particular, the inherent aim); (b) relatively fixed rules governing the attempts to reach the aims; and (c) coherence.

Before explaining these ingredients further, it is important to clarify the epistemological status of Chang's three-level decomposition. First, the decomposition is relative to the aims of the analyst: The decision to categorize a given reality as, say, a *system of practice* rather than an *activity* or an *operation*, depends on the project of the analyst. It is not imposed by some inherent property of the reality under scrutiny. Consequently, one and the same targeted reality might legitimately be categorized as a system of practice, or as an activity, or as an operation.[3] This does not mean that all categorizations are equivalent. It simply means that the relevance and adequacy of the analyst's categorization must be assessed against his aims.

Second, Chang's choice to introduce exactly *three*, and not more or less than three, hierarchic levels and specific names for each, is just a conventional, convenient artifice of presentation. In fact, an indefinite number of levels of analysis could be introduced. Any unit could in principle be decomposed into sub-units, and this indefinitely, without any "objective end" or ultimate fundamental level.[4] Taking that for granted, we could say that there is no "quantum of action" in the realm of human activity. Reciprocally, any unit can be thought of as part of an encompassing higher-level unit. So all in all, we have a sort of fractal scheme.[5]

(a) **Aims.** Each action-type unit is primarily defined by its "inherent purpose" (p. 74). The inherent purpose is, in Chang's framework, the most immediate or proximate aim of a given unit of action. Beside its primary aim, an action-type unit is also associated, in Chang's framework, with multiple more distant aims (or "ultimate desiderata" [p. 70]), called "external functions (Chapter 2, this volume, p. 73)". All in all, an action-type unit must be thought of as being involved in a "teleological structure" (Schatzki, 1996), in relation to other action-type units.

(b) **Rules.** Chang's "rules" of an activity refer to sufficiently stabilized and shared ways of doing through which a given aim is typically achieved by members of a given community.[6] The rules are described as "discernible" (p. 72), "reasonably fixed" (p. 73), and possibly "unarticulated" (p. 72). As an example, Chang decomposes the movements one typically executes in order to light a match ("one hand holds the matchbox with the rough strip facing the other hand" [p. 72], etc.), and re-conceptualizes them as a set of rules for successful action (in the present case: the match indeed lights). Extrapolating to a system of practices associated to a given scientific field during a period of "normal science", we could identify a "disciplinary matrix", as defined by Thomas Kuhn, as a set of rules in Chang's sense—in this case, rules typically followed by scientific practitioners in order to achieve the aim of acquiring new scientific knowledge.

(c) **Coherence.** "Coherence" names, in Chang's characterization, a measure of the quality and efficiency of an action-type unit: a measure of the synergy of different actions in achieving the aims of an activity or system (p. 72–73). Intuitively, coherence refers to a good fit between various items—among which are non-propositional ones—constitutive of the action-type unit. The more a unit is coherent in this sense, the more the aims are systematically and efficiently achieved. This notion of coherence is vaguer and more difficult to analyze than the logical notion of consistency between propositions. Several authors of the practice turn have stressed the need for an enlarged notion of coherence of this kind.[7]

In Chang's proposal, the coherence of an action-type unit is the result of a complex equilibrium involving aims and rules. To assess the coherence of a system of practice—and hence its quality and efficiency—the analyst has to assess and balance multiple types of fits: fits between *various aims,* fits between *different rules,* and fits between *aims and rules.* To be more precise, part of the coherence of a system of practice lies in the good "hanging together" of the multiple aims involved at all levels of this system—including the fit between the different inherent aims *of the different component activities,* and the fit between the *inherent* aim and the *distant* aims *of each activity.* Another part of the coherence of the same system of practice lies in the good fit between the different rules constitutive of *each set* through which the inherent aim of each component activity is pursued. The multiple rules constitutive of the set through which a given aim is pursued—understood as interrelated means involved in the successful achievement of this aim—should at least "not counteract each other" (p. 73). Beyond, they can exhibit a *more or less good* synergy. The more a set of rules enables the achievement of a given aim systematically and efficiently, the better the coherence *between the set of rules and the aim*. The *overall* coherence of the *system* of practice considered *as a unit* lies in these relations *taken together*.

(d) **The ultimate aims of Chang's framework.** What kind of work is Chang's framework supposed to accomplish? Let us answer the question by applying reflexively the categories of Chang's framework to his own research—given that Chang's philosophy of science might be viewed as one system of practice among others.

Above, we quote Chang writing that he wanted to articulate "a *structured* and *precise* philosophical framework for thinking and talking about scientific practices". This can be identified with the *inherent purpose* of *one of the activities* which compose Chang's system of philosophical practice.

What are the "external functions" of Chang's framework? In other words, what are the "ultimate" aims valued by Chang, against which the coherence—the quality and efficiency—of his framework should be assessed? Chang claims that his framework is helpful, and provides new insights, at least with respect to the four following distant aims: (i) to describe bits of past or present science; (ii) to elaborate "philosophical views on the nature of scientific knowledge" (p. 76); (iii) to understand the various kinds of normative evaluations made by actors of science; and (iv) to develop independent normative judgments about past science, for example—to mention a very traditional question—normative judgments of the analyst about the comparative merits of competing past theories such as the phlogiston versus the oxygen theory (see Chang, 2012).

III.   SOME INHERENT DIFFICULTIES OF CHANG'S PROJECT AND ANY ACTIVITY-BASED ANALYSIS OF SCIENCE

As all units of analysis of Chang's framework fall under the rubric of "action", this leaves us with the task to investigate a fundamental issue, namely: What is an action? Chang's insistence on actions as the relevant unit of analysis might perhaps appear obvious—if not tautological—once the need of a turn to *practice* has been recognized. But looking at the works commonly associated with the practice turn, it becomes less obvious.[8] Certainly, most actors of the practice turn would agree with Chang's claim that "scientific work consists in actions carried out by agents" (p. 69). However, not so many researchers of the practice turn stress explicitly that *consequently*, one of the new tasks of philosophers of science should be an investigation of what an action is, and not so many researchers of the practice turn directly address this issue. Most of the time, the notion of practice or activity involved in the works issued from the practice turn has to be reconstructed from what the author is doing and saying. Given that, Chang's project really fills a gap and is highly welcome. Chang's first "few preliminary attempts" (p. 71) provide useful conceptual tools, benchmarks, and promising programmatic suggestions.[9]

That being said, we are not sure that Chang's framework as it stands, and more generally Chang's project, can provide, as Chang hopes, a "relief" (p. 76) to those who take the practice-based approach of science seriously. This is because the realm of action is a highly difficult and complex subject. It involves tremendously hard issues, well-known by philosophers and social scientists who have been interested in human actions in general. Chang excels in the art of making the complex appear simple, but the questions we have to face are all but simple. Let us mention some of these difficult questions.

(a) On what basis is an investigator going to individuate an empirical token *as an action* and to categorize a token of action as *this type* of activity *rather than another type* of activity?

(b) How is the analyst going to identify the motives, aims, values, desires, or affects that shape human behavior, which are possibly tacit, unarticulated, unconscious, if not even undetermined? Will the analyst ask the actors to give their own account of the action, and if so, what method will be used?[10] To what extent will the analyst rely on and take for granted the actors' own account of what happened?

(c) How are we going to think about the relations between the individual level and the collective level of an activity?

(d) And what about still harder and perennial issues related to human freedom? For example, should we accept as genuine actions only free actions? How are we going to define a free action? In practice, how are we going to decide whether a given human behavior is a free action or not? And so on.

Chang's position regarding question (a) shows the importance of aims in his framework. To the first part of question (a), Chang provides a definite, classical, although possibly controversial, answer: "The presence of an identifiable aim . . . is what distinguishes activities from mere physical happenings involving human bodies" (p. 72). The purposive dimension is *the* necessary and sufficient criterion to categorize an observable physical sequence *as an action*. Moreover, the inherent purpose is a crucial element for the definition of an action-type unit as *this type* of action *rather than another type*. On questions (b) and (c), Chang says nothing, as far as we know, neither in his contribution to the present book, nor in other writings. As for issues related to (d), Chang stresses their importance and the need to face them,[11] but does not elaborate positive proposals to cope with them.

The previous list of questions (a) to (d) is just a small sample of the hard problems we must face if we take seriously the idea that science must be studied as an activity. These problems are not just Chang's problems: they are—or should be—problems raised by the very perspective of the turn to practice.

Having pictured the main elements of Chang's framework and stressed general difficulties of projects of this type, let us turn to more proximate and modest aims, and discuss some aspects of Chang's characterization of an action-type unit. First we make a brief suggestion about Chang's "rules" of an activity, and then we consider some issues about aims.

IV.   A BRIEF SUGGESTION IN RELATION TO CHANG'S "RULES" OF AN ACTIVITY

For an activity such as lighting a match, it is not tremendously difficult to identify and articulate "a reasonably fixed set of rules governing the attempts to achieve the aim of the activity" (p. 73), which is to light the match. However, consider the category of an "epistemic activity" (p. 73), also put forward by Chang. What is the "reasonably fixed set of rules" which is supposed to govern the agents' attempts to achieve their aim, defined as the "production or improvement of knowledge" (p. 72)? Even if we let aside the difficulties related to the identification of these rules—which would take us back to the traditional issue of scientific method, possibly in the plural— we feel that there is a *significant difference* between match-lighting and epistemic activities. To generalize, we suggest that a distinction should be elaborated between very standardized, rigidly codified activities (like match-lighting) and by-nature bound-to-be-open, flexible, evolving and creative activities (such as knowledge production).

Scientific *research* is of the second sort, especially if we consider situations akin to Kuhn's "revolutionary science" or if we consider innovative epistemic projects. Of course, what is at stake is not a binary distinction but, rather, a whole spectrum of possibilities. Nevertheless, the "rules" of scientific research do not seem to be well-described as "fixed"—at least not if we want, as Chang does, to characterize science at a sufficiently concrete level. In any case, the "rules" of scientific research are not fixed in the same sense, and at the same degree, than the rules of match-lightning are: they are *bound to evolve* with the advance of science, in close correlation with the historical evolution of the concrete aims of particular scientific activities and the associated evolution of what it means for scientific activities to be successful. In his latest book, Chang (2012) seems to be perfectly aware of these kinds of evolutions. Our suggestion is that it would be helpful to elaborate the distinction between *more or less rigid and more or less open* activities in terms of rules, aims, and criteria of success.

V.   THE INSTANTIATED AIMS OF A SINGLE ACTOR INVOLVED IN A PARTICULAR ONGOING ACTION? OR THE ABSTRACT AIMS OF A TYPE OF ACTION?

We agree with Chang that aims—or something broader akin to aims: teleology, intentionality, what is valued and desired . . .—should be a pivotal ingredient of the characterization of an action-type unit. But further reflection is needed about aims: about their nature; about the methods by means of which the analyst can identify them; about their relation to the identity of a *type* of action; and about the relation between aims at the *individual* level and aims at the *collective* level. In this section and the following, we point to a selection of problematic points.

Chang sometimes talks as if he is concerned with the specific instantiated aims of well-identified epistemic agents—for example, when he writes: "An important part of my proposal is to keep in mind the aims that scientists are trying to achieve in each situation" (p. 72). But Chang also sometimes talks as if he is concerned with general, *impersonal types* of aims—in particular, when he introduces the *inherent* aim *of an activity,* without any explicit reference to any particular agent.

Chang switches from the personal to the impersonal idiom, without addressing explicitly the issue of the differences and possible relations between the two. We think the issue should be addressed, because the analyst is not engaged in exactly the same kind of enterprise, and has to face partially different methodological problems, depending on whether he is concerned with the characterization of a *particular ongoing action involving well-identified individual actors,* or with the characterization of a (more or less general) schematic *type* of action.[12] Let us stress some important differences between these two perspectives, first focusing on the case of *individual* activities.

(i) **When the analyst intends to characterize a particular ongoing action involving a well-identified actor,** the analyst's task is to identify the aims of a particular individual in a given historical episode. For example, the task is to identify the inherent and ultimate aims of someone who, here and now, is scratching a match, or, say, the aims of Einstein as a physicist, when he introduced light quanta in physics in 1905.

At this personal and concrete level, the answer to the question "the aims *of who?"* is clear. What is not clear at all, however, is the *ontological status* of objects like "aims", "purposes", "intentions", "desires", or "values": Are aims definite, sufficiently stable realities that are supposed to be "in the head" of individuals? What is moreover equally unclear is the kind of influence that aims exert on the actions of particular individuals: How do often unarticulated, opaque, or unconscious aims exert a real, definite influence on the verbal and non-verbal behaviors of an individual?

If we turn from the ontological to the *methodological* point of view, the question reads: How is the analyst going to access the aims of concrete epistemic agents? Common answers after the practice turn are, typically: for *past* epistemic agents, rely on the available traces of their doings and sayings, and build a proposal that maximizes the coherence—this is the stance of historians of science; for *living* actors, also interact with them, observe them and ask them direct or indirect questions in order to identify their aims—this is the stance of ethnographers of science. With these common answers, we open the Pandora's box of well-known but enduring methodological problems in the social sciences, such as the influence of the analyst on the analyzed people and activities, the risk of injecting an artificial coherence in the picture because of the application of something akin to the so-called "principle of charity", under-determination, and so on.

(ii) **When the analyst intends to build a general, schematic characterization of a *type* of action,** the task and the difficulties do not coincide with the ones stressed above in (i).

In this second perspective, the analyst's task is, first of all, to identify the *impersonal, inherent* aim of a *type* of activity. For example, the inherent aim of "match-lighting", or "defining", or "doing physics". At this impersonal level, the answer to the question "the aims *of who?*" is far from being obvious. What kind of thing is the *impersonal* aim *of an activity*? Is there an impersonal *subject* of this aim? And if so, what is the epistemological status of such a subject? Clearly, the aims and the subject involved are abstract entities, idealizations. But what kinds of idealizations are they? Are they exemplars in a Kuhnian sense? Or are they supposed to describe "average behaviors"—that is, what happens "most of the time" in a vague or possibly statistically quantified sense? Are they ideal-types *à la* Max Weber, nowhere instantiated in the real world? What is their relation to "what real people really do"? These are issues that anyone interested in scientific practices as sets of actions should face.

VI.   AIMS OF *INDIVIDUAL* ACTIVITIES AND AIMS OF *COLLECTIVE* ACTIVITIES

So far, we restricted ourselves to *individual* activities. Additional problems arise if we turn to *essentially collective* activities such as, say, football, or scientific research as it is practiced today.[13]

A first problem concerns the identification of the *inherent* aim *of an activity* when this activity is *collective*. Let us illustrate the problem. Take the relatively simple, immediately graspable case of soccer mentioned by Chang. Chang writes: "The game of soccer does not have an inherent purpose, so it is not an activity in the strict sense" (p. 75). Such a claim cannot be equated with the expression of a common shared intuition which could be claimed without justification because of its widespread obviousness. The two authors of this commentary provide counterexamples. Reading the latter quotation, Léna Soler's intuition strongly protests: The inherent purpose of football is, according to her, to score and prevent the other team from scoring. Régis Catinaud's intuition disagreed both with Chang's and Soler's (and fluctuated during the elaboration of this commentary). Generalizing, we conclude that the identification of the inherent aim of a collective activity is sustained by intuitions that are susceptible to varying from one analyst to another.

Taking that into account, it *is all the more important, for the analyst, to clarify the basis on which the identification of the inherent aim of an activity has taken place.* Note, moreover, in relation to the remarks made in Section IV, that football is a highly standardized, rigidly codified activity. Things will not get easier for more creative and less rigidly codified collective activities such as scientific research, especially in revolutionary periods. But when dealing with such still more problematic cases, Chang does not provide more arguments. For example, Chang writes about chemical practices of the Lavoisierian revolution: "the Lavoisierian system [of practice] as a whole did not have anything that can be identified as the inherent purpose of the whole system" (p. 75); but no clue is given about the basis of this type of judgment. We think the basis of such judgments should be explicitly discussed and motivated.

A second problem concerns the relation between aims of *concrete individual actors* on the one hand, and the inherent aim and type of a *collective* activity on the other. In Chang's framework, the inherent aim is a crucial feature with respect to the definition of the *type* of an activity. For example, the intention to contribute to the production or improvement of knowledge is what primarily defines an activity as an *epistemic* activity. But how should this idea apply to an *essentially collective* activity? Consider a situation in which *some* members of a *collective* scientific research commit a fraud deliberately. Is this collective activity still an *epistemic* activity? More generally, the issue is the relation between the diverse interrelated aims of individual epistemic agents—which, to make things harder, are frequently unarticulated, often fluctuate in time, and may vary from one agent to another—and *the* inherent aim of a *collective* activity such as science.

Taking the two previous problems together, the question becomes: What are the criteria, or at least the clues, for the attribution of a determined inherent purpose to a *collective* action-type unit? As far as we know, this issue has not yet been considered by Chang in his writings. Chang seems to attribute inherent aims on an intuitive basis. But at least in cases where intuitions might diverge, this is problematic. In his contribution to the present book, Chang dedicates only one brief sketchy passage to considerations related to justification of inherent purpose attributions. The passage applies to the case of soccer. Chang relates the absence of an inherent purpose for soccer to the fact that *different* inherent purposes could be attributed to soccer *according to different points of view.*[14] If, for example, we ask what is the aim of soccer from the point of view of different members of our society, or from the point of view of different players or different teams who practice football, or from the point of view of the "whole institution of soccer", or from the point of view of "the whole society that supports this institution" (p. 75), we will possibly obtain different answers—for example, to win, or to participate, or to get exercise, or to build teamwork. Given that, Chang concludes, soccer has no inherent purpose.

We do not think that Chang's sketchy justification for inherent aims attribution provides the principle of a viable solution. The solution is not viable, to the extent that the multiplicity and variability of aims attribution according to different points of view is not at all a specificity of soccer, but potentially applies to *any* action-type unit—including individual activities. For example, Chang writes, as if it is obvious: "This is not to deny that some particular aspects of soccer have inherent purposes: for example, the inherent purpose of goal-keeping is to prevent the other team from scoring" (p. 75). However, according to some points of view, the inherent purpose of goal-keeping could be to get exercise, build teamwork, or even (imagine some corrupted goal-keeper) to let the ball go through. Hence the multiplicity and variability of aims attribution according to different perspectives cannot, in itself, be the reason why an action-type unit is characterized as deprived of inherent purpose.

We conclude that analysts of science for whom aims are important should explicitly elaborate some minimal criteria for inherent purpose attributions, both for *collective* and *individual* activities, and both with respect to *negative* attribution—no inherent purpose—and to *positive* attribution—existence and specification of a well-defined inherent purpose. With respect to this task, our suggestion is to distinguish between two levels. On the one hand, the level of an institutionalized, socially stabilized, and at least partially standardized activity, the characterization of which corresponds to an idealized and "normal"—in the double sense of "in use" and "normative"—scheme of activity, possibly associated with the mention of a specified group. On the other hand, the level of the actual performance of activities by particular individual actors. Particular individual actors usually are somewhat aware of the normal schemes of actions. For example, they know that normally, the primary aim, when playing soccer, is to score and prevent the other team from scoring. However, their actions might, in fact, display significant shifts from normal schemes. For example, in a given particular situation, one actor or one group of actors can, for some reason, pursue other, possibly incompatible aims—one team might aim to let the other team win, and so on.

VII.   BRIEF CONCLUSION

In previous writings and in his contribution to the present book, Chang takes a first helpful step in the direction of an activity-based characterization of science. No doubt, an activity-based analysis of science, although more promising, is much more difficult than a product-based analysis of science. In this commentary, we have attempted to specify some directions in which Chang's "few preliminary attempts" to build an activity-based framework could be elaborated and completed. To conclude, we would like to express our gratitude to Chang for having given a new impulse to our own attempts to elaborate a theory of practice suited for the analysis of science.

NOTES

1. For example, the framework will "highlight various connections between different strands of scientific work that may not have been obvious so far", and "give us an entirely different view on a whole range of issues in the philosophy of science, including theory-choice, scientific change and progress, realism, and pluralism" (p. 77).

2. The scientific agent has been reduced to an entity "that either believes or doesn't believe certain descriptive statements, fixing his beliefs following some rules of rational thinking that remove any need for real judgment" (p. 70), to "a passive receiver of facts, or an algorithmic processor of propositions" (p. 70).

3. "The activity–system distinction is . . . relative and context-dependent" (p. 74); "a system of practice, if it has a clear inherent purpose, can be taken as a single activity that may form part of a larger system" (Chapter 2, this volume, p. 74).

4. "Each epistemic activity can be analyzed into components . . . and the analysis can go on indefinitely, as those component activities themselves consist of other activities"; there is "no lowest level of description, and no clear end to the process of activity-analysis" (p. 75).

5. This seems to be a general feature of human practices. Several analysts of scientific practices have noticed something akin to the fractal character in different contexts. See, for example, Pickering (2012), where the author formulates the point in terms of "scale-invariant" patterns, or Soler (2012b, section 10.14.2).

6. "[T]o call something an 'activity' implies a routinized and repeated performance of the act, which the agents involved carry out according to *a reasonably fixed set of rules governing their attempts to achieve the aim of the activity*" (p. 73, italics added).

7. Hacking (1992a) spells out what coherence applies to, and gives a coherentist explanation of scientific "justification", in the case of laboratory science. The author introduces a notion of "enlarged coherence" between "thoughts, actions, materials and marks" and describes the resulting configuration as "self-vindicating". Pickering develops a coherentist conception of science according to which elements of heterogeneous types (material, behavioral, social, propositional. . .) co-mature, and in which a mutual stabilization is sometimes achieved. The elements then stand in a relation of reciprocal reinforcement, and this corresponds to a "scientific symbiosis", characterized as a "self-consistent", "self-contained", and "self-referential" complex package (Pickering, 1984a, p. 411).

8. On this point, see J. M. Salanskis' chapter in the present book and the commentary by J. L. Gangloff and C. Allamel-Raffin.

9. About the analysis of the concepts of action and activity, some very interesting insights can be found in the writings of activity theorists concerned by psycho-social and educational issues, especially Vygotski (1993) and his successor Leontyev (2006), who have tried to address the issue of the constitutive components of action.

10. This is of course only possible for living actors. On that point, Vermersch (1994) gives interesting insights. Pierre Vermersch claims that all the dimensions of an action actually experienced by an actor in the course of the ongoing process of the action are not immediately accessible to him, but can be rendered accessible by what Vermersch calls an "entretien d'explicitation" (which is an interview and discussion with the actor that aims to make the content of the action process explicit). Vermersch conceives the corresponding explication as a co-reconstruction of the situation by the actor and the analyst.

11. "We also need to find better ways of describing agents with genuine *freedom*, giving real meaning to words like 'choice' in the phrase 'theory-choice' and 'decision' in 'decision theory'" (Chapter 2, this volume, p. 70).

12. On that point, we recommend the reading of Baudouin and Friedrich (2001), which helps to understand what the necessary requirements are for the construction of the notion of action. Many contributions to this book emphasize the need to go beyond the traditional conception of rational action, which is almost exclusively interested in the articulation of means, ends, and aims from an idealized and a priori point of view. Baudouin and Friedrich favor instead an analysis of the constitutive features of the unfolding action (for instance, the unpredictable shifts of intentions—and consequently of the course of action—while the action is actually happening). In particular, Hans Joas shows that rational ends and means may change considerably during the course of action, to the point that the initial intended action is significantly transformed. He talks about the "creativity of action" to refer to such kinds of transformations.

13. By "essentially collective", we mean cases in which the very identity of the activity implies that a group is involved: To play football alone is not football in the usual sense, and science as we know it today could not be practiced and produced by one single individual.

14. "[I]sn't winning the inherent purpose of the game? That depends on one's philosophy (some might say the real point is to participate, and to build teamwork or character, or to get exercise, rather than to win). And winning each game may be the main purpose for each team involved, but it is not clear that it is the main purpose for an individual player, or for the collective of everyone who is playing each game, or for the whole institution of soccer, or for the whole society that supports this institution" (p. 75).

3 From Normative to Descriptive and Back
Science and Technology Studies and the Practice Turn

Michael Lynch

I. INTRODUCTION

While serving as Editor of the journal *Social Studies of Science* from 2002 to 2012, I read nearly 2000 submissions. I also was active with the Society for Social Studies of Science, and with one of the leading departments in the field.[1] These experiences, as well as my own research, provide an opportunity to reflect on the question that contributors to this volume were asked to address: *How the so-called "practice turn" in philosophy of science has modified both our conception of science and the way in which we analyze it.*[2]

Although I have been privileged to read a substantial amount of recent work in Science and Technology Studies (STS) that virtually nobody else has read—or would want to read—my perspective on the question about the practice turn is no doubt skewed in certain respects. The journal *Social Studies of Science,* and STS generally, is not closely tied to current trends in philosophy of science, though STS itself has developed its own philosophical orientations and undergone its own version of the practice turn. A further caveat is that, as a journal editor, I read one paper at a time and not as part of a systematic theoretical survey of trends in the field. As a gatekeeper who ended up rejecting most of what passed through the gate, I could be accused (and occasionally was accused) of resisting emergent trends rather than facilitating them. For the most part, this paper relies upon my unsystematic impressions, though it mentions some aggregate trends available from the journal's log of download and citation statistics. Finally, aside from my editorial experience, my reflections on the theme of practice have a particular slant, due to the fact that I came to STS from the sociological field of ethnomethodology—a field that did not need to undergo a turn toward practice, because practical action and practical reasoning had been ethnomethodology's central phenomena for decades, going back to the 1960s, when Harold Garfinkel publicly launched the field.[3]

I think many of my colleagues in STS would view the practice turn, which was prominent in the 1990s, as an important phase in the history of the field. Two edited volumes bracket the peak of this phase: Pickering (1992) and Schatzki et al. (2001). During that decade, the practice turn was marked as a development from social constructionism.[4] Since then, *practice* and *practices* have remained on the agenda, but they are no longer emblematic of the latest trends and turns. Various candidates for successor turns have presented themselves—turns toward culture, ontology, and performativity, among others. There is also a widespread move towards political engagement, sometimes in opposition to "neoliberalism" at both local and global levels (see, for example, Lave, Mirowski, & Randalls, 2010), and more generally there has been an increase in enthusiasm for normative engagement in controversies about science, technology, and medicine. This normative turn is supported by criticisms of the (apparently) disinterested historical and social orientation to technical controversies associated with the sociology of scientific knowledge (Winner, 1993; Woodhouse, Hess, Breyman, & Martin, 2002). Such turns often seem to produce old (and not necessarily well-aged) wine in new bottles, but almost invariably they are introduced as bold moves to succeed, or partly undo, allegedly dominant orientations in the field—particularly the orientations toward relativist, constructionist, descriptivist, "micro" treatments of scientific practice which became notable, and widely debated, starting in the 1970s—exemplified by works such as Bloor (1976/1991, 1981), Latour and Woolgar (1979), and Knorr-Cetina and Mulkay (1983b). Calls for normative and activist engagement have been quite successful in attracting adherents among the PhD students and younger scholars who submit most of the articles received by *Social Studies of Science*. The turn to practice is by no means complete, however, and *practice* remains on the agenda, although another keyword—*expertise*—has absorbed much of the more recent interest and debate in STS.

In what follows, I'll start with my impression of the current state of the art in STS. I'll then address the normative turn in STS, and specifically the influential conceptions of expertise and tacit knowledge offered by Harry Collins and Robert Evans. Although there are other normative programs and conceptions of expertise that differ profoundly from Collins and Evans' program, the aggregate download and citation statistics for *Social Studies of Science* make clear that a paper Collins and Evans published a decade ago (Collins & Evans, 2002) helped launch what turned out to be a very successful enterprise. After critically examining Collins and Evans' treatment of expertise, I'll close with a discussion of an alternative way to address the themes of expertise and tacit knowledge. To adumbrate my argument, I'll insist that we should recognize that the terms "expertise" and "tacit knowledge" not only name real phenomena, they also come into play in polemics that establish and defend the authority of professions. These terms are prominent in efforts to establish the credibility of individuals and defend the autonomy of professions—including the professions that gain, or aspire to, recognition as sciences. In brief, I shall argue that expertise commonly is used as a normative category, and that we should recognize that academic social scientists and philosophers take part in a broader polemical field when they construct their own normative arguments about that phenomenon. In other words, expertise is not simply an objective domain to be defined, classified, and critically re-worked by social scientists; it is bound up with political actions and discourses in which social scientists participate.

II. THE CURRENT STATE OF STS[5]

The list that follows is my unsystematic survey of trends in STS following the practice turn in the 1990s. Although I am limiting my perspective to the relatively small field of STS, I think it is fair to say that many of these trends can be found in other fields of social and cultural study. And, arguably, they are responsive to substantive political and economic changes in Europe, North America, and many other regions of the world.

(1) Studies that led the way to the practice turn often took up what were once called "hard cases" of physics and mathematics (for example, Bloor, 1973; Pickering, 1984a; Pinch, 1986). Such studies now are rare and endangered in STS. In place of studies that focus on those subjects, there has been a proliferation of studies of what Steven Shapin (2007, p. 183) has called "sciences of the particular": scientific research with agendas set by government and corporate initiatives and practical objectives. The incentive for conducting studies of "pure" mathematics and physics seems to have waned in STS, though there remains some excellent work on those topics, and studies of the "hard" sciences and mathematics continue to have a privileged (albeit somewhat diminished) place in history and philosophy of science.[6]

(2) STS has become more isolated from philosophy of science and history of science, while at the same time anthropology and cultural studies of science have largely supplanted earlier sociology of scientific knowledge (SSK) programs—and not without friction.[7] In the 1970s and 1980s, leading figures who identified with the sociology and anthropology of science did not always express strong loyalties to the academic disciplines of sociology, anthropology, or any other established social science. Many early protagonists of sociology and anthropology of science, such as Bloor, Collins, Latour, and Pickering, were not sociologists or anthropologists by training. Often, when they invoked sociology and anthropology, they referred to styles of argument rather than academic disciplines. Others who *were* trained in those disciplines often were at odds with prevailing trends in their "home" social science fields. Their research often was social-historical in character, and their interests borrowed strongly from philosophy of science. Today, there continues to be much discussion of epistemology and ontology in STS circles, but the field has weak links to philosophy and somewhat tenuous links with history of science.

STS has its own established journals and professional societies, and, in some universities, it has gained relative autonomy from other disciplines. Anthropology of science has become a robust part of professional anthropology, and it has increasingly put distance between its cultural orientation and the earlier programs in the sociology of scientific knowledge.

(3) Arguments about relativism and constructionism are still in the background, and they sometimes are explicitly invoked, but the decline in the volume of research that argues the case for constructionism using the "hard cases" of physics and mathematics coincides with a decline in overt contentiousness within the STS field since the late 1990s. There are some exceptions,[8] but these tend to be rather restricted and specialized disputes. Topics or alignments that once were contentious (the realist-constructionist debate; debates about feminist standpoint theory; the inclusion of non-human "actors" in Actor-Network Theory) no longer stir as much overt argument as they did in the 1980s and 1990s. A few critics still assign STS to "postmodernism", but, for the most part, their arguments no longer provoke as much interest or debate.

(4) There is an increasing tendency to invoke STS as a brand, as well as to recite key themes and to cite well-known studies without explicating the themes or arguments in any detail. It is common to assert *that* science is constructed, or *that* science and politics are co-produced, rather than to argue for such a position. And, while it is often presumed that studies published in journals such as *Social Studies of Science* march in lock-step with a constructionist drumbeat, a glance through recent issues of that journal and other leading journals in the field should be enough to indicate that many (arguably most) of what is published in such journals has no necessary connection with constructionism.

(5) Multiple constituencies in STS have advocated "normative" turns, sometimes casting one or another version of such a turn against the apparent neutrality or indifference associated with the symmetry and impartiality postulates of the Strong Programme (Bloor, 1976/1991).[9] The Strong Programme and related developments in the sociology of scientific knowledge were sometimes heralded (or criticized) for instigating a *social* turn, but they also represented a *descriptive* turn in contrast to normative epistemology. The more recent normative turn is not necessarily a re-turn to classic epistemology, and it is not choreographed in the manner of a ballet, as different players turn in different directions, at different speeds, and with different postures, sometimes colliding with one another. Consequent to this turn, there has been tension between STS scholars who treat epistemology, ontology, and ethics as historically and socially situated phenomena to be investigated, and those who treat STS *itself* as a source

of a distinctive epistemology, ontology, or ethics. In other words, this is tension between what Peter Dear (2001) calls "epistemography" (which I suggest can be extended to "ontography" and "ethigraphy"), and a widespread tendency to treat STS as itself a basis for a distinctive professional epistemology, ontology, or ethics.[10] It is not a new tension, though I would say the shift toward the latter (sometimes in the name of the former) has reduced the initiative for pursuing the former.

(6) Big-P Politics, and Big-I Institutions, have become more prominent relative to their lower-case counterparts (see, for example, many of the chapters in Jasanoff, 2005). Previously, the politics of science included controversies and machinations within and between esoteric fields, with vague, if any, connections to publicly recognized political issues in conflicts. It still does include such controversies, but nowadays it is more common to see STS research take up widely recognized political and policy issues. This turn toward conventional politics and policy may have to do with the relative success of "mature" STS in gaining sponsored support, and the increased embeddedness of some STS scholars in policy forums. On the critical side, "neo-liberalism" is frequently targeted both in studies of global trends and in case studies of particular developments toward privatization, commercialization, and entrepreneurship.

(7) There is increased talk about globalization in general and "the global south" in particular. For the most part, however, the symbolic inclusion of "the global south" as a theme in articles published in prominent Anglophone journals is advocated by writers from "the global north". There has been an impressive rise in East Asian, and a small but significant amount of South Asian work reaching *Social Studies of Science,* and some work from Brazil, but very little from other regions outside North America and Northern Europe.

(8) There has been growing interest in public engagement in science, and questions about the relations between expert and lay knowledge. Some of the work (especially in Britain and some northern European nations) was sponsored by programs designed to enlist representatives of "the public" in policies about genetically modified foods, nanotechnology, and synthetic biology. Many of the articles published in *Social Studies of Science* on public engagement exercises are critical of the notions of "public" and "engagement" presumed in these programs (see, for example, Irwin, 2006).

(9) There is growing interest in how "futures" are embedded in present day organizations of science and technology. This trend is partly associated with studies of nanotechnology and biotechnology, but also is associated with work on economic forecasting and business marketing. Especially in Britain, there is a cluster of work on sociology of expectations that examines "performative" relationships between

predictions and promises and subsequent developments in science and technology (and consistent with the tendency to turn terms for occasional expressions into pervasive conceptual principles, J. L. Austin's "performatives" become Judith Butler's "performativity").

(10) Taking a number of trends together, I think it's fair to say that there is an abstract movement toward *unity in hybridity*. Many of us reject the idea that STS is an "inter-discipline": a field that gathers together residents of otherwise established disciplines (sociology, history, anthropology, philosophy, ethics) who come prepared with well-formed theories, perspectives, and methods that they apply to science, technology, and medicine. I think a different term has greater salience: *interconceptuality*. It has to do with a love of hybrids and an aversion to boundaries, dichotomies, and even distinctions. In STS, boundary concepts are taken up with relentless frequency and enthusiasm: trading zones, boundary objects, boundary work, boundary organizations, interactional expertise, coproduction, co-construction, technoscience. There is also the "Bio+" family: biosociality, biocitizenship, biocapitalism, biolegality. Interest in such concepts includes a substantive focus on displacement and respecification. The sense of unity in the STS field is less a product of a coherent, agreed-upon theoretical framework than of a more amorphous aversion to fixed boundaries, an opposition to objectivism, and an emphasis on historical contingency and epistemic uncertainty. This sense of unity is expressed and operationalized through an increasingly distinct professional identity, as STS has become a collective with a sense of its own history and progress. This sense of incorporation, growth, and progress supports an agenda to use STS as a distinct body of knowledge, with profound implications for exposing the historical, practical, and political sources of "objective" knowledge, and revealing the lack of clear boundaries between nature and society. However, such implications lately have become increasingly unclear and troubled.

III.   SCIENCE STUDIES AND THE NEW SCIENCE WAR

At the height of the practice turn in the 1990s, social and cultural constructionism became a subject of vociferous, often crude, criticism. The popular term for that campaign—"science wars"—exemplified the hyperbolic arguments that prevailed during that time. We now face what is sometimes called a "new" science war.[11] In 2004, Bruno Latour wrote an article in which he criticized "critique". Initially, he presented his argument as a *Mea Culpa* in which he counted himself among those who had honed a critique that was now aimed at the wrong target—or, rather, it was aimed at a target that no longer was the most salient or threatening one (he alluded to the reputation of French generals for fighting the previous war).

Latour (2004, p. 226) presented a passage from a *New York Times* editorial that quoted a conservative pro-business political strategist as saying, "[s]hould the public come to believe that the scientific issues are settled . . . their views about global warming will change accordingly. Therefore, you need to continue to make *the lack of scientific certainty* a primary issue".[12] As we now know, this strategist and his allies successfully placed *uncertainty* on the agenda, first with the blessing of the George W. Bush Administration, and later with more widespread support from media, corporate, and political allies. This program (or, better, project) of skepticism was underpinned by an infrastructure of think tanks and news outlets in which a seeming embattled minority assumed the position of the underdog faced by a powerful scientific establishment.

Though it would give STS far too much credit to say that studies of scientific controversies furnished political operatives with polemical tools with which to produce what Latour (2004, p. 227) called "artificially maintained controversies", some veterans of the earlier science wars of the 1990s blamed STS researchers for aiding and abetting the climate skeptics, anti-Darwinian creationists, and other professional skeptics who effectively crafted ways to foster public uncertainty about settled scientific matters.[13] And, as if on cue, Steve Fuller, a social epistemologist with a normative agenda for democratizing science, stepped into the breach and lent his voice to the tribunal about intelligent design in a federal trial in Dover, Pennsylvania.[14] Although a few STS scholars have objected to what historian Kevin Lambert (2006) dubbed "Fuller's folly", and Fuller has accused me of trying to stir up a campaign of denunciation against his effort to bear witness against establishment science in that trial (Fuller, 2009, p. 220), for the most part STS research has continued to pursue critiques of established science and its consensual certainties as though such critiques always place us on the side of the angels. If anything, critique was gaining steam within the field at the very time that Latour said it had run out of steam in the academic world, if not in the political world at large.

Latour goes on to set out an alternative program in the politics of science, which would shift from arguing about facts to examining "matters of concern" in which human and non-human agents participate in Heideggerian "gatherings". However, what is of more immediate concern for the present essay is Latour's identification of general academic arguments about contingency and uncertainty with particular arguments that promote uncertainty and skepticism for political purposes—arguments that allege, for example, that so-called consensus about climate change among experts is based on political machinations.

Faced with such circumstances, we can try to pick our fights wisely and lend our polemical ammunition selectively to undermine "certainties" that we are inclined to doubt or oppose. But, of course, inducements are present to lend our voices to less heroically positioned parties who might be happy to employ us. Despite such inducements, STS writings (at least as represented

by submissions to *Social Studies of Science*) tend to align themselves normatively with epistemic (and political) underdogs, without addressing how
the underdog position (and a limited and pointed relativism that *Doones-
bury* cartoonist Garry Trudeau brilliantly dubbed "situational science")[15] is
often used polemically to promote "controversy" about evolution, climate
change, the effects of tobacco smoke, and so on.

Although often explicitly at odds with Latour, Harry Collins and Rob
Evans (2002, 2007) provide a relatively rare, and yet widely cited, response
to "artificially maintained controversies".[16] Their programmatic response
is to pursue a neo-demarcationist program that aims to fashion conceptual tools that would enable real controversies to be distinguished from
pseudo-controversies.

IV.   WAVES CRASHING ON THE SHORES

Collins and Evans (2002) frame their theory with an account of three "waves"
of science studies. The first wave refers to pre-Kuhnian philosophy and sociology of science, with Sir Karl Popper and Robert K. Merton as the iconic
figures. Popper (1963) developed a normative view of science and sought
to demarcate science on conceptual grounds from non-science and pseudo-
science. Merton (1942) also spoke of norms—institutionalized norms—that
enable an autonomous enterprise to pursue and certify knowledge through
procedures of criticism and testing, and with limited interference from religious and political orthodoxy. Science was viewed as an exceptional institution with a distinctive methodology.

The second wave refers to constructionist studies of scientific knowledge,
exemplified by Collins' own work on gravity waves and other subjects. With
respect to demarcation, an emblematic "Wave Two" formulation is Thomas
Gieryn's line: "Even as sociologists and philosophers argue over the uniqueness
of science among intellectual activities, demarcation is routinely accomplished
in practical, everyday settings" (Gieryn, 1983, p. 781). This line marks a shift
from a normative to a descriptive (historical) way of handling the demarcation
and definition of science. This is quite a profound shift, despite its apparent
simplicity, as it requires the STS analyst to refrain from any explicit endorse-
ment of the special epistemic status of science and encourages a detached orientation to claims and counterclaims about what is, or is not, scientific.

The third wave is a vision of a science of knowledge that would return
to the demarcation problem, but with different criteria informed by constructionist (Wave Two) research. The programmatic ambition is to classify
types of expertise, and to furnish decision makers with criteria for recognizing genuine experts and forms of expertise that come into play within specific controversies. Instead of being based on a theory of true (or certified)
knowledge, it is based on a theory of real expertise in which tacit knowledge
is a crucial element (Collins & Evans, 2002, p. 236–237).

Collins and Evans' proposal to move to a new "Third Wave" of STS research has been heavily criticized by prominent STS colleagues (Jasanoff, 2003; Rip, 2003; Wynne, 2003). There are many points of criticism, and I shall only emphasize a few of them. One point has to do with the very idea of a progression of waves. There are good reasons to doubt that a field as amorphous as STS could have progressed in clearly defined waves. However, such schemes are commonplace in internal histories that express ambitions to bring about change in academic fields. Indeed, the "practice turn" in STS can be identified with the transition from Wave One to Wave Two in Collins and Evans' scheme. What is more critical, in my view, is how their explicitly normative orientation to scientific controversy relates to the constructionist position they identify with Wave Two.

As noted earlier, a signal feature of Wave Two was the abandonment of normative conceptions of science (and of the demarcation of science from non-science) in favor of descriptions of "tacit" features of scientific practice that were obscured by edifying conceptions of scientific method. Proponents of the Strong Programme self-consciously initiated this shift in the 1970s. A major source of inspiration for them was Thomas Kuhn's (1970) criticism of present-centered histories of science that treated currently accepted textbook accounts as a basis for reconstructing past developments. Sociologists of science developed Kuhn's argument into a research program—often dubbed the sociology of scientific knowledge (SSK)—that drew a sharp distinction between idealized, normative accounts of science derived from textbooks and general formulations of methods and a dramatically different picture of situated scientific practice derived from historical and ethnographic research (Barnes, Bloor, & Henry, 1996; Collins, 1985/1992). Such a perspective is articulated very well in the following quotation "on the rock-bound difference between the finished versions of scientific work as they appear in print and the actual course of inquiry followed by the inquirer":

> . . . books on method present ideal patterns: how scientists ought to think, feel and act, but these tidy normative patterns, as everyone who has engaged in inquiry knows, do not reproduce the typically untidy, opportunistic adaptation that scientists make in the course of their inquiries. Typically, the scientific paper or monograph presents an immaculate appearance which reproduces little or nothing of the intuitive leaps, false starts, mistakes, loose ends, and happy accidents that actually cluttered up the inquiry. The public record of science therefore fails to provide many of the source materials needed to reconstruct the actual course of scientific developments.

Perhaps it will come as a surprise that this passage was written by quintessential Wave One theorist Robert K. Merton (1968, p. 4). It neatly summarizes what often is taken as the principal lesson drawn from Wave Two ethnographic and social-historical studies of scientific practice (notably,

Latour & Woolgar, 1979), to the effect that actual scientific practices do not develop in linear fashion and are quite "messy" when encountered in the midst of their production. Such studies were undertaken more than a decade after Merton wrote this passage, and they often set themselves in opposition to the idealized, normative conception of science attributed to Merton (1942).[17] However, while Merton's passage anticipates and articulates what often is attributed to ethnographic and social-historical case studies, Merton and the Mertonians did not follow through with such studies (though see Barber & Fox, 1958, for a partial exception), nor did they use the distinction between ideal and actual practices as a cornerstone for developing a constructionist view of scientific facts and theories.

Unlike Popper, who offered falsifiability as an explicit criterion for distinguishing science from metaphysics and pseudoscience, Merton (1942) treated the norms of science descriptively. However, the norms he highlighted—commun(al)ism, universalism, disinterestedness, and organized skepticism—present an edifying picture of a very special social institution. While the above quotation (like many others that can be drawn from Merton's voluminous writings) indicates that he was well aware of the difference between the edifying portrait and the actual practice of scientific investigation, he did not endeavor to describe actual practice in close detail or to develop a general conception of science based on such description. Instead, by highlighting the idealized norms, and attributing the distinctive institutional character of science to their sanctioned adherence, he reified as the essential core of the institution what, at the time, was a common view of the epistemic and normative virtues of science.

Social constructionism did not simply bubble up from historical and ethnographic descriptions of actual practice—it was as much a motivation for producing and framing such descriptions. Philosophical arguments, which were adopted from Popper's critique of inductive empiricism and then turned against Popper's own views of falsification resulting from crucial experiments, set up the forms of social explanation characteristic of SSK (exemplified with great clarity in Collins, 1985/1992). However, empirical studies provided a strong impetus for declaring that, contrary to what Merton or Popper would lead us to believe, "scientific knowledge is like other forms of knowledge" (Collins & Evans, 2002, p. 239). Accordingly, like other forms of knowledge, science was said to be inflected by vested interests, culturally located, and open to political influence, so that controversies are closed, not because of crucial experiments that falsify contending theories, but because persistent dissenters are discredited on "social" grounds and marginalized from the core groups that dominate particular research fields at particular times.

Social constructionism has been criticized many times over, both by rival theorists within the STS fold and by others who express sweeping hostility to STS (often placing it under the rubric of "postmodernism"). There are many different lines of critique, but the one I shall emphasize is articulated by Collins and Evans themselves, and is used to motivate a turn from a descriptive orientation to practice to a (new and improved) normative theory of science. It is less of a critique, in the sense of rejecting the grounds of a position, than a call to move on to the next stage in a linear progression.

For Collins and Evans, like many others who advocate a normative turn, the practice turn was a developmental phase rather than a destination. It marked a transition from one normative era to another, better one, in which analysts should be able to support their criticisms and policy recommendations with empirical knowledge of actual scientific practices, instead of grounding them in ideals that have never had more than a tenuous hold on what scientists actually do. Perhaps the most common normative approach in STS credits constructionist studies with having dismantled the epistemological distinctions that privilege expert knowledge. Such epistemological leveling is linked explicitly to democratic politics, and to programs designed to give greater voice to those who, for want of expert credentials, are assumed to be passive beneficiaries of techno-scientific innovations.[18] By raising familiar political questions about who benefits from such arrangements, scholarly work promises to redress the imbalance between state and corporate actors who control "sound science" and the rest of us who are on the other side of the great demarcation. By relativizing (alleged) scientific certainties with research that identifies vested interests, closed policy networks, and strategies for closing off controversy, STS research continues the tradition of what is sometimes called "underdog sociology". As noted earlier, however, the practical and rhetorical production of "artificially maintained controversies" aligns unquestionably powerful actors with the weaker side in a controversy—weaker in terms of official scientific support or consensus—so that a relativistic leveling strategy works to their advantage. It does so under political and legal circumstances in which the burden of proof is placed on the opponents of the actors who promote the controversy. When faced with such circumstances, both as citizens and academic analysts, we may find it necessary to distinguish when controversies are, or are not, genuine:

> . . . should you eat British beef, prefer nuclear power to coal-fired power stations, want a quarry in your village, accept the safety of anti-misting kerosene as an airplane fuel, vote for politicians who believe in human cloning, support the Kyoto agreement, and so forth. These are areas where both the public and the scientific and technical community have contributions to make to what might once have been thought to be purely technical issues. (Collins & Evans, 2002, p. 236)

The promise of their theory is to yield sound advice about the limits of expert knowledge and public participation in debates about technical matters of public concern, and about which fields of expertise should guide, if not determine, decisions that have broader public consequences. In some of these cases, it is necessary to decide which experts and bodies of expertise are trustworthy. Related to this are questions about when, and to what extent, expert advice should be sought. Collins and Evans propose to begin with the assumption that expertise is a real quality possessed by some individuals and social groups but not by others. They also propose to supply criteria for demarcating genuine expertise from pseudo- or non-expertise, and to equip us with an ability to judge how much expertise is needed in order to participate in policy decisions involving expert judgments. They note that "interactional expertise"—the ability to talk *about* a technical subject in an informed and convincing way but without the tacit knowledge required to "contribute" to it as an expert—often can be sufficient for participation in decision-making forums in which technical matters are crucially important.

Instead of counseling unremitting skepticism toward the role of experts in public life, Collins and Evans propose that STS should itself become a source of expertise about how to distinguish genuine from artificial expert knowledge. In order to secure such a position, they argue, it is necessary to understand that "expert" is not simply a social or relational construct—a social "attribution" with no stable, non-arbitrary attributes. Instead, in their view, real expertise confers genuine authority over particular domains of knowledge. Indeed, *we* in STS should claim our expertise on the matter of what expertise is, how it is conferred, and when it is authentic.

> The emphasis on the "social construction" of science has meant, however, that when expertise is discussed, the focus is often on the attribution of the label "expert", and on the way the locus of legitimated expertise is made to move between institutions. . . . We believe, however, that sociologists of knowledge should not be afraid of *their* expertise, and must be ready to claim their place as experts in the field of knowledge itself. (Collins & Evans, 2002, p. 239)

In other words, instead of merely *describing* "boundary work" in historical cases, we should not be afraid to stipulate boundaries prospectively on the basis of our cumulative knowledge about knowledge and expertise. When advancing to Third Wave studies, in Collins and Evans' progressive scheme, it is necessary to go beyond a "relational" approach and to assume a realist orientation:

> To treat expertise as real and substantive is to treat it as something other than *relational*. Relational approaches take expertise to be a matter of experts' relations with others. The notion that it is only an "attribution"—the, often retrospective, assignment of a label—is an example of a relational theory. The realist approach adopted here is different. It starts with the view that expertise is the real and substantive possession of groups of experts and that individuals acquire real and substantive expertise through their membership of those groups. . . . Under our

treatment, then, individuals may or may not possess expertise independently of whether others think they possess expertise. (2007, pp. 2–3)

In some respects, it is hard to disagree with what they say in this passage. Surely, expertise is not *only* an attribution—a nominal category. There must be more to it than that. But why should anyone suppose that "real and substantive" is incompatible with "relational"; or that retrospective attributions would be the only way to identify the relations in which expertise is embedded? If the assertion that expertise is "real" meant nothing more than that there are people who merit being called experts, and groups of people who successfully cultivate expertise in one or another practice, it would be an unremarkable claim. The problem has to do with how Collins and Evans purport to settle the reality question *themselves* on the basis of their analytical judgments, rather than leaving it open to determination by participants and procedures in the institutions that social scientists investigate. They acknowledge that there are different modes and degrees of expertise, and that not everyone who claims to be, or is designated as, an expert really is an expert, but they propose that they can determine what counts as a "real" expert, independent of the social occasions and relations in which the use of that social category is embedded.

V. EXPERTISE AND TACIT KNOWLEDGE AS ACTORS' AND ANALYSTS' CATEGORIES

When outlining their normative approach to expertise, Collins and Evans propose to go beyond the description of expertise as an "actor's category":

> [We] must emphasize the role of expertise as an *analyst's category* as well as an *actor's category,* and this will allow *prescriptive,* rather than merely *descriptive,* statements about the role of expertise in the public sphere. (2002, p. 240)

However, when we consider a historically developed, institutionalized discourse in which "expertise" features as an important actors' category, the distinction becomes untenable: the actor's category is also, and importantly, an analyst's category; indeed, the actors themselves perform analyses of, and with, that category. Moreover, the category is used to do organizational work; among other things, it is used tendentiously to gain advantage or to counteract the advantages of others. In other words, the term already is used prescriptively and normatively.

Collins and Evans recognize that "expert" and "expertise" are commonplace categories, but they assert that sociologists should feel free to stipulate what these terms mean, independent of their everyday uses and associations. However, I believe that before doing so, we should remind ourselves how we commonly use these terms. As J. L. Austin (1961, p. 133) famously said, although "ordinary language is *not* the last word . . . it *is* the *first word*".[19] This doesn't mean that we should become complacent about ordinary usage. However, efforts to bypass this starting point in the interest of a more stable, context-independent, and methodologically precise theoretical conceptualization has well-known pitfalls: overgeneralization, insensitivity to contextual differences, academic arrogance, category errors, and many others. Such efforts also can miss what, arguably, is most sociologically relevant about institutionalized as well as so-called "ordinary" uses of key concepts.[20]

To understand why I say this, consider a key concept in Collins and Evans' treatment of expertise: tacit knowledge—skill acquired through long apprenticeship in an acknowledged practice that enables the practitioner to make technically informed judgments. For Collins and Evans, tacit knowledge is exemplified by a science, such as a branch of experimental physics, in which practical skill (and not just credentials and abstract knowledge) enables a member to contribute to research and argumentation among the relatively few highly skilled participants in a "core set".[21] Collins and Evans explicitly assume that there is no serious question that the field (astrophysics, as opposed to astrology, for example) is recognized as an authentic field of science, and that the most genuine basis for attaining "contributory expertise" in that field is informal peer-recognition by other members of the "core set". Others—managers, science journalists, specialists in neighboring fields, and ethnographers—can attain variable degrees of abstract, second-hand, verbal knowledge of the technical issues ("interactional expertise"), but the baseline of expertise is tacit knowledge (and informal recognition that a given individual possesses tacit knowledge).

What Collins and Evans refuse to countenance, however, is that public disputes often involve the primary issue of *what counts* as *relevant* expertise. This is more than a preliminary attribution, because questions can arise throughout such disputes about whether or not "experts" are necessary and, if so, what they are necessary *for*. Moreover, decisions about the relevance and limits of expertise aren't necessarily made by qualified "experts"; designated or self-appointed experts may be consulted, and may offer their advice, but such advice is not necessarily given much weight. By disregarding the particular relations in which the very salience and content of expertise is embedded, Collins and Evans launch their theory in a purified space that severely limits their ability to address the demarcationist questions they propose to resolve.

"Expert" is not only a descriptive category. In public domains, "expert" is a valued and contested term. So is "tacit knowledge". Michael Polanyi, Collins, and others who have written on the topic address tacit knowledge with great respect for its role in the acquisition of technical authority. Polanyi (1967), especially, treats tacit knowledge as a basis for skepticism about centralized planning and efforts to "manage" science. Collins (2010) carries his conception of tacit knowledge over to arguments that draw a line between what can or cannot completely be simulated with expert systems. For both Polanyi and Collins, tacit knowledge is an unmitigated good in labor conflicts and professionalization struggles, but for Frederick Winslow Taylor, the tacit knowledge of manual workers presented a *practical* challenge to his efforts to manage labor and improve its efficiency (Braverman, 1976). Unlike Collins (2010), who, as a matter of definition, distinguishes tacit knowledge from mere secrecy, Taylor initially faced a seamless alliance between workers' secrecy and their possession of incommunicable skills: He aimed to break the workers' "soldiering" (collective guild-like maintenance of a barrier between insiders' and outsiders' knowledge of work-practices) in order to formalize, standardize, and control the labor process. In a different way, the rhetorical and political significance of tacit (or "incommunicable") knowledge comes up in Chris Lawrence's (1985) account of efforts by Victorian gentlemen-doctors to resist the introduction of diagnostic instruments such as the sphygmomanometer and stethoscope. Fearful that these items of equipment would open up the medical profession to mere technicians, the gentlemen-doctors emphasized the "incommunicable" skill necessary to perform diagnosis at the patient's bedside. Their arguments highlighted the embodied knowledge that accumulates with long experience, and they even suggested that the refined nerves of the gentle classes gave gentleman physicians an inherent advantage over their low-born colleagues. Warwick Anderson (1992) identified similar arguments (though not, in this case, supplemented by the pseudo-neurology) in his study of a more recent conflict in an Australian hospital over medical technologies. Considering this kind of case produces a gestalt switch on the virtues of tacit knowledge. Instead of being a substantive possession, tacit knowledge becomes a normative concept and rhetorical resource that is deployed in specific historical and contemporary situations. We can gain insight into such usage through descriptive studies such as Lawrence's and Anderson's, but Collins and Evans propose to transcend such Wave Two accounts in favor of a normative theory in which tacit knowledge is the substantive core of contributory expertise. The possession of tacit knowledge was their demarcation criterion for distinguishing real expertise from its artificially constructed double, but then tacit knowledge itself is subject to games of hide and seek (after all, it is "tacit" by definition). We have what might be called an analytical regress.

Collins and Evans thus face two, related, problems with their normative program. One is that it they set a task for themselves and their followers that may be impossible: to treat tacit knowledge as a "real thing" that is transparent to a general form of social analysis.[22] By itself, this is not such a serious problem—if it turns out to be intractable to analysis, then, at worst, they end up wasting a lot of time. The more serious problem is that by valorizing the tacit knowledge of a core set of experts, they fail to take into account *how* the concept is used to claim and allocate authority and to promote domains of expertise in contested situations.

Collins and Evans are more interested in presenting their own analytical classification than in explicating how the category of "expert" is administered in specific social circumstances, such as legal cases. However, there is a large literature on the subject, including typologies of expert witnesses (Gross & Mnookin, 2004; Risinger, 2000; Turner, 2001). A particularly large amount of work has been done in reference to the U.S. Supreme Court ruling in *Daubert v. Merrell Dow Pharmaceuticals, Inc.* (1993) about the admissibility of expert evidence in Federal courts. A brief acquaintance with the topic may be sufficient to elucidate some of the difficulties with Collins and Evans' proposals.

Expert witnesses have a privileged role in the Anglo-American courts: When testifying in a trial, they are permitted to draw inferences on the basis of their specialized experience and education, which would be excluded under "hearsay" rules for ordinary fact witnesses. At the time of *Daubert*, there was widespread agitation about the possibility that juries were being persuaded by "junk science" in the courtroom to award large rewards to plaintiffs suing corporations for alleged damage to their health and well-being. The case was one such lawsuit. The question before the court in *Daubert* was about which standards trial judges should use when deciding if the expert evidence a litigant is prepared to submit is relevant and sufficiently reliable for presentation to the jury. In other words, the courts addressed a situated variant of the philosophical problem of demarcation.

Most of the literature on the Daubert case is in law journals, but some discussions are in STS sources, as well as in law journal articles that take an STS perspective.[23] There is no clear separation in this instance between actors' categories and analysts' categories. The legal actors perform analyses, and once in a while they even cite and draw upon writings by philosophers of science and STS scholars. One way to preserve a distinction between the actors' analyses and those of STS scholars is to say that the actors *perform* boundary work, while the scholars describe the rhetorical strategies and organizational machinations through which they do so. However, that is the very sort of "Wave Two" descriptive approach that Collins and Evans intend to supersede with their new and improved normative program. The problem is not that they need to supplement actors' categories with analysts' categories, but that they are committed to doing a better job of developing prescriptive and normative analyses than do the actors who deploy the same categories. These categories not only include "expert" and "expertise", but also "consensus" and "controversy" and various familiar demarcation criteria.

There is precedent for proposing to instruct actors with analysts' categories. A year before the Daubert decision, an essay of Sheila Jasanoff's (1992) published in a law journal declared in its title: "What judges should know about the sociology of science". The lessons Jasanoff conveyed to judges were largely couched in "Second Wave" terms, but the possibility of advising legal actors with STS knowledge was not beyond imagination. Indeed, Jasanoff and a number of other STS scholars and philosophers were cited in the Daubert decision, though the citations exhibited a highly truncated understanding of the cited sources (Edmond & Mercer, 2002). Following the decision, numerous books and articles were written by Jasanoff (1995) and others to critically discuss the way the Daubert majority discussed science, expertise, and particular aspects of scientific methodology. One of the more notable articles was by philosopher Susan Haack (2005), who titled her paper "The Supreme Court's philosophy of science", and criticized how the court conflated the philosophies of Popper and Hempel.[24]

Describing and criticizing the court's actions was a matter of participating in a normative discourse. The extent to which the court was doing philosophy—and especially bad philosophy—provided a basis for critiques such as Haack's. However, as Haack also points out in her analysis, there was argument among the justices about the very relevance of philosophical criteria for distinguishing between reliable and unreliable expert testimony (see Justice Rehnquist's dissent in *Daubert*). To a large extent, the Supreme Court settled for granting broad discretion for trial judges to exercise their "gatekeeping" function on a case-by-case basis, and the Court refrained from providing a closed set of criteria for demarcating expertise.

Although there is no reason not to criticize U.S. Supreme Court decisions (doing so has become a widespread pastime with the current Supreme Court), when we consider the resources (scholarly, financial, etc.) that feed into big court decisions, as well as the consequences of such decisions, we may want to invert Jasanoff's formulation: Instead of proposing what judges should know about the sociology of science, we may instead hope to enrich the sociology of science by examining how judges articulate phenomena of interest to our field. I should add that the reasoning of individual judges, or even entire panels of judges, is not the phenomenon of interest; instead, it is the entire nexus of arguments, written briefs, and deliberations that feed into a key decision. I should add further that such an interest is not a matter of deferring to the way judges articulate what the word "expert" means, but of examining *how* they articulate the concept and of delving into the implications and consequences of their actions and interactions.

VI. SUMMARY AND CONCLUSION

Permit me to present a summary of the "relational" conception of expertise that my argument adumbrates:

(1) "Expert" is a membership category.[25] Like the word "science", in many contexts of use it is a valued category that implies epistemic and personal virtues (Haack, 2005, p. 51). Experts are not always venerated, of course, as it is fairly common to hear complaints about

experts, but often such complaints distinguish "so-called experts" from "real experts". "Expert" also can be a name for a formal status—there are registries of experts and formal procedures for vetting experts in particular institutions, such as the courts.

(2) Expertise is circumscribed. There are no general experts in any serious sense, just as there are no general scientists. One can teach general science at an elementary level, and a doctor can be a general practitioner, but I know of no general experts. To appreciate this point about the folly of the generalized expert, consider the following characterization of a well-known parody on U.S. radio: "'Ask Dr. Science' is a daily 90-second radio parody of deep-voiced, all-knowing experts with white lab coats, Bunsen burners and test tubes. It opens with a ditsy, 1950-ish theme song, and then Dr. Science's trusty sidekick declares: 'Remember, he knows more than you do. He has a master's degree . . . in science!'" (Barron, 1991). To say that expertise is circumscribed is not the same as saying that modern scientists are specialized; it is that they are specialized in *something*. This also is not to say that experts are only experts in one narrow specialty; a person can claim expertise at more than one activity, and an expert in one domain often can find inroads into other domains, or parlay a reputation in one area to gain authority over a broad range of subjects.

(3) What Collins and Evans call "issues [that] are of visible relevance to the public" often (perhaps always) involve complex relations among multiple technical specialties, no single one of which owns the issue; these specialties often are integrated with, and take initiative from, government agencies, corporations, and other organizations that include non-scientific personnel (some of whom wield considerable authority) and have agendas besides those of scientific research.

(4) Qualified and acknowledged experts and expert bodies may have unquestioned input into public decisions, but often their input is regarded, disregarded, and weighed alongside other parties' input, or otherwise is "digested" by a larger organization according to priorities controlled by no single expert or field of expertise.

(5) How much esoteric knowledge or technical experience is necessary for making sound decisions on matters of concern is a highly variable, relative, and contested question.

I mention these points not to claim anything new, but to recall how we ordinarily use the terms "expert" and "expertise". Perhaps I am wrong, but I don't think any of these points should be controversial.

So, where does this take us? In line with the practice turn, key conceptual issues (issues identified by what Ian Hacking [1999, p. 22] has called "elevator words") are brought down to earth in a deflationary engagement with specific cases. Contrary to Collins and Evans' ambitions to initiate a science of expertise, what I have in mind is more of a form of casuistry than of an aspiring science. It is an intensive effort to come to terms with cases in which key concepts are situated in particular practices. Initially, and predominately, this is a descriptive approach—indeed, it is a descriptive approach that focuses on actions and discourses that often are explicitly normative. However, when a case study is pursued in depth, it can provide a basis for particularistic judgments about the credibility of particular participants and the plausibility of their arguments. Further, it can motivate normative critique and advice of a rather down-to-earth sort. This differs from "expertise about expertise", and is more akin to expertise about *something* in particular.

STS research may lead us to the point where we take normative positions that are embedded in or directed towards the settings under study, but those positions are not grounded in generalized STS concepts, typologies, or theories. They cannot be divorced from an intimate understanding of the practices studied. Consequently, the practice turn should remain integral to normative as well as descriptive orientations in STS.

NOTES

1. Other activities that involved exposure to a range of recent research in the STS field included being co-editor of the major handbook in the field (Hackett, Amsterdamska, Lynch, & Wajcman, 2008), editor of a four-volume set of previously published articles in STS (Lynch, 2012), and President of the Society for Social Studies of Science in 2008 and 2009.
2. The question was suggested by the organizers of the workshop, "Rethinking Science after the Practice Turn", held at the PratiScienS Group, Université de Lorraine, Nancy, France, 19–20 June, 2012.
3. Garfinkel coined the term "ethnomethodology"—the study of practical action and practical reasoning in ordinary as well as specialized activities—prior to the publication of his seminal work, *Studies in Ethnomethodology* (1967), but the publication of that book introduced that field to a broader readership in the social sciences and philosophy.
4. The term "constructivism" is more commonly used in STS than "constructionism", but I use the latter term throughout this paper, consistent with Hacking's (1999, p. 48) suggestion that it is more easily distinguished from constructivist programs in mathematics and other fields that are quite unlike the various social constructionisms that run through current humanities and social science programs.
5. This section draws upon a presentation I gave at Harvard University for a meeting organized by Sheila Jasanoff on "STS the next 20 [years]" (April, 2011).
6. A search through recent session and paper titles at the annual meetings of the Society for Social Studies of Science (4S) with the keywords "physics" and "mathematics" turns up very few entries, compared with "politics" or "policy", "medicine" or "medical", and "technology".
7. Pickering's (1992) *Science as Practice and Culture,* while exemplifying the turn to practice, included a series of chapters and debates that exhibited tensions and outright antagonisms between proponents of various sociological, feminist, cultural, and post-structural programs. More recently, historian

Lorraine Daston (2009) announced a split between history of science and STS, which drew a response from Peter Dear and Sheila Jasanoff (2010).

8. See, for example, the exchange between Trevor Pinch (2011) and Karen Barad (2011).

9. In an often-cited argument, Bloor (1976/1991, p. 7) proposes that his program in the sociology of knowledge would be "strong" because its mode of sociological explanation would cover even the most robust scientific and mathematical knowledge: It "would be impartial with respect to truth or falsity, rationality or irrationality, success or failure. Both sides of these dichotomies will require explanation". In addition, "it would be symmetrical in its style of explanation, the same types of cause would explain, say, true and false beliefs". Both impartiality and symmetry express the aim to move the sociology of knowledge beyond what Bloor called the sociology of error (or, more broadly, the sociology of variable belief), in order to encompass "true" as well as "false" knowledge/belief.

10. The Actor-Network Theory developed by Bruno Latour and Michel Callon, which has been Anglicized by John Law and feminized by Annemarie Mol, is the prime example of an STS approach that has attained the status of an alternative ontology that refuses standard distinctions between facts and artifacts and between humans and non-humans.

11. See Turner (2003) for an essay that chronicles successive variations on the theme of "science war".

12. "Environmental Word Games", *New York Times*, 15 March, 2003. Available at http://www.nytimes.com/2003/03/15/opinion/environmental-word-games.html

13. Abundant examples can be found in the blogosphere. One example—a brief article originally published in the *Los Angeles Times*, by Chris Mooney, author of *The Republican War on Science*, and physicist and veteran science warrior Alan Sokal—links "postmodern" science studies to climate skepticism and the like, and advocates a realist agenda (Mooney & Sokal 2007).

14. *Tammy Kitzmiller, et al. v. Dover Area School District, et al.* (400 F. Supp. 2d 707, Docket no. 4cv2688, 2005)

15. The cartoon appeared on 5 March 2006, and is available at http://www.gocomics.com/doonesbury/2006/03/05/

16. I should add that, while both Collins and Latour recognize a common phenomenon and problem for science studies, they do so independently, and especially from Collins' side, there is a rift between how they go about addressing that phenomenon.

17. Merton (1993, p. xxii) later quotes this passage, and further points out that he later realized that Medawar (1964) anticipated it in his famous essay "Is the scientific paper fraudulent?", and that he (Merton) had made a similar point in his doctoral dissertation a half-century earlier.

18. Brian Wynne's (1996) account of the conflict between the local knowledge of Cumbrian sheep farmers and the (dubious but domineering) scientific knowledge of government ministry scientists has attained the status of a parable in STS.

19. Austin is, of course, identified with the "linguistic turn" in philosophy, which usually is treated as pre-dating the practice turn. However, there is an affinity between the two, especially when we consider that for Austin and others associated with the earlier turn, language is understood as practical action. Incidentally, there is a parallel in Anglo-American jurisprudence to Austin's "first word": this is the doctrine that, when interpreting legislation, judges should interpret the words of statutes "in their ordinary and usual" meaning, unless the words are given a more specific definition.

20. In a response to a book review in which I made a similar argument about the concept of "expertise", Collins and Evans (2008) dismiss my perspective by associating it with "a very small school called 'ethnomethodologists'". Although ethnomethodologists have argued along such lines, many others also have done so. Mills (1940) invokes American pragmatism, and Winch (1958) invokes Wittgenstein's later philosophy, to critically treat "motives" not as an explanatory concept for a social science theory, but as a vernacular term whose ordinary and institutionalized uses invite explication. Does it not seem reasonable to suppose that similar arguments apply to the terms "expert" and "expertise"?

21. Compare this to Shapin's account of research on matters of public concern: "The heterogeneity of expertise on a very large number of occasions of public concern is a notable feature of our culture . . ." (2007, p. 175). In reference to expert advice on dietary matters, Shapin observes that the members of the general public persistently turn to experts with great hopes and expectations, while at the same time they maintain skepticism about specific bits of advice. "The laity assert their freedom to pick and choose which expertise is credible, while giving few signs that they find the whole domain of dietary expertise wanting" (p. 176).

22. Turner (1994) provides a critique of abstract, generalized treatments of "tacit knowledge" in philosophy and social theory.

23. See, for a few examples, Jasanoff (1995), Edmond and Mercer (2004), and Caudill and LaRue (2006). Also see the special issue of *Villanova Law Review*, 54(4), 2007.

24. For an earlier round of criticism of how philosophy of science was deployed in the 1982 Federal case *McLean v. Arkansas*, see the exchange between Ruse (1992, 1996) and Laudan (1992).

25. See Sacks (1972) for a conception and analysis of membership categories. Coulter (2001), Hester and Eglin (1997), and many other ethnomethodological studies make use of the idea. In this paper, I am invoking the idea without going into a detailed analysis, though such an analysis could prove to be interesting and informative. As I use it here, a membership category means, simply, a conventional term used to categorize people and social activities. In many cases, individuals actively seek inclusion, while in others they attempt to avoid or evade inclusion. Some membership categories count as terms of praise, while others imply disrepute.

Normative Engagement and Descriptive Accuracy in Science Studies

Commentary on "From Normative to Descriptive and Back: Science and Technology Studies and the Practice Turn", by Michael Lynch

Vincent Israel-Jost and Katherina Kinzel

I. INTRODUCTION

In his paper, Michael Lynch critically examines recent trends in STS that encourage scholars to turn *from* the description of scientific practices to a normative engagement with science. Lynch recognizes the need to include normative perspectives in STS research, but criticizes the sort of generalist and essentialist theorizing that underwrites the normative account of expertise given by Collins and Evans. His own stance on the question of normativity is casuistic in spirit and dissolves the boundaries between actors' and analysts' (normative) categories.

In this commentary, our goal is to reconstruct how the normative engagement with science emerges as a response to problems associated with purely descriptive accounts of scientific practice and to assess whether Lynch's normative approach is satisfactory in escaping these problems. In Section II, we take a look at the different ways in which descriptivism in science studies highlights the constructed and uncertain character of scientific knowledge, and we discuss the difficulty that in doing so, science studies may provide argumentative ammunition to groups and actors with questionable political objectives, such as creationists or climate change skeptics—to which we will refer as "the problem of wrong allies". In the third section, we present why a normative engagement with science is indeed necessary to avoid this difficulty. We argue, with Lynch, that such a renewed normative engagement needs to incorporate the central lessons of the practice turn. Two challenges result: First, STS scholars need to find the right balance between descriptive and contextual approaches on the one hand, and normative proposals formulated in more general terms on the other. Second, STS scholars need to find ways to ground and justify their own normative proposals in the absence of stable and certain normative standards. In Section IV, we then present how Collins and Evans' conception of expertise addresses the problem of wrong allies and point to several shortcomings of their work. In the last part of our paper, we focus on Lynch's own stance toward normativity in science studies. We point out important differences in the way Lynch uses the terms "expert" and "expertise" compared to Collins and Evans, and we highlight how this permits him to respond to the challenges resulting from the practice turn. We conclude, however, that Lynch's own conception of what a normative engagement with science may look like remains too crude to provide a clear answer to the problem that STS's emphasis on uncertainty can serve the wrong forces.

II. UNCERTAINTY AND WRONG ALLIES

In recent years, the lack of scientific certainty has turned from a topic of primarily philosophical interest into a political issue. Pro-business strategists seeking to promote climate change skepticism and religious organizations pushing to put creationism on school curricula have pointed to the uncertain and contested character of scientific results to promote their objectives. Bruno Latour (2004) links the science skeptics' arguments to the research practices of science studies. He expresses the worry that when analyzing the practices of scientific consensus formation in such a way as to highlight their historically contingent and epistemically uncertain character, science studies provide ammunition to social and political actors with questionable intentions: If these groups and actors can convincingly argue that what appear to be settled scientific matters actually are uncertain and open to revision, they have an easy job of it. To the degree that STS research calls into question the consensual certainties of established science, it finds itself aligned with wrong allies. If it is the emphasis on uncertainty that helps to create such problematic alliances, we need to understand better what are the arguments that lead science studies to challenge the certainty of science.

Latour claims that the highlighting of the constructed character of scientific facts in science studies has been too closely associated with criticism of the established science, and with skepticism about scientific results. The identification of sociological accounts of knowledge with skepticism has a long history. On the one hand, the identification is partly motivated by early theories of ideology that sought to uncover the social causes that systematically distort ideas and perceptions of reality (e.g., Mannheim, 1929/1995). On the other hand, in the infamous *Science Wars*, opponents of social constructivism found fault with social explanations of scientific knowledge which, in their eyes, set out to debunk scientific knowledge and undermine the authority of science (e.g., Gross & Levitt, 1994). If social causes are primarily conceived of as biasing and distorting factors, revealing the social roots of accepted knowledge claims is identical to criticizing them.

Still, many sociologists have denied that social explanations and debunking go hand in hand: Pointing to the social roots of scientific knowledge does not necessarily amount to criticism, they argue. The *locus classicus* of this argument is to be found in the Strong Programme's symmetry principle, according to which the "same types of causes would explain true and false beliefs" (Bloor, 1976/1991, p. 7). The symmetry principle presupposes that social factors are not conceived as biasing factors, but as factors constitutive of knowledge that provide the basis for the explanation of beliefs, whether they are true or false, rational or irrational, successful or unsuccessful. This *ipso facto* implies that social factors can be causes of true beliefs, and hence do not constitute distorting, but rather constituting, factors.

However, even if sociology can describe and explain scientific knowledge without directly debunking it, social explanations of knowledge have also been found to bear problematic consequences for questions of theory appraisal and theory choice. The central idea is that if all knowledge claims, and even the standards that we use to justify these knowledge claims, are considered a contingent product of local and specific historical conditions and social negotiations, our decisions between different knowledge claims are to some degree arbitrary, groundless, contingent, or at least fundamentally uncertain. The fear that highlighting the relevance of social (rather than epistemic) causes in scientific knowledge processes destroys the rational basis for theory appraisal is one prominent subgenre of this more general concern (see, e.g., Boghossian, 2006; Franklin, 1994; Laudan, 1981; Roth & Barrett, 1990).

The problem of theory appraisal is often linked to the notion of relativism, but relativism comes in a variety of forms and degrees of strength, and it is not entirely clear how drastic the relativist consequences of different forms of social constructivism are. Barry Barnes and David Bloor argue that relativism is compatible with at least some realist commitments (Barnes, 1992; Bloor, 1976/1991, p. 34; Bloor, 1999), and in the "Epistemological Chicken" debate, Harry Collins and Steven Yearley present relativism as a rather unproblematic version of anti-foundationalism. Relativism, then, neither implies that "anything goes" nor that one cannot make reasoned judgments about which knowledge claims to accept; it only means that these judgments are never without presuppositions. Sociological relativism means "the rejection of any kind of foundationalism and its replacement . . . by permanent insecurity" (Collins & Yearley, 1992, p. 308). The upshot of these arguments is that relativism, even if it indeed follows from social constructivism, need not be considered very threatening in itself.

Maybe Collins and Yearley are right. But the problem of wrong allies remains. In order to reach the verdict of uncertainty that provides argumentative ammunition for those wishing to attack accepted scientific consensus, neither a sociological debunking of science nor radical relativism is required.

The mere highlighting of the historical contingency and contested nature of scientific results is enough to be used to promote "artificially maintained controversies" (Latour, 2004, p. 227) about such issues as climate change, evolution, and the harmful effects of tobacco smoke.

To make matters worse, meta-levels of analysis are affected by uncertainty as well. If the standards against which competing knowledge claims are evaluated are socially situated and never without presuppositions, even the methodological canons of SSK and science studies are to be regarded as contingent, historically situated, and socially construed. This issue has been discussed under the label of reflexivity (Ashmore, 1989; Woolgar, 1988). This implies that no ultimately compelling viewpoint on science can be found by "going meta". Each layer of analysis is found to face the same problems of uncertainty, which, as we will see later, complicates the task of developing normative considerations that would allow STS to respond to the problem of wrong allies.

III. FROM THE PRACTICE TURN TO THE PROBLEM OF NORMATIVITY

The practice turn shifted the attention of science studies from theories as static products of scientific activity to knowledge practices in open-ended research processes, and it emphasized the importance of skills and tacit knowledge in the production of scientific knowledge, at the expense of abstract methodological rules (see the introduction of this volume, Section III.5). It at the same time was a turn to the social dimensions of science and to the local historical contexts of scientific activity and, as such, fostered precisely those types of arguments about contingency and uncertainty that are at issue when discussing how science studies may give voice to the wrong political forces. The practice turn also constituted an effort for greater descriptive adequacy and aimed at correcting for the overgeneralizations and idealizations of earlier normative accounts of science.

But the mere descriptivist approach to science associated with the practice turn cannot answer the problem of wrong allies. Purely descriptive accounts of how scientific results are achieved in concrete and local practices or of how science is entangled in its social context do not provide the argumentative resources that would allow STS researchers to take sides in scientific controversies or even to discriminate between genuine and "artificially maintained" controversies. This is where the need for a renewed normative engagement with science, as identified by Lynch, originates. But what type of normative engagement with science is it that is being called for, and how could it be realized?

Being normative about science can mean a series of different things. It can, for example, refer to judgments about how to demarcate science from pseudo-science or non-science. Demarcationist arguments usually imply that by following certain general methodological procedures and by adhering to specified epistemic values, science gains a greater degree of epistemic authority than other activities. Traditional demarcationist proposals, however, were unsuccessful in providing the criteria to clearly distinguish science from non-science, as the distinctive methodology they thought characteristic of scientific research proved to be partly imaginary. But normative approaches to science can also be situated at the level of theory-choice (rather than the level of demarcation), referring to the assessment of particular scientific knowledge claims and to decisions between competing knowledge claims in situations in which the conflicting parties do already adhere to accepted epistemic standards. It is normative judgments of this latter sort that science studies practitioners need to engage in, in order to circumvent being aligned with the wrong allies—they need to be able to discriminate between more and less credible, more and less justified scientific knowledge claims, promoting the closure of "artificially maintained" scientific controversies.

Reconciling this demand with the lessons drawn from the practice turn leads to two distinct challenges, the challenge of descriptive adequacy and the challenge of contingent standards. The first challenge results because one lesson from the practice turn was that close attention to how science is actually performed in local contexts is necessary to avoid the overgeneralizations and idealizations connected to the blind normativism of the early demarcationist projects. STS scholars thus need to make sure that the general considerations that seem to be necessary in order to get a normative perspective on scientific knowledge off the ground are not detached from local and particular contexts and that they remain descriptively accurate.

The second lesson of the practice turn has been not only to focus on scientific endeavor in its concrete and local historical contexts, but also that when analyzing science in this way, this reveals the historical contingency and epistemic uncertainty of science. As described in the first section, contingency and uncertainty can arise at different levels, such that not only scientific knowledge claims, but also the norms and standards on the basis of which those claims are assessed are affected.

This leads back to the difficulties of theory appraisal and reflexivity that we discussed before. The worry of relativism connected to the problem of theory appraisal is exactly that no normative perspective on scientific theory choice can be formulated once we grant that the outcomes of scientific processes are historically contingent and subject to contextual influences. Relativism so understood precludes the sort of normative engagement STS scholars call for, as it thwarts our capacity to ultimately justify decisions between competing knowledge claims.

It also relates back to the reflexivity problem introduced above. If our own standpoint of analysis is as contingent and uncertain as our object of analysis, the grounds on the basis of which we put forward normative proposals about which scientific knowledge claims to accept are too. Attempts at intervening normatively in scientific controversies can thus be challenged on the basis of them being as much subject to uncertainty as the positions in the controversies they seek to settle.

The challenge for STS scholars is thus to put forward and justify their own normative proposals in the absence of stable and certain normative standards.

In the following section, we will discuss Collins and Evans' account of expertise, as well as Lynch's own casuistic approach to normativity. We investigate their responses to the problem of wrong allies and evaluate whether they answer satisfactorily to the two challenges just outlined.

IV. EXPERTS OF EXPERTISE—COLLINS AND EVANS' NEO-DEMARCATIONIST PROPOSAL

Collins and Evans diagnose an additional normativity problem, the "problem of extension" (Collins & Evans, 2002, p. 235–237) that arises from dissolving the boundaries between experts and the public. Their proposed account of expertise seeks to regulate technical decision-making rights in processes at the intersection of science, technology, and politics. If successful, this account would respond to the problem of wrong allies by substituting to the certainty of science the trust in the judgment of experts. The role of STS would then be to provide expertise about expertise—that is, STS would serve as a basis for prescriptions about what contributions to technical decision-making are to be considered legitimate. This way, it would establish STS as a normative, rather than merely descriptive enterprise, and it would guarantee that STS forges alliances only with those forces identified as real experts. One could also argue that their account of expertise constitutes a response to the problem of relativism as it provides relatively stable ground for our judgments about whose and which knowledge claims to accept. However, Collins and Evans' proposal is anything but beyond dispute. For example, critics have argued that their essentialist conception of expertise downplays the extent to which what counts as relevant expertise depends on how certain issues are framed in the first place, and on how the meaning and relevance of seemingly technical problems are constructed—questions which are settled in broader social and political processes (Jasanoff, 2003; Wynne, 2003). In the present context, however, we will focus on whether and how Collins and Evans answer to the challenge of descriptive adequacy and the challenge of contingent standards.

How does the "expertise about expertise" strategy laid out by Collins and Evans perform with respect to descriptive adequacy? On several occasions in his paper, Lynch addresses and criticizes the generality of their programme. Collins and Evans wish to establish a general "science of expertise", which for Lynch is not sufficiently focused on details, specificities, and contexts. He therefore accuses them of disregarding the particular relations in which expertise is embedded, overgeneralizing and losing track of the practical details of the circumstances in which their theory is supposed to apply. His conclusion is that Collins and Evans "launch their theory in a purified space that severely limits their ability to address the very demarcationist questions they propose to resolve" (Lynch, p. 106). Lynch's criticism shows that Collins and Evans fail to respond to the challenge of descriptive adequacy when offering their science of expertise and, thus, do not take into account a central lesson of the practice turn.

Assessing Collins and Evans' strategy with respect to the second challenge that normativism faces—the contingent standards—we make a logical point. As Lynch points out, Collins and Evans' framework depends on a strict separation between analysts and actors, in which analysts keep a privileged point of view: It is analysts (STS scholars) who are to sort out which actors (scientific practitioners) qualify as real experts. In these judgments, however, the analyst has to rely on her own expertise about what constitutes expertise. And this pushes the question of how to identify expertise one step higher up. A regress occurs. How can we secure that the analyst really does possess the relevant expertise to identify real experts? In other words, how can Collins and Evans claim to escape the reflexivity problem described in the second section? Collins and Evans reply to this worry by stating that the analysts who seek to determine expertise can simply "proceed with an imperfect set of classifications, just as other experts proceed" (Collins & Evans, 2002, p. 255). But this answer already indicates that there might be difficulties in cases where two experts working with different, possibly contradictory, sets of classifications, arrive at different conclusions. Who's to count as the "real expert"? Do we need another meta-expert to decide who's the expert of expertise?

V.   LYNCH'S CASUISTRY—NORMATIVITY WITHOUT PRIVILEGE?

Lynch's opposition to Collins and Evans' approach does not amount to a plain rejection of the notions of "expert" or "expertise" but to a conception of those notions that is framed in a much more particularistic agenda. Lynch attempts to answer to the concerns raised by the problem of wrong allies while avoiding the objections to Collins and Evans' account of expertise that we identified in the previous section. Since his conception of "expertise" is for the most part developed by contrast to Collins and Evans', let us summarize the main differences that Lynch identifies.

First, according to Lynch, one crucial premise of Collins and Evans' account of expertise proves to be unsound. In order to provide expertise with some prescriptive power, Collins and Evans want to rely on a clear separation between two categories: that of actors and that of analysts. Only analysts would then be able to perform a normative role in the attribution of expertise, because they can take a bird's eye view and enjoy a privileged perspective based on their own "expertise about expertise". But if the respective normative powers of actors and analysts are to be contrasted, one has to make sure that the two categories are actually well separated. This is what Lynch opposes to, arguing that the two categories are conflated or, at least, not clearly demarcated. Indeed, because actors (scientists) constantly perform normative analyses within their own field, they are not just actors, but also analysts. Conversely, we could add that analysts (STS scholars) that provide normative prescriptions in a given field may also become active participants in this field by doing so. The conclusion is then that a clear demarcation between the categories of actors and analysts seems impossible to obtain and that this is an important charge against Collins and Evans' attempt to address problems of normativity.

Another way that Lynch's conception of expertise differs from Collins and Evans' is by avoiding any sort of general or essentialist account of expertise. This is where we are getting at Lynch's main desiderata regarding expertise, which is to avoid the related pitfalls of overgeneralization and insensitivity to contextual differences. One place of particular interest for him is the courtroom, where both questions of expertise and of the role of STS analysts in determining expertise can be at issue. In the Daubert case, for example, the debated question was whether the drug Bendectin, produced by Merrell Dow Pharmaceuticals Inc., could have caused serious birth defects to Jason Daubert and Eric Schuller. During the trial, a number of STS sources were cited in order to cast light on what constitutes good scientific evidence, but, apart from a questionable reading of those sources, what stood out was that the court was not willing to ground their appreciation of reliable and unreliable expert testimony on philosophical criteria. For Lynch, this points precisely to the problem of general accounts of expertise such as Collins and Evans'. Such accounts are too coarse to serve in an actual deliberation where they can be, and will be, easily contested. On the other hand, the ways judges proceed in their evaluations by considering various sources of evidence, listening to actors and groups with opposing interests, and deliberating, is a phenomenon of the greatest interest for Lynch, as it appears to be the road to a much more contextual and comprehensive approach to normativity that conceives of normative analyses as constitutive features of particular practices.

Lynch's approach to normativity not only differs strongly from Collins and Evans' generalist account of expertise, it is also capable of escaping the problems that this account faced, identified in the previous section. Indeed, Lynch's detailed focus on local and contextual judgments permits him to take into account the multi-factorial aspects of the determination of expertise, including, for instance, the fact that it is a political process. The same interest for details permits him not to take the privileged bird's eye view on expertise that is threatened by a regress. But then, the question is whether Lynch can solve the main problem that motivates his paper: the problem of wrong allies.

In order to do so, Lynch must be able to take some sort of normative stance on the basis of local studies, not towards science conceived in general terms, as this would conflict with his particularistic agenda, but towards specific issues at the intersection of scientific and technical matters and politics. His proposal, however, is hardly more than the sketch of a proposal. What is postulated here is a gradual emergence of normative judgments in a specific situation when enough descriptive efforts have been made in the first place. But we lack details regarding why, how, and at which point one can claim to have reached the possibility to express normative conclusions. There remains a strong tension in this enterprise between Lynch's avoidance of any privileged position for the analyst and the necessity to be normative to respond to the problem of wrong allies. It seems almost impossible to make a normative claim without at the same time occupying an epistemically or morally privileged perspective from which the claim receives its authority and commands assent. Whether Lynch's careful and more modest approach can be developed into a position that solves this tension certainly is the main question that follows from reading his paper.

4 Chemistry's Periodic Law
Rethinking Representation and Explanation After the Turn to Practice

Andrea I. Woody

I. THE TURN TO PRACTICE

Contemporary philosophy of science makes frequent mention of the "turn to practice", but what exactly does this signify? The phrase refers not to a single shift, but rather a cluster of related, and potentially interdependent, changes, each consisting in a retreat from a certain sort of abstraction:

- **Conception to representation:** Theories originally treated as abstract conceptual objects characterized predominantly by their logical structure are replaced by theories as artifacts, rendered as particular representations, whether linguistic, diagrammatic, or other, and associated with distinct sets of models. In important respects, the relation here is analogous to the one between propositions, taken as abstract objects, and the specific linguistic statements used to express them.
- **A priori to empirical:** Concepts such as evidence and explanation, long considered crucial for articulating the distinctive nature of science, were traditionally cast at the highest level of generality and analyzed in terms of syntactic and semantic conditions generated through a priori analysis. The turn to practice has replaced these analyses with accounts based on examination of the reasoning invoked by scientists in particular contexts that arise in their ongoing work. Here the methodological shift is at once from a priori principles whose normative weight is presumed to empirical investigations that must extract, and justify judgments of, the normative, and also from a top-down assumption of generality to a bottom-up construction out of the local and the contextual. In addition, the intention to situate philosophical analysis meaningfully within practice entails being respectful of, and accountable to, the categories of practice operative within the relevant scientific communities.
- **Ideal agent to human practitioner:** Drawing on empirical studies of reasoning, the turn to practice has replaced the ideal agent with something closer to an actual practitioner, considering what types of training and background knowledge such practitioners bring to their

investigations and asking in self-conscious ways what impact knowledge of human cognitive capacities should have on our analyses of the central concepts of practice.

- **Knowing subject to social epistemology:** The inherently social nature of contemporary scientific practice introduces a variety of issues surrounding how knowledge is generated and transmitted, how knowledge is intertwined with issues of authority, expertise, trust, and divisions of labor, and how social structure is established and perpetuated to coordinate community deliberation and action. We shift from the perspective of the individual scientist to that of particular disciplinary communities and their interactions with other such communities.

Arguably the first work in philosophy of science to advocate explicitly and self-consciously for a turn to practice was the self-labeled "New Experimentalism" led by Ian Hacking and Allan Franklin in the 1980s, and conditioned in part by scholarship in science studies by Pickering, Latour, Woolgar, Collins, Galison, and others (Hacking, 1983, 1988; Tiles, 1992). This research sought to examine experimental practice in ways that would counterbalance the theory-centric bias within traditional 20th-century philosophy of science. Experimental practice no longer was assumed to be nothing more than theory testing, and abstract concepts of evidence and data were traded in for discussions of measurement techniques, instrumentation, embodied skills, and the social organization of large-scale laboratory science.

Hacking's popular book *Representing and Intervening* (1983) laid out the agenda for a turn to practice, in large part by articulating the position of experimental realism, but in doing so it further entrenched a dualism between theory and experiment—the realm of the conceptual/linguistic versus the realm of practical action—which effectively perpetuated a high level of abstraction with respect to theoretical science, despite the fact that the turn to practice aimed to reject such abstraction. Whereas experiment was conceptualized as practice, theory essentially was not. In the next section, I will discuss the periodic law in chemistry with the aim of advocating for a parallel conception of theoretical practice, one engaged with the active construction and manipulation of representational artifacts and shaped by practical concerns and contingent, contextually determined goals. From this vantage point, representing *is* intervening of a conceptual sort that is significant for our understanding of science.

I.1. Challenges Faced by Any Turn to Practice

As a methodological stance, the turn to practice, no less than earlier traditions to which it is a response, faces distinct and substantial challenges. The first concerns how to avoid getting stuck at a level of particularity that evades any reasonable effort to generalize. While we do not want to assume uniformity in science nor essentialize in ways that cannot be justified and are potentially distorting, case studies typically would be of little value if they shed no light on activities and communities beyond those directly considered. I have written for several years now on the representational practices of chemistry, and while I believe that some of the issues I have explored are perhaps most vivid and perspicuous in the chemical sciences (and thus are capable of shedding light on the very nature of that science), I have always assumed, and in some cases know, that these same issues arise in other corners of science (see, especially, Woody, 2000, 2004a, 2012). The hope has always been to make claims of some broad sweep. But the induction, of course, is tricky.

The second challenge is perhaps the more vexing. When we build philosophy on strong intuitions and a priori reasoning, in effect, the normativity of the analysis is built in from the get-go. But when we turn to practice and take it seriously, there is a possibility that we will all become social scientists of a more thoroughly descriptive stripe, that is, more akin to certain anthropologists and sociologists, only without the proper training and skill set. What is required, instead, is a framework for analyzing practice that allows us to make assessments regarding the effectiveness or appropriateness of particular practices or actions situated within those practices. When we abandon the logical analysis of traditional confirmation theory to ground our concept of evidence, for example, we still require a notion of evidence that can underwrite judgments of justified belief—that is, a concept that does not deflate to nothing more than social consensus or acceptance.[1] The proper sorts of normativity must be retained by alternate means, but it is not clear how this may be achieved.

II.   THE PERIODIC LAW IN CHEMISTRY

I have characterized the turn to practice in broadest terms as a retreat from abstraction. Philosophers of science have long made use of scientific episodes and achievements in their reasoning. The issue at hand is not whether philosophers have traditionally looked to actual science to inform their work, for undoubtedly they have, but how exactly these events and achievements *have been characterized* within philosophical discourse. Chemistry's periodic law is interesting in this regard. It has served as stock example in contemporary discussions of the relative evidential weight of accommodation versus prediction even while its status as a law has seldom been scrutinized. In addition, there is a rich history within chemistry of alternative representations of periodicity, accompanied by healthy debates regarding their respective merits, that continues to this day (see, especially, Mazurs, 1974, and Van Spronsen, 1969).

In its 19th-century formulation, the periodic law expresses the empirical generalization that trends in various chemical and physical properties of the elements tend to follow a single repeating pattern with respect to increasing atomic weight.[2] According to historical lore, the validity of the law was demonstrated originally by a series of remarkable predictions made by the discoverer, the now immortal Dmitri Mendeleev. Philosophical debates about the relative evidential weight of prediction of unknown facts versus accommodation of known facts make frequent use of this episode (Maher, 1988; Howson & Franklin, 1991; Lipton, 1991; Scerri & Worrall, 2001; Akeroyd, 2003; Brush, 2007; Scerri, 2007).[3] Yet as we will see, it is not clear the example can serve its intended function, as vindication of a predictivist stance, in part because it is not clear in what respect exactly the periodic law was predictive. Closer attention to representational practice, I contend, suggests that the content of the law remained indeterminate through these early years in ways that complicate any notion of the law's predictive capacity.

Such complications have been overlooked, perhaps, because of an ungrounded assumption that this generalization could be made to fit the standard philosophical mold for scientific laws. During most of the 20th century, philosophical analysis of scientific laws focused primarily on issues relevant to larger philosophical projects, specifically those related to evidence and explanation. Appropriate analysis aimed to fill in the characterization of Hempel's (1965a) nomological account of scientific explanation and to serve as a cornerstone for traditional confirmation theory and projects aimed at justifying inductive inference. A satisfactory account was needed to provide a non-question-begging way to differentiate genuine laws from accidental generalizations, and also to resolve Goodman's (1954) new riddle of induction by identifying, out of the set of infinite possibilities, which candidate generalizations were legitimately law-like. In analyses that take on these challenges, laws are typically rendered quite differently from how they are expressed in practice, almost exclusively taking the form of universal conditionals.[4]

Chemistry's periodic law is an odd bird from this perspective. It is never explicitly cast as a logical conditional, and only seldom, in the earliest years of its development, were efforts made to generate a precise mathematical expression; indeed, most of the time the law is not rendered in words at all. Also, as far as I am aware, only the structuralists have attempted to make sense of the law in ways that would mesh with traditional philosophical analysis (Hettema & Kuipers, 1988). Even so, philosophical literature often seems implicitly to accept the periodic law as a genuine law. Discussions of accommodation versus prediction, for example, appear to assimilate it to traditional conceptions of law, for which the notion of prediction is generally well defined.[5]

I want to re-examine this particular example, highlighting an aspect of practice that has been curiously neglected: namely, the explicit representational choices made in expressing the periodic law. My discussion aims to exemplify the turn to practice first as a means of making intelligible why the periodic table was entrenched in the first place and in what respects it can be a powerful inferential tool for chemists, and consequently, why we should consider it to be explanatory. Additionally, this analysis offers resources for re-engaging with the accommodation versus prediction debate and confronting more explicitly the status of the periodic law *as a law*. I begin with the original establishment of the law and subsequently turn to its role in contemporary practice.

II.1. Establishing the Periodic Law: How Representing *is* Intervening

A flurry of mid-19th-century activity in the emerging science of chemistry paved the path that led to the "discovery" of the periodic law. Recognition of periodicity relied upon—indeed required—different types of background knowledge, which themselves relied upon the development of various supporting practices:

1. A large number of elements had to be identified and distinguished.[6]
2. The physical and chemical properties of these elements had to be established in considerable detail. In particular, elemental (atomic) weights had to be determined (i) with sufficient accuracy and (ii) according to a consistent set of conventions.

In essence, periodicity is a pattern that could emerge only after the community accumulated a sufficient number of data points, determined individually with adequate accuracy and precision and collectively by commensurable methods, standards, and conventions.

The state of such knowledge evolved rapidly throughout the 19th century. Many elements were isolated for the first time as a result of tremendous refinements in analytic laboratory techniques coupled with chemistry's blossoming professionalization and the accompanying swell in the populations of its practitioners. Achieving the requisite accuracy, for atomic weight as well as other quantitative properties, required the development of novel measurement procedures. These, in turn, demanded the tandem development of technologies and cultivation of skills by chemists in multiple communities scattered throughout Europe and England (see, e.g., Chang, 2007).[7] Until the mid-1860s, the *facts,* upon which the *relations* identified as the "periodic law" rested, were not available to practitioners. (It is worth noting, however, that several individuals did surmise various patterns of more limited extent that, in hindsight, might be taken as precursors of the periodic law.[8])

Establishment of a widely endorsed set of conventions for assigning atomic weights materialized only in the aftermath of Cannizzaro's pleas for a standard based on Avogadro's hypothesis at the Karlsruhe Congress in 1860, which Mendeleev attended while holding a postdoctoral position at Heidelberg.[9] By setting uniform standards for determining atomic weights, the Congress conventions removed a crucial obstacle. Periodicity does not emerge from every possible internally consistent set of conventions for determining the weights of elemental substances. The history of the meeting's negotiations provides a rich illustration of the demands for social coordination within and among scientific communities and the diverse factors that can pull against such coordination. After establishment of uniform standards, however, and given the ready availability of chemical information (a product of existing communication structures, including well-established professional journals), simultaneous or near simultaneous discovery of the periodic law is not overly surprising.

These early efforts encompassed a diverse set of representations, but within a short time Mendeleev's table was firmly entrenched as the dominant expression of periodicity.[10] Why was this the case? Textbook history asserts that it was Mendeleev's bold predictions, with their remarkable precision, that paved the road to the table's hegemony (see Table 4.1). Stephen Brush (1996), as part of a series of investigations into the role of predictive power in the acceptance of scientific theories, largely vindicates this traditional viewpoint, at least with respect to public statements published in textbooks and professional journals. But given the fact that within a relatively short time frame several men publicly articulated some version of the law of periodicity, we might ask why no one else displayed predictive capabilities comparable to those of Mendeleev.[11]

There were likely significant differences of opinion about the appropriateness of such predictions, which could appear overly speculative and thus not properly empiricist. Indeed, reticence is not hard to imagine given the surface similarities between various codifications of periodicity in the 1860s and 1870s and numerical relations suggested in earlier decades that had been ridiculed as pieces of Pythagorean numerology. More generally, there were likely relevant differences in interests and intellectual style among the individuals involved and across the communities they inhabited.[12]

While acknowledging all this, the response I will explore is of a different sort. Looking at the tables of Dmitri Mendeleev and the graphs of Julius Lothar Meyer, the other individual with the most compelling claim to have discovered the periodic law, I will argue that differences in the representational resources of these two sets of artifacts correspond to fundamental differences in the sort and range of predictions, and inferences more generally, these representations could support. To understand why the periodic table became entrenched, we must understand, furthermore, what forms of practice it could enable and sustain and how this potential meshed with the skills, interests, aims, and background knowledge of a burgeoning chemical community. Framing things in this way allows us also to see in what respects the choice was not altogether obvious.

Table 4.1   Mendeleev's predictions

Predictions of unknown elements:

	Prediction	Experimental results
	eka-aluminum (1871)	gallium (1875)
atomic weight	68	69.9
specific gravity	6.0	5.935 (4.7)
atomic volume	11.5	11.7
	eka-boron (1871)	scandium (1879)
atomic weight	44	43.79
	eka-silicon (1871)	germanium (1886)
atomic weight	72	72.3
specific gravity	5.5	5.47
atomic volume	13	

Additional predictions of atomic weight:

eka-manganese	100
eka-niobium	146
eka caesium	175
tri-manganese	190
dvi-caesium	220
eka-tantalum	235

Corrections to atomic weight:

	Accepted	Mendeleev
beryllium	9 or 14	9
yttrium	60	88
cerium	92	138
uranium	120	240
tellurium	128	125

The Structure of the Representations

Less than a year apart, Julius Lothar Meyer and Dmitri Mendeleev each published representations displaying periodicity, but the representations are strikingly different. Meyer published a set of graphs of physical properties, such as atomic volume versus atomic weight (see Figure 4.1), and Mendeleev

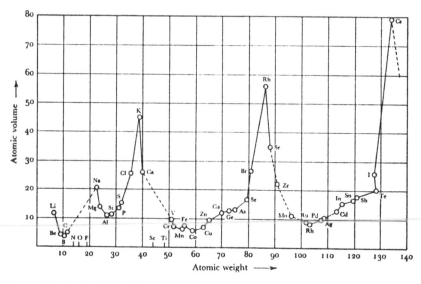

Figure 4.1   Lothar Meyer's graphical representation of periodicity

published a vertical form periodic table (see Figure 4.2), followed soon after by a horizontal form table (see Figure 4.3).

Meyer's Graphs

In Meyer's (1870) graphs, each dimension displays a quantitative measure of a property. Some particular physical property serves as the dependent variable, with atomic weight as the independent variable. There is an explicit metric, which in principle is dense and continuous.[13] The two dimensions are measured, again in principle, independently of one another. Thus the graph can be as precise as (i) relevant laboratory procedures for measuring the required quantities, and (ii) techniques for generating the graphs, will allow. In this representational format, the periodic law is identified as a characteristic *shape,* a repeating series of peaks. Identification is not based on strict repetition, however, because the particulars of the shape change from period to period. It is rather some generalized shape that constitutes periodicity.

Mendeleev's Table

Mendeleev's table, in comparison, is a two-dimensional array, or matrix, of discrete elements. The representation is based on two significant relations: ordering with respect to atomic weight and similarity with respect to chemical

ОПЫТЪ СИСТЕМЫ ЭЛЕМЕНТОВЪ.

ОСНОВАННОЙ НА ИХЪ АТОМНОМЪ ВѢСѢ И ХИМИЧЕСКОМЪ СХОДСТВѢ.

```
                    Ti = 50    Zr = 90    ? = 180.
                    V = 51     Nb = 94    Ta = 182.
                    Cr = 52    Mo = 96    W = 186.
                    Mn = 55    Rh = 104,4  Pt = 197,1.
                    Fe = 56    Rn = 104,4  Ir = 198.
                  Ni = Co = 59  Pl = 106,6  O· = 199.
H = 1                Cu = 63,4  Ag = 108   Hg = 200.
        Be = 9,4 Mg = 24  Zn = 65,2  Cd = 112
        B = 11   Al = 27,4  ? = 68   Ur = 116   Au = 197?
        C = 12   Si = 28   ? = 70    Sn = 118
        N = 14   P = 31   As = 75    Sb = 122   Bi = 210?
        O = 16   S = 32   Se = 79,4  Te = 128?
        F = 19   Cl = 35,5 Br = 80   I = 127
Li = 7 Na = 23    K = 39   Rb = 85,4  Cs = 133   Tl = 204.
                  Ca = 40  Sr = 87,6  Ba = 137   Pb = 207.
                  ? = 45   Ce = 92
                ?Er = 56   La = 94
                ?Yt = 60   Di = 95
                ?In = 75,6 Th = 118?
```

Д. Менделѣевъ

Figure 4.2 Mendeleev's tabular representation of periodicity (vertical form, 1869)

behavior in the aggregate. It is important to note that periodicity emerges only if there is sufficient fidelity in the ordering of elements with respect to weight. This is because the dependence between the two relations encapsulated in the table—a dependence that constitutes the higher order relation of periodicity—is only visible if the lower order relations are determined with relative accuracy. Combining the two relations amounts to transforming a one-dimensional array, the sequence of weights, into a helical structure. This

Reihen	Gruppo I. — R^2O	Gruppo II. — RO	Gruppo III. — R^2O^3	Gruppo IV. RH^4 RO^2	Gruppo V. RH^3 R^2O^5	Gruppo VI. RH^2 RO^3	Gruppo VII. RH R^2O^7	Gruppo VIII. — RO^4
1	H=1							
2	Li=7	Be=9,4	B=11	C=12	N=14	O=16	F=19	
3	Na=23	Mg=24	Al=27,3	Si=28	P=31	S=32	Cl=35,5	
4	K=39	Ca=40	—=44	Ti=48	V=51	Cr=52	Mn=55	Fe=56, Co=59, Ni=59, Cu=63.
5	(Cu=63)	Zn=65	—=68	—=72	As=75	Se=78	Br=80	
6	Rb=85	Sr=87	?Yt=88	Zr=90	Nb=94	Mo=96	—=100	Ru=104, Rh=104, Pd=106, Ag=108.
7	(Ag=108)	Cd=112	In=113	Sn=118	Sb=122	Te=125	J=127	
8	Cs=133	Ba=137	?Di=138	?Ce=140	—	—	—	
9	(—)	—	—	—	—	—	—	
10	—	—	?Er=178	?La=180	Ta=182	W=184	—	Os=195, Ir=197, Pt=198, Au=199.
11	(Au=199)	Hg=200	Tl=204	Pb=207	Bi=208	—	—	
12	—	—	—	Th=231	—	U=240	—	—

Figure 4.3 Mendeleev's tabular representation of periodicity (horizontal form, 1871)

in turn can be manipulated in various ways to produce a two-dimensional image for the printed page. The helix can be flattened into a spiral, something in fact suggested by Heinrich Baumhauer (1870), or alternatively, sliced along an edge and flattened into a table, as we find in Mendeleev's table, as well as the "telluric screw" of Beguyer de Chancourtois (1862).

Undoubtedly, one could argue that these two sets of artifacts, Meyer's graphs and Mendeleev's table, do not have identical content.[14] But for our purposes, it is not whether they are identical in content that is pertinent but whether they were genuinely competing contenders for representing the chemical phenomenon of periodicity. For this, the judgments of practitioners seem definitive, and there is ample evidence that both were taken seriously. Official announcement in 1882 of the Royal Society's Davy Medal, awarded jointly to Mendeleev and Meyer for the discovery of the periodic law, for example, makes explicit mention of both formats. To guard against the possibility that mere lack of familiarity with Meyer's graphs will inhibit us, in our contemporary context, from seeing them as contenders, I would offer Mendeleev's own linguistic expression of the law: "I define the law of periodicity as the mutual relations between the properties of the elements and their atomic weights which can be applied to all the elements. These relations have the form of a periodic function" (Mendeleev, 1871). It seems undeniable that Meyer's graph is as legitimate and direct a representation of the law, as expressed by Mendeleev, as the table is. Indeed, perhaps the most intriguing issue here is that while both representations were legitimate contenders, and were treated as such by practitioners, one could hardly deny that what these artifacts represented was not equivalent (nor even straightforwardly inter-translatable).

How should we make sense of this? At the time of its introduction, the content of the periodic law was neither obvious nor settled. There were standard under-determination issues; from finite data, chemists had to *project* the law, and there were multiple ways to do so. It was not even clear what *sort* of relations grounded the periodic law, as evidenced in part by the widely varying formats. Yet because it would condition practitioners' thought and direct research efforts, the particular representation scheme adopted would likely influence the subsequent specification of the law's content. In this regard, such representational choices are substantive and significant for the future elaboration of the content these representations aim to capture. In this respect, representing *is* intervening.

We can perhaps see this point most vividly by returning to the historical story, with its stress on predictions, and posing a question. Are predictions of the sort Mendeleev made more readily supported by one representational format than the other? To approach this question, we must distinguish two types of predictions: (i) predicting the *existence* of new elements, and (ii) predicting the *physical and chemical properties* of elements, whether previously identified or not.

Predicting the Existence of New Elements

In Mendeleev's table, establishing the length of the period is sufficient for gaps to appear. This requires some nontrivial assumptions, of course, concerning the relative completeness of data, the accuracy of measured properties, and the constancy of period length. Depending on the precision with which other properties, including weight, are determined for the surrounding elements, the exact *position* of a gap may be more or less determinate, but the mere *existence* of the gap is a necessary consequence of the establishment of the period. The table makes this absence starkly visible, and this is all that is required for the prediction of missing elements.

The format of Meyer's graphs implicitly suggests continuity, because one tends to extrapolate a continuous function from the data points (see Meyer's own graph in Figure 4.1). This characteristic could lead one to infer that any point on the function is compatible with the law. But contrary to what the graphical format suggests, not all atomic weights are possible, nor did anyone at the time assume they were. Moreover, because the representation holds an explicit metric, locating a missing element would require determining a quantitative pattern for increasing atomic weights (the spacing, in effect, of x-coordinates for the data). Clearly, the graphical format does not indicate the presence of missing elements in any straightforward sense. It does not reveal "holes".

Predicting Properties of Elements

Turning now to consider the prediction of specific properties, let us start with Meyer's graphs. Without a means of locating points corresponding to particular elements, there can be little hope of predicting any given element's properties. Even if one could determine the atomic weight of a given element (the x-coordinate mentioned previously), one also needs to codify explicitly the shape corresponding to the periodic function in that region of the graph. If this could be done, however, one could in principle generate precise and accurate predictions merely by reading off the corresponding y-coordinate for the data point.

For Mendeleev, any prediction of quantitative properties must rely on some calculation of ratios, something not explicitly supported by the table. Since the table has no explicit metric, but only ordering relations, the table reliably supports only interval predictions that result from this ordering. In other words, it can provide something we might loosely call "ballpark" predictions, if any predictions at all.

Now recall the precision of Mendeleev's actual predictions of various properties (Table 4.1). Does it seem reasonable to assume that Mendeleev relied on the table in making these predictions, or that in some meaningful way the table has the capacity to support such predictions? I do not see how; the very format of the representation rules out the possibility. The point is relevant for those philosophical projects that use the periodic law to consider the role of predictive power in theory assessment. What evidential role should the predictions have played in this case? Should they increase confidence in the periodic law as expressed in the table? Or alternatively, in some abstract, unarticulated, conception of the periodic law? (Or perhaps simply in Mendeleev himself?) We will return to this issue briefly in Section IV, because it exemplifies a concrete benefit of the turn to practice.

Representations as Tools

We have discussed only one very specific use of these representations—the generation of particular sorts of predictions—and this focus has been imposed somewhat artificially because of the presumed role of prediction in acceptance of the law. Perhaps further insight may be gained by considering the roles Meyer and Mendeleev intended for their representations. Here we can draw on detailed historical work, most recently by Gordin (2004, 2012) and Scerri (2007), and earlier by Bensaude-Vincent (1986). Both scientists produced their representations in the context of writing textbooks, but the two scientists' basic conceptions of chemistry and their priorities for the discipline were different. Meyer had strong empiricist leanings and tended to focus on measurement and quantitative analysis of substances' physical properties. Meyer's work also was explicitly sympathetic to Prout's Hypothesis, which postulated that all elements were composed of some more fundamental primary matter. Here, the quantitative specificity of the graphical format is advantageous. Highly precise quantitative data would be required to probe underlying constitution. Mendeleev's work, on the other hand, included disavowals of any Proutian hypotheses. He tended to foreground chemical properties over physical properties, his overarching aims were predominantly classificatory, and the guiding force for the table was concern with an orderly presentation, a framework, for the entire domain of chemistry. Ultimately, Mendeleev desired a structure both useful in pedagogical contexts and capable of providing directives for further research. He articulated "element" as the core concept of chemistry and used the periodic table to express the lawful regularities of these elements.

Recourse to the specific aims of individuals certainly helps to rationalize their representational choices. But if our goal is more generally to understand acceptance of the periodic law and entrenchment of the tabular format, we must expand the circle further to consider a wider chemical community alongside a more comprehensive rendering of practice. The value of particular representations, as artifacts or tools employed in inferential practices, is determined by several pragmatic features, including the general reasoning practices available to the community and the other intellectual tools that support these practices, as well as the types of questions practitioners wish to address or consider compelling. Let me briefly compare the two representations with regard to dominant characteristics of 19th-century chemical practice. This quick sketch cannot hope to capture all the relevant details, especially the diversity, of chemical practices across Europe at this time, but I believe the factors outlined here are indeed central ones and thus provide a good place to start.

On first consideration, it seems plausible that chemists would prefer the graphical format. Efforts to legitimate the new science created pressures both to establish chemistry's empirical grounding and to align the discipline with the more mature science of physics. The graphical format, with its emphasis on measurement and quantitative specificity, could facilitate both. Graphical representation of data, in fact, gained predominance in chemistry during this period. But these are reasons for general representational preferences, not reasons specifically for the community to prefer a particular representation of periodicity. The quantitative precision of Meyer's graphical format hinders many of the inferences that the periodic law promised to facilitate. Because periodicity is essentially a topological feature of the graph that is not strictly repeated, recognition of the pattern requires indifference to variations of shape. At the same time, using the periodic function either to direct ongoing research, including the search for new elements, or to provide explanation for chemical reactivities, valences, and trends in physical and chemical properties requires responsiveness to details to a degree that could hardly be made robust given the perceptual capacities of practitioners and the tools at their disposal. As we saw earlier, these graphs could not plausibly support predictions of the sort made by Mendeleev. More generally, because chemically similar elements are spatially separated in the graphical format, comparative perceptual judgments that might otherwise reveal informative trends or patterns are obscured. In essence, there is a tension here between demands for indifference and responsiveness that makes the representation challenging to use in the desired ways. These graphs display periodicity without making the implications of periodicity perspicuous.

Mendeleev's table imposes discreteness by eliminating or suppressing (or perhaps we should say transforming) certain information. Thus, in contrast to Meyer's graph, requirements of indifference and responsiveness are *internalized within* the structure of the tabular format. The creative work necessary to produce the table far outstrips that required to use it. The resulting structure allows gaps to appear, something which itself provided directives for research, even in the absence of explicit quantitative predictions. In addition, classification is highly efficient as a result of two-dimensional indexing. Many inferences require attention only to limited portions of the table—one need only move along a row or column, or be mindful of nearest neighbor relations—and effectively involve judgments of set membership rather than quantitative comparison. The table, taking advantage of the perceptual capacities of its users, employs spatial arrangement to heighten inferential utility.

But the table's value was, and still is, contingent upon the sorts of reasoning the community needed to do, and this in turn is dependent upon the discipline's overarching aims, interests, and current state of knowledge. In the late 19th century these included (i) imposing order on a daunting and diverse set of facts (something that Mendeleev saw as a significant challenge to successful pedagogy), (ii) investigating valence properties and their consequences, and (iii) discovering new elements. As the community's knowledge base and interests evolved, the table faced significant challenges, including the incorporation of the noble gases, the location of helium, and the inclusion of ether. In each case, the structure of the table arguably made the proper judgments more difficult to recognize and accept (Stewart, 2007). Mendeleev's own reticence to acknowledge the existence of noble gases exemplifies how representational format conditions the pathways of our thoughts and conceptualizations.[15] Certain possibilities seem almost preordained, while to recognize others, even as possibilities, strains tremendously against intuitions cultivated by the representational tools we habitually employ.

It is noteworthy that in the face of such challenges, the chemical community held fast to the periodic table, amending it to accommodate missing elements and erroneous placements. The entrenchment was solid. I have argued that the unique representational resources of the table fruitfully supported a wide range of 19th-century chemical reasoning in ways alternative representations could not. The table's power was a consequence of the way content was represented, specifically as a discrete structure of relations that did not depend on the absolute magnitude of atomic weights. (With the discovery of isotopes and the identification of atomic number, a more direct ordering relation would replace the one originally provided by atomic weight and vindicate, by appeal to atomic theory, the representation's imposed discreteness.)

The practices of late-19th-century chemistry established the discipline and significantly shaped the path of its development. Patterns of reasoning invoked then still thrive today, and the desire for control over a large, diverse set of facts has become only more pressing. A brief discussion of contemporary uses of the table will reveal the continuing productivity of the representation, something that could not be evident in its early days and yet is partially a consequence of the original entrenchment.

II.2.   The Periodic Table in Contemporary Context: Linking Representation to Explanation[16]

The periodic table is now the single most iconic representation in contemporary chemistry (see Figure 4.4). Its centrality to chemical education and practice is easily discerned from the table's ubiquitous presence in classrooms, laboratories, and textbooks. But why exactly is it there? What role does it play? No doubt the periodic table is a primary tool for classification. Descriptive chemistry constitutes an enormous domain of inquiry, and the periodic table aims rather heroically to impose structure, and intelligibility, on this vast landscape.

A well-known university chemistry textbook makes the point in this way:

> With the help of this *periodic law,* it is possible to organize and to systematize the chemistry of the elements into a manageable subject.

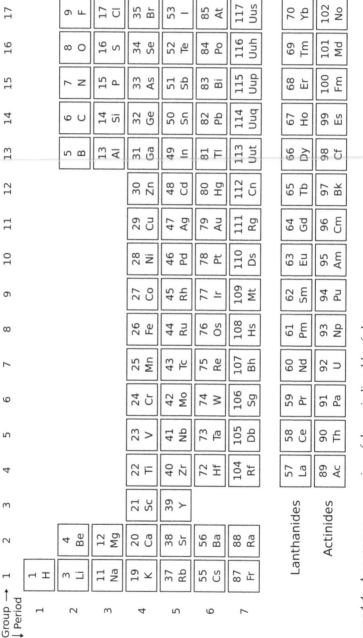

Figure 4.4 A contemporary version of the periodic table of elements

Learning descriptive chemistry then becomes a process of discovery and assessment of facts, prediction and verification of chemical behavior, and evaluation of correlations and explanations. All of this leads to an understanding of why elements have the properties they do. (Mahan, 1975, p. 569, italics in original)

This statement is telling because it suggests that the ultimate goal is not organization *per se* but rather what such organization facilitates. According to Mahan, the periodic law makes certain information manageable and, as a result, facilitates "discovery", "assessment", "prediction", "verification", and "evaluation". And here is the punch line: "All of this leads to *an understanding of why elements have the properties they do*" (emphasis added). In other words, the periodic law, by facilitating central aspects of inquiry, plays a significant role in *explaining* the properties of the elements.[17] I want to take Mahan's assessment, as that of a professional practitioner, seriously. While this is just one statement extracted from one textbook, I believe it to be representative of a common view, which often remains implicit.

Mahan's statement, like those of many 19th-century chemists, refers to periodicity as a law, and just like Mendeleev, Mahan characterizes the law in terms of periodic functions: "the properties of the elements are periodic functions of their atomic numbers" (Mahan, 1975, p. 569). Yet this generalization, as stated, seems too imprecise to organize and systematize in the requisite ways. Most obviously, the wording makes no mention of the periodicity for different properties being the same, even though this is a crucial feature of chemical periodicity, the very one that enables a single representation like the table to suffice for a wide range of physical and chemical properties. Thus it seems implausible that this particular generalization could be considered explanatory in the specific sense suggested by Mahan. Moreover, as mentioned previously, there is no explicit logical or mathematical rendering of the law that might serve as a more precise substitute. Having argued that, in comparison to Lothar Meyer's graphs, there are specific reasons why a tabular format was particularly well suited for 19th-century practice, I now suggest that the very same representational features that helped to entrench the periodic table also grant the capacities that Mahan sees as vital to the law's current explanatory power.[18] In a nutshell, it is the periodic *table* that is explanatory in Mahan's sense.

As we have seen, two structural features enable the table to reveal relations among chemical properties effectively: (i) suppression of quantitative magnitudes to produce simple ordering relations, and (ii) a two-dimensional arrangement that embodies periodicity in the representation scheme itself, effectively projecting the *content* of the periodic law unto the *format* by which it is displayed. These structural features highlight patterns that otherwise would be obscured. Two-dimensional display of the ordering relation creates vertical and horizontal, as well as diagonal and other "close neighbor", relations. The simultaneous presentation of many such relations facilitates

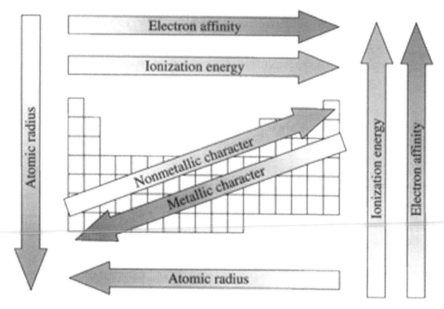

Figure 4.5 Trends for some chemical and physical properties of the elements as organized by the modern periodic table

the recognition of higher-order patterns. Categories such as "halide" and "transition metal" emerge and trends in electronegativity, ionization energy, and oxidation states can be rationalized, in part because they become readily identifiable. Patterns of atomic volumes and ionic radii become conspicuous and consequently make a multitude of facts concerning chemical bonding and reactivity comprehensible and predictable (see Figure 4.5).

The periodic table's ability to guide reasoning, and inquiry more generally, flows from this particular way of aggressively projecting order onto its complicated domain. The nature of this guidance is suggestive, however, rather than explicitly inferential. The periodic table displays patterns and implicitly represents them to be far-reaching and robust. The prediction of a new element, for instance, follows from recognition that something is absent from the imposed structure. The table reveals a "hole" and insinuates that something is missing. But there is no formal structure to generate quantitatively precise predictions.

It is also significant that much of the reasoning supported by the periodic table is analogical. A category such as "halide" is important within the practice precisely because it allows us to reason, "If compound *X* displays certain properties, then the related compound *Y* will display analogous properties". In a domain such as descriptive chemistry, analogical reasoning is an important tool for gaining control over complexity; a multitude of facts can be managed if they fall into conspicuous patterns rather than remain independent and singular. As I have argued, a tabular format, in particular, carves out such patterns. But precisely because the reasoning is analogical, conclusions must be accepted tentatively and recognized as defeasible. To assume similarity in the chemical reactivities of chlorine and bromine requires, on the surface, simply skimming down a column of the table, while beneath this surface, warrant for the projectibility of the pattern is complicated; chlorine is like bromine only in certain respects and to certain degrees, and the two elements behave similarly only under certain conditions.

Although many will agree that the periodic table organizes, and even classifies, there is no similarly robust intuition that it explains.[19] Yet the insight implicit in Mahan's statement is that an understanding of the chemical elements and their properties consists, in large part, in the recognition of patterns among these elements and a capacity to use these patterns to guide chemical inquiry. My discussion aims to establish that the representational format of the periodic table makes these patterns evident, and partly as a consequence, enables types of reasoning that facilitate dominant forms of inquiry. In this respect, from the perspective voiced by Mahan, the table has explanatory power. This is not to claim that the periodic table, taken in isolation, supplies complete explanations of the properties of elements, or substances more generally. Rather it is to claim that in contemporary practice the periodic table has an essential role in such explanations.

Today, of course, we tend to view the periodic table, ordered as it is with respect to atomic number, as representing the electronic structure of atoms. Is it possible that all the productivity, and explanatory power, I am claiming for the periodic table is more accurately located in the theory of electronic structure? Obviously the periodic table does not tell us why the patterns it displays exist, and a theory of atomic electronic structure may indeed help with this (though the claim is contentious according to many, including Van Brakel [2000] and Scerri [1997, 2007]). The issue at hand, however, is whether the table may be considered explanatory of the properties of elements and substances. What explains the patterns themselves is a separate issue, in principle, and one that raises familiar questions concerning levels of explanation and threats of regress.[20] Even if we assume every trend and pattern identified by the periodic table could be captured through electronic structure, unless we also presuppose a reductionist account of explanation, it does not follow that this electronic theory necessarily gets explanatory credit with respect to particular chemical properties.

And let us be clear about exactly what electronic theory needs to be invoked. A general scheme for electronic configuration based on assignment of the four quantum numbers provides theoretical foundation for the maximal number of electrons in each electron shell, which underwrites the lengths of the periods, a fact that is admittedly arbitrary and contingent given the bare structure of the periodic law. But as Scerri (2007) rightly points out, accounting for periodicity requires more than this. It requires a way to assign electronic configurations to particular elements—in essence, to determine the order in which electron shells are filled. Yet assignment of electronic configurations, especially for the tricky transition metals, arguably still cannot be derived from underlying theory in a principled manner. (See Scerri [2007] for an extended argument.) Moreover, the extra information in a full-blown quantum mechanical treatment, which in principle could generate quantitative values for a range of measured observables, would obscure many of the patterns clearly revealed by the periodic table. This issue is simply a variant of the one we already confronted with respect to Meyer's graphical format. A certain indifference to quantitative specificity is required to make these chemical patterns perspicuous. The periodic law gives us "halide" while the Schrödinger equation does not (Woody, 2000). At best we may read "halide" into the formalism *post hoc*—something akin to seeing objects in drifting clouds.

While our theoretical understanding of atoms has been fundamentally transformed since periodicity was first articulated, the periodic table has not been eliminated as an outdated relic, providing further evidence of continuing utility. The heuristic value of the table is a primary reason for this longevity. So is the fact that the table does a wonderful job of capturing many of the central categories and concepts of chemical practice (e.g., noble gas, halide, transition metal, electronegativity, ionization energy). Instead of eliminating the periodic table, a quantum theoretic basis for the ordering of elements, to whatever extent that we have one, only further entrenches its centrality by providing theoretical legitimacy without threatening, in any serious way, its utility. Thus, I would argue we may recognize the periodic law as explanatory even while granting that some ideal atomic theory should in turn explain periodicity.

This is precisely the perspective voiced by Mahan in the textbook quotation with which we began. The periodic table is an intelligent map of the domain of chemistry. Like all good maps, it helps one navigate effectively not by presenting the most detailed representation possible but by highlighting features of the landscape that are taken to be significant. As Mahan proclaims, the explanatory power of the periodic table stems from its organizational capacity, which has granted both great productivity and cognitive intelligibility. It is not, however, that organization per se is valuable or necessarily explanatory. Rather, it is that in this particular domain, classification and organization are prerequisites for productivity. The table's achievements stand in relation to the vast set of empirical facts that it organizes and classifies. In effect, because the aims of chemistry demand control over large and diverse sets of information, as well as a foundation for analogical reasoning, the periodic table is explanatory. And as our original comparison of graphical and tabular formats reveals, it is not the periodic law, in the abstract, that facilitates this reasoning, but the specific representation embodied by the periodic table.

III. REORIENTING DISCUSSIONS OF SCIENTIFIC EXPLANATION: A FUNCTIONAL PERSPECTIVE

In contrast to my assessment, built upon a practitioner's judgment, the periodic table seems a rather poor candidate for explanatory status under any of the traditional philosophical accounts of scientific explanation. There is no explicit logical statement of the law; it does not reveal causal structure, identify mechanisms, serve as premises in articulated argument patterns, or in any obvious manner aid the attainment of empirical adequacy.[21] Rather, the table is valuable because of its ability to reveal, or make perspicuous, certain relations. This being the case, I might attempt to wield the periodic table as a counterexample to some of these accounts. Or I might use it as Exhibit A in introducing a new, alternative account. My diagnosis of the problem, however, rests at the level of the general enterprise rather than the specifics with which it has been articulated.

Traditional accounts of scientific explanation focus on one general question, "What is the nature of scientific explanation?" and—importantly—interpret that question in a particular way by assuming that an appropriate answer will reveal the common *logical and/or conceptual structure of individual scientific explanations.*[22] Each of the standard accounts—inferential, causal, mechanist, erotetic, and unificationist—draws on deep intuitions to offer a detailed response to this question. Yet the intuitions grounding each account are distinct, and such differences reveal the problem most persistent in the lineage of philosophical analyses of explanation: The explanations that are generated and endorsed across modern scientific communities are diverse and pluralistic, rather than homogeneous, in kind. An account of the features of individual explanations that is both rich in detail and genuinely unified would seem to be out of reach. It simply would not fit the multiplicity of practices we observe.

More importantly, this focus neglects other questions we could ask about explanation in science, questions that I would argue are equally important to consider, though they have received considerably less philosophical attention. One question seems to me a direct product of the turn to practice: What role does explanatory discourse play in scientific practice? Descriptive fit may entail that there is no uniform answer to this question either, but in any given case, whatever the answer may be would seem to form the core of any possible response to another related question—namely, why should we desire any particular theory (or representation) to be explanatory? Or in other words, why should we consider explanatory power a theoretical virtue? In posing these questions in this way, we displace from center the project of articulating necessary and sufficient conditions for individual explanations, while still assuming we can identify a considerable range of explanations robustly, and turn our attention to the reasons why, and in what regards, explanation might be crucial to particular scientific practices.

This issue requires, effectively, a functional analysis along the lines delineated, for different purposes, by Hempel (1965b), and drawing on the work of sociologist R. K. Merton (1957):

> The kind of phenomenon that a functional analysis is invoked to explain is typically some recurrent activity or some pattern of behavior in an individual or a group . . . And the principal objective of the analysis is to exhibit the contribution which the behavior pattern makes to the preservation or the development of the individual or the group in which it occurs. Thus, functional analysis seeks to understand a behavior pattern or a sociocultural institution by determining the role it plays in keeping the given system in proper working order. (Hempel, 1965b, pp. 304–305)

In short, what I call *the functional perspective* aims to reveal how the practice of explanatory discourse functions within science. In comparison to traditional accounts of scientific explanation, there is a shift in focus away from explanations as *achievements* and toward explaining as a coordinated *activity* of communities.

I discuss the functional perspective in greater detail elsewhere, arguing that explanatory discourse inculcates particular patterns of reasoning generally, and thereby functions to sculpt and subsequently perpetuate communal norms of intelligibility (Woody, 2003, 2011; Dear, 2006). By doing so, explanatory discourse frequently plays a methodological role for scientific communities by enabling coherent practice across practitioners and subcommunities. Here I will mention only a few issues that are especially relevant to our discussion of the periodic law. The periodic table's centrality in chemical education and everyday reasoning by chemists reflects the nature of its subject matter, the phenomena of chemistry. Precisely because elements, the fundamental chemical kinds, are so numerous (and the resulting compounds even more so), classification and pattern building are essential to maintaining coherent and productive cognitive practice. Identifying elements by their location within the periodic table has been a useful way to condition reasoning in the discipline in ways that contribute to this coherence. At the same time, entrenchment of the table reflects something about the overarching aims of the discipline. More than any other modern science, chemistry is oriented toward the production of novelty, in the form of new substances with specified properties, in ways that align it strongly with engineering and design work. The periodic table provides a framework for keeping track of diversity at the phenomenal level while also facilitating, largely through analogical reasoning, the compositional, building block structure that has enabled chemistry to generate cancer drugs, household plastics, agricultural pesticides, and nanoparticle-laced sunscreens.

The value of the functional perspective lies in its potential to reveal how explanatory discourse may support the particular epistemic and practical aims of given scientific communities. Considerably more detail is required to substantiate this value in the case of the periodic law but the outlines of the project should now be clear.

IV.   TAKING STOCK: IMPACT OF THE TURN TO PRACTICE

In closing, I want to circle back to the issue with which I began: the turn to practice within philosophy of science. Our discussion of the periodic law embodies each of the four shifts away from abstraction mentioned earlier:

1) **Conception to representation:** I have stressed the significance of the particular representational format embodied by the periodic table in discussing the initial acceptance of the periodic law as well as the entrenchment of the law in practice up to the current day, and I have argued that we need recourse to the particular representational format of the periodic table to properly assess both the role of prediction in the law's acceptance and the law's explanatory status.

2) A priori **to empirical:** My argument that the periodic table should be considered explanatory is grounded by appeal first to a prominent practitioner's public assessment and subsequently to contemporary disciplinary practice in chemistry. I have also given reasons, more briefly, why existing philosophical analyses, based on a priori intuitions, appear incapable of recognizing the table as genuinely explanatory.

3) **Ideal agent to human practitioner:** Analysis of the graphical and tabular formats makes crucial reference to the general perceptual capacities of practitioners, the kinds of background knowledge available to them, and the sorts of reasoning commonly employed within the relevant communities.

4) **Knowing subject to social epistemology:** In explaining both the original entrenchment of the periodic table and its role in contemporary practice, I have appealed to the aims of the relevant chemical communities. Moreover, the functional perspective focuses on explanatory discourse as a social phenomenon directed toward achieving and maintaining coherence within and across communal scientific activities.

My discussion here has been schematic in certain regards. A maximally compelling argument for the explanatory status of the periodic table built upon practice would require richer description of the relevant chemical communities, whether historical or contemporary, as well as grappling with nontrivial questions regarding how to differentiate communities, their practices, and their aims. There are many sub-disciplines within chemistry today, and distinct communities within each of these, and no doubt their practices, aims, and even explanations differ in telling ways. Even so, the status of the periodic table extends like an umbrella across these communities, helping to bind them together as a distinct scientific discipline. It is important to recognize, furthermore, that even this preliminary discussion of the periodic law brings to the surface ways in which the turn to practice can provide a necessary corrective to some existing discussions within philosophy of science. I consider four issues briefly below: accounts of scientific explanation, general analyses of scientific representation, the role of accommodation versus prediction in theory acceptance, and analysis of the concept of scientific law.

In advocating the functional perspective, I am suggesting one way the turn to practice helps us avoid unproductive cycles of research concerning scientific explanation. As discussed in Section III, there are a few distinct questions we could pursue with regard to explanation. If we focus exclusively on the structure of individual explanations, we bottom out with competing intuitions and no resources for adjudicating among them. Furthermore, a priori assumption of a monolithic account of individual explanations stands at odds with existing practice. The traditional project has been informative and productive in many ways, but the turn to practice opens up new avenues for exploring the nature of explanation in science and accounting for its epistemic significance, something I argue elsewhere is hard to generate from the traditional line of analysis.

The past decade has seen a regular cottage industry of philosophical research on representation in science. This attention might itself seem a consequence of the turn to practice, but appearances can be deceiving. Much of this research, motivated by the assumption that there is something distinctive about scientific representation, tackles the very nature of the representing relation (e.g., Bailer-Jones, 2003; Hughes, 1997; Suárez, 2004). Most treats representation as effectively, if not essentially, a two-place relation between the representation and its target. The aims of users are often mentioned in passing but do not figure in analysis, and few consider the specifics of representational format or the particulars of representational use. (A noteworthy exception is Perini [2005, 2010].) Callendar and Cohen (2006) argue persuasively against the very idea of a distinct representing relation in science, recommending instead a Gricean framework for making sense of all representation. This seems exactly right, but it misses the real target. If we believe there is something fundamentally distinctive about representation in science, why assume it is located in the representing relation itself rather than in the concrete social practices of representation cultivated by scientific communities? The turn to practice suggests a similar corrective to the fuss over false models and "fictions", where focusing exclusively on the truth-value of a given representation, itself a byproduct of the two-place relation perspective, generates the very problem it attempts to address, in part by suggesting that truthful representation is the norm.

Due attention to practice constantly reminds us that individual representations cohere in codified representational systems (see, e.g., Haugeland, 1998) that are themselves embedded in sophisticated practices for storing, organizing, and reasoning with information and constrained by the values, aims, and capacities of individuals situated in particular communities. If there is something *distinctive* about representation in science, my bet is firmly on it being located in the sophistication of the representational and inferential practices of the relevant communities. It is not the representations themselves, but how scientists develop, use, and build upon them that should draw our scrutiny.

Our discussion of the periodic law likewise reveals some naiveté in the literature concerning the role of accommodation versus prediction in theory acceptance. As we have seen, the periodic law could generate no precise quantitative predictions, because at the time it was not articulated in a sufficiently determinate way, and as a generalization, its content was effectively qualitative and comparative.[23] The debate surrounding prediction and accommodation presupposes a certain conception of evidence whereby we can speak meaningfully about predictions following from, or being consequences of, a given theory. This is not the case with the periodic law. Rather, Mendeleev, as a practitioner situated within a landscape of communities and their respective practices, generated precise quantitative predictions. The evidential role of these predictions thus becomes hard to ascertain, and consequently, the example cannot function in the philosophical debate as intended.

Much of this research also takes insufficient care to distinguish sociological reasons for the law's acceptance from factors that are more narrowly evidential. There is little doubt that Mendeleev's vindicated predictions, with their impressive precision, had tremendous influence on the acceptance of the periodic law by the chemical community and entrenchment of the tabular format (Brush, 1996, 2007). But Mendeleev also made numerous erroneous predictions (Scerri, 2007; Scerri & Worrall, 2001). Historical lore, it turns out, suffers from a significant selection bias, and this surely complicates the role of prediction even in the sociological story. I have argued, alternatively, that social acceptance of the law was crucially dependent on its particular representational format and the ways this format meshed with existing aims and practice. From the time of its introduction, the periodic law was productive for the discipline, in ways that far outstrip any narrow conception of prediction, and this is a central reason why it endured.

Finally, a few words about the elephant in the room. In what sense is the periodic law a law? Clearly it fits no standard philosophical mold, leading some to recast it in an attempt to preserve its status (Hettema & Kuipers, 1988), others to deny that it is in fact a law (Scerri & Worrall, 2001; Shapere, 1977), and others still to avoid the issue by shifting locution to "periodic system" or "periodicity" (Scerri, 2007).[24] But the turn to practice, as a methodological stance, perhaps suggests a way to take the law designation seriously, just as we know Mendeleev did. One virtue of traditional accounts of scientific laws is that they make the normative status of laws clear by highlighting their role in explanation and prediction. Laws that fit the traditional mold should be productive in practice; they offer guidance for research, and they answer questions. Yet this is exactly the sort of productivity provided by periodicity. It codified chemical types through its classification scheme, introducing a fundamental conceptual framework that could guide reasoning and generate research questions for the community. In other words, perhaps the normative status of the generalization, earned through its utility in practice, marks it as law-like (and—connecting back to our earlier discussion—reveals its explanatory power). There is no space to pursue the issue here but others, including Sandra Mitchell (2000), have argued for a more pragmatic account of laws in an attempt to make sense of generalizations in biology. Though the periodic law turns out to be a poor choice of example for the prediction versus accommodation debate, it might be a fine example for exploring alternative analyses of the nature of scientific laws.

V.   CONCLUDING REMARKS

Even in this single case study, the diverse advantages of the turn to practice are apparent. We gain a better understanding of the history of the periodic law in chemistry; the function that it serves within the discipline; and, consequently, an awareness of various shortcomings, some coupled with tentative insights, in philosophical projects concerning representation, explanation, prediction, and the nature of scientific laws. Challenges, no doubt, remain, but the potential pay-offs suggest these challenges are well worth pursuit.

NOTES

1. Of course consensus and acceptance may be significant components of genuinely normative accounts of evidence.
2. Properties that display periodicity include atomic volume, specific gravity, melting point, malleability, compressibility, latent heat of fusion, thermal conductivity, coefficients of expansion, among others. After general acceptance of an atomic theory in the 20th century, the periodic law was redefined in terms of atomic number rather than atomic weight.
3. There are two distinct issues we could discuss here: one is the relative evidential strength of prediction versus accommodation (an issue of evidence) and the other is the role of prediction versus accommodation in theory acceptance, either ideally or in particular historical episodes (an issue of theory acceptance). The philosophical literature frequently fails to differentiate clearly between these two related, but distinct, projects.
4. These analyses also typically interpret the conditional as a material conditional, an assumption that generates its own set of issues (and potential problems).

5. There are noteworthy exceptions. Scerri (2007) and Scerri and Worrall (2001) both give careful attention to the notion of prediction and discuss in detail problems inherent in attributing predictive power to the periodic system, which they deny is a genuine law. Shapere (1977) argues that the periodic table is an "ordered domain", something distinct from a traditional law or theory.

6. By the 1860s, more than 60 elements had been identified.

7. Perhaps most notable among them is the Belgian chemist Jean Servais Stas, whose careful measurements of atomic weights Mendeleev mentions in his (1891) textbook *Principles of Chemistry*.

8. Here is a partial list (with year of relevant publication) of scientists who postulated relations arguably concerned with periodicity prior to 1860: Johann Döbereiner (1817, 1829), John Hall Gladstone (1853), Peter Kremers (1856), Jean-Baptiste Dumas (1857), and Ernst Lenssen (1857).

9. Mendeleev himself changed conventions several times in the period between 1858 and 1868.

10. There were spiral and cylindrical representations, graphical as well as alternative tabular formats, and even several attempts at algebraic formulation. Excellent sources on this diversity are Van Spronsen (1969), Mazurs (1974), and for the earliest history, Venable (1896).

11. Those who wrote about periodicity after the 1860 Karlsruhe Congress included Beguyer de Chancourtois, William Odling, John Newlands, Gustav D. Hinrichs, Julius Lothar Meyer, and Dmitri Mendeleev.

12. Michael Gordin (2012) has detailed such intellectual differences between Lothar Meyer and Mendeleev, as well as differences in their general conceptions of chemistry as a science.

13. That is, in principle it could be mapped onto the real numbers. Of course measurement capabilities will determine how fine-grained quantitative determinations can be in practice.

14. A *set* of graphs of various chemical and physical properties measured in relation to atomic weight would be the closest analogue within Meyer's format to the information captured by Mendeleev's periodic table. However, the two sets of representations, one fully quantitative and the other essentially qualitative, are so different in kind that it is admittedly hard to know how to approach the issue of content identity.

15. For example, rather than acknowledge argon as a new element, Mendeleev argued that the new substance was actually a nitrogen compound, N_3.

16. The discussion in this section draws on Woody (2004b).

17. Note that my claim here is relatively weak, being neither that periodicity is sufficient nor necessary for explanation of the properties of chemical elements, but only that periodicity plays a significant role in such explanation within contemporary chemistry.

18. Although I do not state as much in Section II.1, I believe Lothar Meyer's graphs can serve as proxy for a wide range of alternative representations in my original argument for the representational advantages of the tabular format.

19. Scerri (1997, p. 239) expresses the more common viewpoint eloquently: "The periodic systems, both naïve and sophisticated, are systems of classification which are devoid of theoretical status in much the same way as the Linnaean system of biological classification or the Dewey decimal system of library classification. None of these systems can be regarded as theories since they do not seek to explain the facts but merely to classify them".

20. Hempel (1965a) considers this issue, arguing that a true universal empirical generalization explains an instance of that generalization, while explaining the generalization itself requires a separate argument.

21. It may be possible to incorporate the periodic table within Kitcher's (1989) unification account of explanation, and it would be worthwhile to attempt doing so. Even so, I would argue that taking the functional perspective is a more productive means of grasping what we mean when we attribute explanatory power to the law.

22. The inferential, causal, mechanistic, and erotetic accounts each provide a direct characterization of individual explanations (Hempel & Oppenheim, 1965; Hempel, 1965a; Salmon, 1984; Woodward, 2003; Bechtel, 2005; Craver, 2007; Van Fraassen, 1980). The unification account concerns itself, somewhat differently, with determining the overarching theoretical system that is most unifying and, as a consequence, confers explanatory status on the explanations provided by that structure (Friedman, 1974; Kitcher, 1989). But even here there is an assumption that the general desiderata for unified theoretical systems are invariant in ways that implicitly emphasize the structure of individual explanations. Consider, for example, Kitcher's characterization of argument patterns.

23. Scerri and Worrall (2001) and Scerri (2007) make essentially the same point, in somewhat different guise.

24. I have generally availed myself of the last option.

Theoretical Artifacts and Theorizing Practices

Commentary on "Chemistry's Periodic Law: Rethinking Representation and Explanation After the Turn to Practice", by Andrea I. Woody

Régis Catinaud and Frédéric Wieber

I. INTRODUCTION

In her chapter, Andrea Woody addresses the central question of this volume (*How has the practice turn modified both our conception of science and the way in which we analyze it?*) by showing us how an interest in scientists' practices can open up new ways of analyzing *theoretical* activity in empirical science as well as general and classical philosophical issues concerning scientific representation and scientific explanation.

Woody's interest in theoretical practices is particularly welcome because, as she emphasizes, the turn to practice has been chiefly, and was originally, focused on experiments. Theoretical sciences have for the most part been left aside, probably because the practice turn was above all a reaction against "the theory-centric bias within traditional 20th-century philosophy of science" (Woody, Chapter 4, this volume, p. 124).[1] In the founding works of the practice turn, what was perhaps the main obstacle to the development of a full practice-based view of scientific activity (where theory is also conceptualized as practice) was a misleading dualism between experiment as *concrete practices* and theory as *abstract concepts*. Woody's works (in this essay as well as in many others; see Woody, 2000, 2004a, 2004b, 2011) contribute to the development of a renewed conception of theory in science, a conception of theory *as practice*. Within such a conception, an analysis of the theoretical tools and models constructed and used by scientists becomes central. Woody's specific object of inquiry consists, more specifically, in the representational formats scientists construct and use. Her approach is then inscribed in a conception of theories as artifacts, that is, a conception in which the understanding of theories demands that one takes into account the way they are concretely represented by particular material objects or representational formats and the way they are used.[2]

In this commentary, we will first underline how Woody's contribution exemplifies the virtues of a study of science in practice (see Section II). This will also give us the opportunity to outline her main arguments concerning the role of representational formats in scientific theorizing and to point out their links to what she calls the "functional perspective" (p. 144) on scientific explanation. Our second goal is to draw attention to two challenging points emerging from our reading of this chapter. The first one questions the way the "productivity" (understood as the possible effects and outcomes resulting from the use of a tool) of Mendeleev's table is discussed (Section III). We find Woody's demonstration concerning the way the table has *perpetuated* dominant reasoning practices in chemistry convincing. We suggest, however, that it would be interesting to address the further question of how this theoretical artifact could have led as well to *new and unforeseen achievements* and to *innovative practices*. Further discussing this second aspect of the productivity of the table would lead, we think, to a more complete understanding of the place of Mendeleev's table in the historical evolution of chemical practices. Our second point is related to the conceptualization of the chemists' practices offered by Woody in her chapter (Section IV). We think that this conceptualization could be refined, by specifying how the components of practices identified by Woody relate to each other, what their comparative powers are, and how they concretely act on artifacts. In that regard, it would be interesting to construct a comprehensive picture of human practice against which it will probably be easier to understand even more precisely the power and performativity of the table.

II.   PRACTICE TURN, REPRESENTATIONAL FORMATS, AND SCIENTIFIC EXPLANATION

In her account of the practice turn, Woody identifies four shifts that allow for a more concrete conception of science, and dictate a different way of analyzing it. The first one is related to a shift in the conception of theories, from abstract logical structures to artifacts with particular representational formats. The second concerns the way we study science, from a priori analysis to empirical studies centered on the local context within which scientists work and on the concrete ways scientists reason, argue and categorize their actions. The third shift is related to the way we conceive of scientists, from ideal agents to actual practitioners with limited capabilities, bounded rationality and particular skills and background knowledge. And the fourth concerns the way we recognize science as a collective activity, moving from the perspective of individual scientists to that of disciplinary communities.

In accordance with the second shift, which recommends studying science on empirical grounding from a descriptive perspective, Woody supports her discussion of general questions concerning scientific representation and explanation with a rich case study. In her analysis of the periodic law in chemistry, she states that the explanatory power of that law for chemists cannot really be understood if we don't take into account the way the law is represented as a table (i.e., Mendeleev's periodic *table*). This makes clear the fact that it is not so much the abstract periodic *law* which is explanatory in practice, but the periodic *table,* that is, a particular representational format of the law. And this format of the law has explanatory power because its particular structural features impose order on a vast set of empirical facts, and because one constitutive aim of chemistry is to classify numerous elements with regard to their properties. Then, the explanatory power of the table depends on the "epistemic and practical aims" (p. 144–145) of the chemists' community. In turn, it also supports these aims once the table has become one central element of the community's toolbox.

This two-way relationship between the power of the table and the collective shared aims and practices of the community is, we think, at the heart of Woody's argument. As the first part of the case study demonstrates, the success and entrenchment in the chemical community of Mendeleev's table has to be understood, historically, by considering how the resources of that particular representational format fit with the practices and aims of the 19th-century chemical community. For this reason, the power of the table *depends on* the collective shared practices of this community, but the resources of the table in turn *support (and reinforce)* these practices, especially because, as the author says, "representational format conditions the pathways of our thoughts and conceptualizations" (p. 137). This support is central within what Woody calls the "functional perspective" on explanation, which put an emphasis on the function of explanatory discourses, more specifically on their ability to play "a methodological role for scientific communities by enabling coherent practice across practitioners" (p. 144) and by "sculpt[ing] and subsequently perpetuat[ing] communal norms of intelligibility" (p. 144).

Woody's case study shows how the four shifts of the practice turn that she identifies can transform the way we analyze science and our conception of scientific activity. By choosing an empirical perspective (second shift), it is possible to show how a concrete conception of theories as representational artifacts (first shift) is necessary in order to fully understand the way a law has become central in a scientific field. Such an understanding relies on the comprehension that we have of the properties of a representational format, knowing that this representational format is used by non-ideal agents (third shift). This understanding also relies on the way the resources of such a format fit with collective practices and aims of a community (fourth shift). This case study thus highlights some of the virtues of the practice turn, by showing how our understanding of science can be transformed by a practice-based perspective.

Woody's discussion underlines another trend within the practice turn, which is the particular attention paid to scientific pluralism. On the one hand, the interest of studying the multiplicity of representational formats of theoretical contents in science is highlighted. Two different representational formats of the same law are discussed and compared in the case study. Such an interest for this multiplicity seems a necessary prerequisite for considering, as is the case within a conception of theories as artifacts, that the way theories are expressed by means of particular representational artifacts is of pivotal importance. Moreover, this interest leads one to analyze and discuss the specific virtues of some kinds of formats (such as tables, graphs or other diagrams), which were not considered potentially legitimate objects of study before the practice turn. On the other hand, concerning scientific explanations, the "functional perspective" is rooted in the recognition of the pluralistic nature of explanations that we can observe across scientific communities. Recognizing their pluralistic nature can lead us to change our philosophical perspective on explanation by wondering "how the practice of explanatory discourse functions within science" (p. 144) instead of trying to "reveal the common logical and/or conceptual structure of individual scientific explanations" (p. 143). But while the project of devising a unified conception of the nature of explanation would then be abandoned, pluralism as regards explanation is nevertheless, in Woody's project, associated with what seems to be a unified conception of the *function* of explanatory discourses within scientific communities. As previously seen, Woody considers that explanatory discourses have a methodological role in scientific groups: They generate and perpetuate "communal norms of intelligibility" (p. 144).

While Woody's chapter shows some virtues of the practice turn, it also underlines two challenges faced by that turn. The first one is: How can working on case studies nevertheless lead to general claims concerning scientific activities? The second one is: How can we try to ground our normative judgments concerning scientific practices? The author concludes by stating that these "challenges, no doubt, remain" (p. 148). But some directions are given throughout her chapter. Concerning the first challenge, Woody claims that, although difficult, it is nevertheless possible to reach more general and normative conclusions through local analyses. For instance, she shows that well-chosen case studies can lead us to criticize the way general questions are discussed in the philosophy of science. In doing so, new perspectives on these questions emerge, as is the case with the "functional perspective" on explanation and the central role of representational formats in scientific theorizing.

With regard to the second challenge, the "functional perspective" on explanation shows how norms of intelligibility are sculpted and perpetuated within scientific communities by explanatory discourses. Mendeleev's table can then be seen as a means of perpetuating such norms. Disclosing in this way how norms function within scientific communities is an interesting first step, for us, as analysts of science, in constructing a "framework for analyzing practice that allows us to make assessments regarding the effectiveness or appropriateness of particular practices or actions situated within those practices" (p. 125).

III. PRODUCTIVITY OF MENDELEEV'S TABLE AND HISTORICAL TRANSFORMATION OF CHEMICAL PRACTICES

As seen in the previous section, the link between the power of Mendeleev's table and the reasoning practices and aims of the 19th-century chemical community is central in Woody's analysis. It enables us to understand why the table became entrenched in the 19th-century chemical community and how the representational resources of the table supported and reinforced that community's aims and reasoning practices. Moreover, the same representational resources of the table account for the explanatory power of the table in contemporary chemistry: This underlines "the continuing productivity of the representation" (p. 137). The *productivity* of the table, in the 19th-century chemical community as well as in contemporary chemistry, is then one of the main features of Woody's discussion. The table is seen as being productive because it enables, sustains, supports, conditions and guides reasoning practices in chemistry. And this leads, more generally, to a guidance of research efforts and to the emergence of research questions.[3] In this sense, the table is conceived as an artifact whose use is a driving force in the historical transformation of chemistry. Nevertheless, if our reading is correct, one of these dimensions of the productivity of Mendeleev's table within this historical transformation could be discussed more extensively.

We understand clearly, in Woody's chapter, how the table has supported and reinforced the reasoning practices and the aims of the 19th-century community, which chose this format of the periodic law. We also understand that the table has an explanatory power in contemporary chemistry because it "enables types of reasoning that facilitate dominant forms of inquiry" (p. 141). Moreover, we agree with Woody when she writes that "explanatory discourse inculcates particular patterns of reasoning generally, and thereby functions to sculpt and subsequently perpetuate communal norms of intelligibility" (p. 144). We are then convinced by her demonstration concerning the way the table has been (and still is) productive because its use has instituted and perpetuated norms of reasoning which have contributed to the coherence of cognitive practices in chemistry. We believe, however, that it would be interesting to further explore another dimension of the productivity of this theoretical artifact. While one of the functions of this artifact has been to shape chemists' reasoning practices and to generate norms of reasoning within the community over its history, we believe that it would be interesting to show more thoroughly how the table could also have generated *new and unforeseen* achievements and *innovative* practices. Of course, the table has enabled the prediction of the existence of new elements, but is this the single innovation it has produced? In sum, we think that once the productive role of the table has been demonstrated in its *perpetuating* and then perhaps *conservative character,* it would be worth trying to determine more precisely what the role of the table is with regard to *innovation.*

Klein (2003), in her "pragmatic account"[4] of the introduction and use of Berzelian formulas in organic chemistry between the 1820s and the 1840s, has interestingly discussed what we mean, here, by the innovative character of an artifact. She demonstrates how such "paper tools" (as she labels these artifacts) have been a central "agent of change"[5] in the transformation of organic chemistry, from animal and plant chemistry to carbon chemistry. What is particularly interesting in Klein's discussion is the way she demonstrates how the manipulation on paper of Berzelian formulas led the chemists Jean-Baptiste Dumas and Auguste Laurent to develop the new concept of chemical substitution, which was not supported by existing theories and not suggested by empirical results, and which was, more importantly for our present discussion, in contradiction with the theory of the constitution of organic substances that Dumas defended at this time. Klein then claims that the actual manipulation of the formulas caused Dumas "to do something completely new that served neither his individual goals nor the collectively shared goals of European chemists" (Klein, 2003, p. 195). She concludes by stressing that the "new concept of substitution was not the consequence of an atomistic research program, of social interests, or of any other form of deliberate intention on the part of chemists; rather, it was an originally unforeseen and unintended result of work on paper with chemical formulas that was coordinated with a new type of experiments performed with organic substances" (Klein, 2003, p. 199). There is then, for Klein, a dialectic of tools and goals: Tools have to be aligned with existing goals of the actors, but by being used, they tacitly alter such goals and generate new goals.

Woody's "pragmatic account" of the entrenchment of Mendeleev's table in the 19th-century chemical community and of the explanatory power of this artifact no doubt shares many epistemological concerns with Klein's account of Berzelian formulas. So, it would be interesting to know if more emphasis could be put on the possible innovations that the use of the table could have generated. Perhaps such types of innovation, like the emergence of the concept of substitution in the case of the manipulation of Berzelian formulas, could not be disclosed in the same manner for the periodic table. Putting more emphasis on this possible dimension would lead, we think, to a more complete way of trying to understand how the table has been a central "agent of change" in the historical transformation of chemistry. This, of course, would require a more detailed historical narrative of the uses and transformations of the table according to the different contexts in which it has been used by chemists.[6] In the concluding chapter of her book, Klein considers that "tools are an amalgam of working stability, adaptive flexibility and creative openness" (Klein, 2003, p. 246). The "working stability" and "adaptive flexibility" of Mendeleev's table are well underlined in Woody's analysis. We think that an even greater concern for the "creative openness" of the table would be interesting.

IV. ARTIFACTS, MATERIAL AGENCY, AND PRACTICES

In this last section, we discuss the way the role and evolution of theoretical artifacts in specific communities can be analyzed. It seems to us that in order to address these issues we must keep a careful balance between the focus on the characteristics of artifacts and the focus on the practices of the community from which these artifacts emerged and in which they evolve. We consider that Woody's case study could be interestingly extended by refining this second aspect—that is, the way chemists' practices are conceived.

As already mentioned (see Section I), Woody's approach is inscribed in a conception of theories as artifacts. In such a conception, the analyst focuses on the material objects in which theoretical contents are embodied, such as paper tools (as Berzelian chemical formulas),[7] representational formats (as the periodic *table*), three-dimensional models (as Watson and Crick double helix model of DNA),[8] or any kind of substantive "form" (*"formulaire"*).[9] These artifacts have a function of representation and, at the same time, they carry and transmit the practices that led to their construction. The conception of theories as artifacts is especially interesting for the unique way in which it analyzes the place and the role of theoretical objects in relation to scientific practices. Particular attention is paid to the materiality of theories, to their actual inscription in a concrete delimited object (artifact), to the structural qualities (resources) of the artifacts, and to their position in a social field of practices. The basic idea, as stated by Kaiser (2005), is to "follo[w] a nonhuman scientific object around as an organizing principle" for scientific practices.[10] The study of the action and of the trajectories of these objects constitutes what is known as the *material agency* of practice, which is the capacity of an instantiated object to act in the world (notably on humans). Studying these objects provides a useful perspective to understand the material culture of scientific communities and to have access—although indirect—to the evolving practices of these communities. But in order to account for the evolution and viability of these objects, we think that such a study has to be carried out in strong connection with the study of human agency. Without this association, it might be difficult, if not impossible, to understand the historical trajectories, and the potential deviation of these objects, to explain their emergence, their establishment, and their transformation or conservation over time.

Woody's chapter provides some elements for understanding both the material agency of the table and its connection to human agency. In her account, formats are capable of certain kinds of "actions".[11] For instance, the periodic table "carves out [chemical compounds] patterns", "highlight[s] features of the [chemical] landscape", "capture[s] many of the central categories and concepts of chemical practice", has an "organizational capacity", and so on (pp. 141–42). The table has then, as already pointed out, a "suggestive" (p. 140) and "heuristic" (p. 142) power. This power is, however, tightly linked to human agency, inasmuch as it is dependent upon the practices of the chemical community.

The components of the practices that constitute human agency—or more precisely, in this particular case, the *disciplinary agency* of the chemist community—are revealed, in Woody's case study, through the way they interact with the table. Woody's description includes elements such as: "background knowledge" (p. 128), "consistent set of conventions" (p. 127), "procedures" (p. 127), "skills" (p. 128), "general reasoning" (p. 135), "intellectual tools" (p. 135), "types of question practitioners wish to address or consider compelling" (p. 135), pedagogical preferences, and specific "aims" and "interests" (p. 136), as well as epistemic requirements (like classificatory needs). These are, one might say, the constituents of general practices, of the "matrix" (in a Kuhnian sense) that defines a community; it is what Woody is referring to as the "dominant characteristics of 19th-century chemical practice" (p. 135).

We think that it would be interesting to refine this way of conceiving of the chemists' practices, by specifying more thoroughly how these components of practices relate to each other and what their comparative weights and roles are with regard to representational formats. Indeed, since theoretical artifacts are strongly dependent on human agency, it seems important to have a picture of the structure of human practices and of the arrangements of their elements that is as precise as possible, in order to assess the effectiveness of representational formats. In that regard, a better understanding of the structure of chemical practices and of the way its components interact with the format would lead to an even better understanding of the reasons why the tabular format has been chosen. The human and material agencies are indeed intimately connected in Woody's case study—the chemical community having an *active* role in the fate of representational formats—but it seems to us that this role could be further discussed. The human agency and the chemists' practices could be analyzed by using less general and more discriminate functions or elements from which it would perhaps be possible to construct a coherent and comprehensive image of human practice. Such an image seems important to fully analyze the evolution of artifacts. In Woody's case study, the main factor explaining the success of Mendeleev's table over Meyer's graph is its better alignment with the general aims and interests of the community as well as the perceptive capacities of humans in general.[12] Of course, some other factors are mentioned, such as the "reasoning patterns" of the community. But contrary to aims, their role is far less significant in the decision process surrounding the acceptance or rejection of a new artifact, and at the end of the day it is still the overarching aims of the community that are decisive.[13] However, it seems that, in order to assess the success of a new artifact—in order to explain for which reasons a community at this specific moment has chosen the periodic table—we cannot rely only on the relation of what the table provides and suggests to the overall aims and interests of the chemical community. First, as we pointed out, there are other elements in scientific practices that can play a distinctive role regarding the choice of one representational format over another. For example, a more thorough analysis of the integrated and routinized ways of seeing and doing of the community before the table was accepted could have been interesting. This would have required paying greater attention, for instance, to the different patterns of explanation, argumentation, and justification running across the community, on the way they come into play regarding artifacts. But even if we suppose that aims are ultimately a sufficient element in some cases to justify the choice of an artifact, the way they are discussed in Woody's chapter could be further developed. Aims are plural, oriented toward divergent ends and extending over different temporalities. Additionally, the particular understandings of these aims among practitioners can vary significantly depending on their social situation and historical background. We think that it would be interesting to take this diversity into account.

The relation between the *disciplinary agency* of scientific communities and artifacts is complex. It cannot always be assessed in terms of match or mismatch of the artifact's power with the community aims and interests. It might be regarded as a normative or constraining relation (when the "integrated patterns of reasoning" and "sets of conventions" encourage looking at the format and using it in certain ways, or when the uses of the format reinforce the habits and conventions of the community). But it might as well be regarded as a productive relation (when new goals emerge and incite reevaluating the representational formats in use in the community, or when the use of a format makes new goals of research emerge). Therefore, our last suggestion will be that it would be interesting to devise a better identification of the different elements that compose individual and collective practices. Such analysis would need to display the way these elements operate, as well as their comparative impact on practitioners' actions in different situations. This analysis would ultimately have to account for the relation of each of these elements to the characteristics of the artifacts and explain how this relation can be a source of innovation, conflict, or perpetuation (of norms or habits).

NOTES

1. When quoting Woody's chapter in this volume, we will only indicate, in the following, the page number.
2. See especially Klein (2001a, 2001b, 2003), Kaiser (2005), and Latour (1998).

3. In the author's words, the table "could enable and sustain" (p. 128) reasoning practices, it conditions "practitioners' thought and direct[s] research efforts" (p. 133), it provides "directives for further research" (p. 135), it has supported "a wide range of 19th-century chemical reasoning" (p. 137), it has the "ability to [suggestively] guide reasoning, and inquiry more generally" (p. 140), it "enables types of reasoning that facilitate dominant forms of inquiry" (p. 141), it has a heuristic value which "is a primary reason for this [its] longevity" (p. 142), its use "condition[s] reasoning in the discipline [chemistry] in ways that contribute to this coherence [the coherence of cognitive practices in chemistry]" (p. 144); and it "could guide reasoning and generate research questions for the community" (p. 148).
4. As labeled by Grosholz (2007).
5. This expression is used by Grosholz in her discussion of Klein's account (Grosholz, 2007, p. 28).
6. We are well aware that this was not possible within the limited scope of a chapter.
7. See Klein (2003).
8. See De Chadarevian and Hopwood (2004).
9. See Latour (1998).
10. Kaiser (2005, p. 7).
11. Woody never uses the notion of "action" to qualify the effects of artifacts. Perhaps she reserves the notion of action only for humans, because, like for those who are skeptical about the existence of a material agency, actions are conceived as necessarily intentional. What is certain, however, is that in her conception artifacts have more than just mere affordance; they are not only suggestive but perform more concrete "actions".
12. There are three different interactions that are mentioned as criteria to explain the choice of the table. Two of them a related to the general aims of the community: (1) the relation between the individual aims and epistemological preferences of Mendeleev and Meyer and the collective aims of the larger community. On that point, there are good arguments on both sides, both fit with some aims of the community, but those of Mendeleev fit better than Meyer's as they explicitly recognize and focus on the core elements of chemistry. (2) The relation of the features of the format with the aims and interests of the community. (3) The relation of the features of the format with the perceptual capacities of practitioners and the tools that they already use.
13. Woody is aware of the fact that she selected only a few central elements of the chemists' practices:

> This quick sketch cannot hope to capture all the relevant details, especially the diversity, of chemical practices across Europe at this time, but I believe the factors outlined here are indeed central ones and thus provide a good place to start. (p. 135)

Our concern is not that her study has selected only some aspects of the practices, nor that these aspects have been qualified as central. It is that, in the end, she put the emphasis on one of these elements that has a disproportionately decisive power compared to other aspects.

5 Epistemic Dependence in Contemporary Science

Practices and Malpractices

Hanne Andersen

Despite an increased focus on scientific *practice* in the philosophy of science in recent years, there has been relatively little focus on *mal*practices such as intentional fraud or gross negligence.[1] This is the more striking because malpractice in research—both in the form of outright misconduct such as fraud and deceit and in the form of the so-called "grey zone" behavior such as sloppiness and incompetence—has been a topic of growing concern both among scientists themselves and among politicians, administrators, and in the general population (for an overview of this development, see, e.g., Steneck, 1994, 1999).

Most existing philosophical analyses of malpractice in science have centered on intentional deceit and treated the phenomenon primarily as a topic for ethical analyses. However, in this paper I shall go beyond this focus on deceit and discuss intentional, reckless actions as well as negligent ones, and I shall argue that an analysis of these actions goes beyond research ethics and includes important epistemological aspects as well. Hence, one of the aims of this paper is to point to a new area for philosophy of science in practice to address.

I shall start with the notion of epistemic dependence and the necessity for scientists to be able to trust their collaborators and their peers, and reiterate core contributions to the literature on the epistemic and moral components of trustworthiness and how trustworthiness is assessed. Based on this background, I shall examine situations in which scientists have *not* been trustworthy, and I shall discuss how the assessment of trustworthiness compares to the assessment of *untrustworthiness*.

I. THE IMPORTANCE OF EPISTEMIC DEPENDENCE

Scientists become epistemically dependent by drawing on the work of each other in many different settings. Close collaborators often divide labor among the group rather than all working on the same thing at the same time. This division of labor saves time and resources, and it enables specialization among the group members. But dividing the tasks among the collaborators means the group members become epistemic dependent. Often they will have to rely on each other, whether it is for a result which they could in principle also have arrived at themselves (but for the sake of efficiency they do not repeat), or where they have different areas of expertise. As argued by Wray (2002), by sharing resources that are scarce and for which there is strong competition, collaborative scientists can realize the epistemic goals of science more efficiently. Further, the extent and historical development of collaborative practices in science are very well documented. Historians and sociologists of science have produced detailed analysis of the dramatic increase in collaborative practices up through the late 19th and 20th centuries (Beaver, 2001; Beaver & Rosen, 1978, 1979a, 1979b), and recent analysis of scientific literature reveals that by far the majority of all new scientific publications published today are produced by groups in which several scientists collaborate and combine knowledge, staff, materials, and other resources (Wuchty, Jones, & Uzzi, 2007). Similarly, empirical studies of co-authored papers and their influence vindicate Wray's philosophical argument for the epistemic significance of collaboration in science by showing that it has a range of benefits, including increased productivity (Price & Beaver, 1966; Godin & Gingras, 2000; Lee & Bozeman, 2012), higher visibility to the participants (Beaver & Rosen, 1979a), more citations (Wuchty et al., 2007), and even a higher association with Nobel Prizes than scientists working alone (Zuckerman, 1967).

But scientists are also often epistemically dependent on distant peers. Researchers follow the work of their peers in the field in order to be able to build on their results and move quickly on to the next open research puzzle. Similarly, when scientists are in need of knowledge about a particular subject they may find it in the scientific literature and build on it in their own research, while citing the relevant article as an authority that is deferred to. Whenever scientists draw on existing literature, citing another scholar in deference to his or her authority on this particular matter, they are engaging in epistemic dependence.

Given this importance of epistemic dependence, both between close collaborators in a group and between distant peers in the scientific community as such, it is important to understand a) how it is established and maintained, and b) what happens when the preconditions for epistemic dependence are violated. In the following, I shall first review existing philosophical analyses of epistemic dependence. These analyses have all focused on how a scientist establishes that another scientist is trustworthy. However, I shall primarily focus on the opposite situation—namely, how a scientist establishes that another scientist is *not* trustworthy.

I.1.   Analyzing Epistemic Dependence

Hardwig has provided a philosophical analysis of the pervasive epistemic dependence on which much of the scientific endeavor builds in a series of now-seminal papers (Hardwig, 1985, 1988, 1991). This work argues that epistemic dependence in science is rational, and it also goes on to analyze the conditions for the trust in expert testimony that epistemic dependence implies. The basic structure of Hardwig's argument is that, in a domain for which scientist A has limited competence, he or she may appeal to another and more competent scientist B who knows something, *p*, that A does not. In this case, A's belief that *p*, based on the testimony from B, will be superior to a belief that is based on direct, but non-testimonial evidence for *p*, because B's reasons for believing *p* are epistemically better than any that A could come up with alone. However, this means that not only does A not have the reasons necessary to justify *p*, A may not even know or understand what B's reasons for believing *p* are and why they are good reasons.[2] In this situation, Hardwig (1991) claims that for scientist A to trust scientist B implies that

1) A knows that B says that *p*;
2) A believes that B is speaking truthfully;
3) A believes that B is in a position, first, to know what would be good reasons to believe *p* and, second, to have the needed reasons; and
4) A believes that in the case at hand, B also actually has good reasons for believing *p* when saying so.

A's belief that *p* relies both on the *moral* character of B, that B is truthful, and on the *epistemic* character of B, that B is knowledgeable[3] about what constitutes good reasons in the domain of his or her expertise and has kept him- or herself up to date with those reasons, has done his or her work carefully and thoroughly, and is capable of epistemic self-assessment—that is, that B does not have a tendency to deceive him- or herself about the extent of his or her knowledge, its reliability, or its applicability to the question of whether that *p* or not. Further, as also pointed out by Hardwig, A's belief that *p* also relies on A being in a position to assess B's moral and epistemic character. Hence, what is important here is that *both* scientists' moral and epistemic character *and* their practices of assessing each other's moral and epistemic character are key ingredients for the practices of epistemic dependence.

I.2.   Calibration of Trust

In the philosophical literature, the practices of assessing the moral and epistemic character of peers have primarily been analyzed as a calibration of trust where no distinction is made between the moral and the epistemic component of a scientist's trustworthiness. The most prominent analyses of how scientists (and others) calibrate their trust in others have been provided by Kitcher (1992, 1993) and Goldman (2001). While Kitcher has focused primarily on the warrant for one scientist to trust another (expert-expert calibration), Goldman has focused on the warrant for members of the general public (or novices in science) to trust a scientific expert (lay/novice-expert calibration). For expert-expert calibration, an important ingredient is a comparison of the reliability or error rate of the trusting and the trusted scientist. This can be done either by *direct calibration* where the trusting scientist compares the output of the trusted scientist to his or her own on topics where there are overlaps, or, in areas outside the trusting scientist's specialty where there is no such overlap on which to ground direct calibration, then by *indirect calibration* where the trusting scientist has to draw on the judgment of others, but in a chain that starts from a scientist that can be directly calibrated (Kitcher, 1993).[4] Similarly, for lay people who are not in a position to evaluate experts based on their own understandings, their lay/novice-expert calibration has to draw on such indicators as argumentative superiority, agreement from additional experts, appraisal by meta-experts, evidence of possible conflicts of interest or similar forms of bias, and past track records (Goldman, 2001). Obviously, this distinction between experts and lay people/novices is somewhat artificial. Often, scientists assess peers with an area of expertise different from their own, and they will here be in a situation similar to that of a lay person. Consequently, a strict distinction between expert-expert calibration and lay/novice-calibration is untenable, and the calibration practices should not be seen as different in kind but instead as lying on a continuum.[5]

While the described calibration practices may work well for establishing that another scientist *is* trustworthy, they have various shortcomings when it comes to establishing that another scientist is *not* trustworthy. First, because a trustworthy scientist will necessarily be both truthful and knowledgeable, there is no need in the calibration practice to distinguish between moral and epistemic character. In contrast, an untrustworthy scientist may either be untruthful or unknowledgeable (or both), and calibration practices may therefore need to assess moral and epistemic character separately. Admittedly, some elements of the calibration practices described by Kitcher and Goldman do seem to address such differences indirectly. As examples, argumentative superiority seems primarily to address whether the assessed scientists appear epistemically unimpeachable, and conflicts of interest seem primarily to address whether the assessed scientists appear to be morally unimpeachable. However, these differences have not been pursued in any detail.

Further, the calibration practices suggested by Kitcher and Goldman take a comparative or relational approach that calls for additional specification of their applicability. On Goldman's analysis, lay people or novices assess experts by comparing them to each other, and on Kitcher's analysis scientists assess other experts by comparing them to themselves. Hence, in this view, a prerequisite for direct calibration is that the calibrating scientist is him- or herself a very knowledgeable expert who is up to date, who works carefully and conscientiously, and who has an infallible epistemic self-assessment. Consequently, when it comes to assessing the epistemic character of a colleague or of a peer in the field, different standards should be expected from, for example, junior or senior researchers.

To examine this issue of how calibration practices may work for establishing that scientists are not trustworthy, let us turn to cases in which scientists have been fraudulent, sloppy, or incompetent and examine how they have been assessed by collaborators and by peers in the wider community.

II.   WHEN EPISTEMIC DEPENDENCE IS UNDERMINED

Similar to the philosophical analyses and historical and sociological studies that argue for the importance and document the extent of collaborative practices in science today, recent decades have also seen increasing attention paid to scientists who have intentionally attempted to deceive other scientists by reporting manipulated or invented research results. Scientific malpractices became a topic of public attention when a number of spectacular cases were picked up by the media, including the case of the immunologist William Summerlin from the Sloan-Kettering Cancer Center whose drafting of skin from a black to a white mouse turned out to have been painted with a black felt pen (Hixon, 1976), and the publication of the monograph *Betrayers of the Truth* written by two *Science* journalists. This book included a long list of historical and contemporary cases of misconduct (Broad & Wade, 1982). In response to the increasing number of reports on cases that did not seem to have been adequately addressed by the scientific community, the U.S. Congress held hearings in 1981 and 1985 that eventually led to the introduction of explicit regulations regarding scientific misconduct in publicly funded research.[6]

Despite such regulations, new cases have kept emerging.[7] In 2002 and 2006, scientific misconduct was even listed as "breakdown of the year" in *Science's* Breakthrough of the Year sections. In 2002, the "breakdown" was due to reports on fraudulent data in the work on single-molecule semiconductors performed by the nanotechnologist Jan Hendrik Schön from Bell Labs and in the work on element 118 led by Victor Ninov at Lawrence Berkeley National Laboratory (Service, 2000). In 2006, it was due to fraudulent data in the work on human stem cells performed by Woo Suk Hwang from Seoul National University, in the work on cancer research by Jon Sudbø from the Norwegian Radium Hospital, and in the work on obesity by Eric Pohlman from the University of Vermont, Burlington (Couzin, 2006).

These examples are all spectacular cases that have been reported in general news media and often been discussed in detail in general science journals such as *Science* and *Nature,* but several surveys indicate that they are only the tip of the iceberg (De Vries, Anderson, & Martinson, 2006; Martinson, Anderson, & De Vries, 2005). A recent meta-analysis reports that around 2% of scientists admitted to fabricating or manipulating data or results at least once in their careers, while more than 30% admitted other questionable research practices (Fanelli, 2009).

II.1. Identifying Fraud, Sloppiness, and Incompetence

In response to the many reports of fraud in science, concerned scientists, administrators and politicians have requested that measures be taken, first, to prevent misconduct if at all possible and, second, to investigate allegations and penalize misconduct if it occurs. One of the first countries to establish official regulations was the United States, where misconduct was defined as fabrication, falsification, or plagiarism[8] and at the same time explicitly distinguished from so-called "honest error":

> Fabrication, falsification, plagiarism, or other practices that seriously deviate from those that are commonly accepted within the scientific community for proposing, conducting, or reporting research. It does not include honest error or honest differences in interpretations or judgments of data. (Handling Misconduct, 1989)

However, this definition was heavily criticized from the outset due to the difficulties defining "serious derivations from commonly accepted practices", and the "serious deviations" clause was later removed. Nevertheless, as pointed out by Steneck in a review of the regulation of misconduct during the 1980s and 1990s, studies indicate that poor research performed in careless or incompetent ways could well be a larger problem than outright fraud, although it is the latter rather than the former that has remained the focus of debates:

> it is likely that much more error makes its way into the scientific literature by sloppy practices than by deliberate falsification and fabrication, which raises an interesting non-sequitur in current policies. Sloppy science is either not addressed in most misconduct/integrity policies, or, more commonly, specifically excluded from these policies. If the preservation of the integrity of science is a major concern, how does one justify specifically excluding practices that may have more to do with the health of science than major misconduct? (Steneck, 1999)

One of the major problems when addressing the grey zone of poor research is how to distinguish between negligence, incompetence, and so-called honest error (see also Andersen, 2007). One response to this problem has been to argue that since erroneous results would be caught by the self-correcting practices of science such as peer review, regardless of whether error was due to negligence, incompetence, or just accident, the distinction is unimportant.[9]

However, this view seems too simplistic. Science is not as perfectly self-correcting as this argument assumes. Although peer review and peers' critical scrutiny of published works may reveal loci of trouble, such as failure to replicate experiments, contradictions, or incoherence with existing knowledge, these practices are unlikely to catch all erroneous results. Further, relying simply on peer review in the publication process and peer criticism after results have been published is to make the calibration of trust exclusively a *post hoc* activity. Instead, we should examine how trust—and *dis*trust—can be calibrated while science is still in the making.[10]

III. CALIBRATING *DIS*TRUST

One way to analyze calibration practices in situations where scientists come to *dis*trust another scientist is by examining reports on known cases of malpractice, including both misconduct and ignorance or incompetence. These reports often provide retrospective accounts of how the process of science-in-the-making unfolded and can therefore give at least some indication as to how and when the involved scientists started (or should have started) questioning the moral or the epistemic character of their collaborators.

III.1. Relational Calibration of Distrust

From descriptions of some of the first major misconduct cases in the United States, Broad (1981) has argued that often the detection mechanism is the "detective work of young lab assistants or young scientific rivals who have the extra-experimental evidence of cheating, who have some independent reason for suspicion" (p. 140). For example, Summerlin's fraud when improving a graft of black skin onto a white mouse with a black felt tip pen was revealed when a laboratory assistant wondered about the color of the black skin graft, wiped it with alcohol and observed a change in color (Medawar, 1976/1996). Another publicly well-described case concerns Professor Eric T. Poehlman, who worked on metabolism and obesity at the University of Vermont. A junior colleague became suspicious when, during the process of writing a paper based on data he had received in a spreadsheet, he received a new version of the spreadsheet from Poehlman and suddenly got different results in the data analysis. After the junior colleague had turned to university officials, an investigation revealed several cases of fraud in Poehlman's work (see, e.g., Dahlberg & Mahler, 2006 for a detailed account of this case).

In these cases, discrepancies or sudden, unexplained changes prompted collaborators to doubt the truth of presented material and hence the *moral* character of the scientist who had presented it. Admittedly, in many cases of reported misconduct it is clear that this is not necessarily an easy step to take. First, many factors may discourage whistle-blowing, including, as in the Poehlman case, outright threats of discontinued collaboration or of notice of dismissal. Second, some reports indicate that many scientists feel there is a moral barrier to be overcome if one is to question the truthfulness of a peer. For example, the Nobel Laureate Peter Medawar talks of his own hesitation before questioning the truthfulness of Summerlin's reports. In this case, Medawar was presented with a rabbit, which had allegedly received a cornea transplant that extended over the whole dome of the cornea right into the rim where the blood vessels run: "I could not believe that this rabbit had received a graft of any kind . . . because the pattern of blood vessels in the ring around the cornea was in no way disturbed. Nevertheless I simply lacked the moral courage to say at the time that I thought we were the victims of a hoax or confidence trick" (Medawar, 1976/1996).

The situation has been even more complicated in several recent misconduct cases related to the work of, among others, Jan Hendrick Schön, Woo-Suk Hwang, and Diederich Stapel. First, in these cases suspicion of misconduct has originated not with close collaborators, but instead with peers in the wider community. Second, these cases have raised debate as to whether co-authors have special responsibilities for assessing the trustworthiness of each other and, consequently, whether the co-authors of Schön, Hwang, and Stapel *should* have had concerns—and, therefore, whether they did an inadequate job as scientists.

For the purposes of this discussion, the Schön case will serve as the prime example. In this instance, the committee that was charged with investigating the case wrote in their *Report of the Investigation Committee on the Possibility of Scientific Misconduct in the Work of Hendrik Schön and Coauthors* (Lucent Technologies, 2002) that it "hoped to stimulate a discussion" (p. 16) around the question of whether co-authors have a special responsibility. But while the aim was to stimulate a discussion, the committee was also careful not to take a clear stand on these questions. Instead, it summarized its reflections by referring to the lack of a general consensus in the scientific community:[11]

> These are extraordinarily difficult questions, which go to the heart of what we as a community of scientists can expect of one another professionally, in the real world within the context of a collaborative research endeavor. The Committee does not consider itself qualified to make a specific judgment in this case, in the absence of a broader consensus on the nature of responsibilities of participants in collaborative research endeavors. (2002, p. 18)

Importantly, they stressed that they saw this responsibility purely as an issue of "professional responsibility" that was clearly distinguishable from scientific misconduct (p. 3). An interpretation of this position is, according to the analysis of epistemic dependence presented above, that it serves the epistemic aims of science to trust another scientist only insofar as there is good reason to believe in this scientist's moral and epistemic character. Obtaining these good reasons forms part of the *epistemic* practices, and not obtaining them is therefore an epistemic, and not a moral, failure. Further, the committee found that the responsibility of participants in collaborative research endeavors would vary between co-authors according to their expertise, the centrality of their contributions, and their leadership roles (Lucent Technologies, 2002, p. 17). This is in line with the relational accounts of calibration advanced by Kitcher and Goldman, in which the strength of a direct calibration necessarily depends on the epistemic character of the scientist performing the calibration.[12] The committee therefore primarily focused on Schön's superior, Bertram Batlogg, who, as the senior co-author, was considered to be in the strongest position to perform calibration of trust on this relational basis. The committee expressed the concern that it

> felt the need to question whether Bertram Batlogg, as the distinguished leader of the research, took a sufficiently critical stance with regard to the research in question, even in the absence of direct knowledge of the growing concerns. Should Batlogg have insisted on an *exceptional* degree of validation of the data in anticipation of the scrutiny that a senior scientist knows that such *extraordinary* results would surely receive? Such attention need not violate the spirit of trust. To the contrary: Unprecedented and spectacular results are guaranteed to lead to serious critical inquiries from the scientific community at large. A senior coauthor who has paid close attention to the detail of the work, who recognizes the fallibility of the experimental process and the human beings carrying it out, and who has asked searching questions is in a much better position to support his colleagues in defending the work. (Lucent Technologies, 2002, p. 18)

What is particularly interesting here is that what is perceived as the admissible route to calibrating mistrust in the absence of any clear discrepancies indicating untrustworthy reporting (and thus indications of untruthfulness) is to focus on the epistemic character and inquire into whether the work was performed with sufficient conscientiousness, given its extraordinary character. In contrast, when the committee poses the question, "In a similar spirit, but admittedly more difficult, should Batlogg have crossed the line of trust and questioned the integrity of the data?" (2002, p. 18) it seems that calibration of trust explicitly directed at truthfulness is seen as a taboo that requires very special reasons for suspicion to be set aside.[13]

Similar concerns regarding epistemic practice of evaluating collaborators' work have been raised in other recent high-profile cases regarding scientific misconduct,[14] among others, the case of the South Korean stem cell researcher Woo Suk Hwang and his collaboration with Professor Gerhard

Schatten at the University of Pittsburgh. In this case, the University of Pittsburgh appointed an inquiry panel to investigate the possible role of Gerhard Schatten, who was senior co-author to one of the Hwang papers in which misconduct was suspected. Also in this case the inquiry panel focused on whether a senior co-author (Schatten) had provided sufficient oversight to assure himself of the integrity of the data. The committee first described how Schatten had not asked for explanations of why data had changed in a table during the writing process, nor had he commented on the reported contamination that called for the start of a new cell line that would have made part of the reported data impossible to obtain in the noted timeframe (University of Pittsburgh, 2006, p. 7). On the bases of these omissions, which according to the analysis above must be seen as an epistemic rather than a moral failure, as well as his failure to ascertain that all co-authors had read the manuscript before submission and confirmed that data were reported correctly, the committee concluded that Schatten shirked his responsibilities, and added that "while this failure would not strictly constitute research misconduct as narrowly defined by University of Pittsburgh policies, it would be an example of research misbehavior" (University of Pittsburgh, 2006, p. 9).

Similarly, but addressing the responsibilities of junior researchers rather than senior researchers, the investigative committee in the Stapel case also inquired as to why it took so long for fraud to be discovered. With respect to Stapel's collaborators, who were primarily junior researchers, the committee argued that while reciprocal trust is an essential part of all team work, "the junior must be able to trust *absolutely* in the integrity of the 'master'" (University of Tilburg, 2011, sec. 3.2, italics added). Further, since Stapel both developed very close relations to the junior scholars in his group and at the same time shielded his group from outside relations, the committee argued that "it was precisely because of the isolated approach that the young researchers were unaware that this was not a normal state of affairs in social psychology research" (University of Tilburg, 2011, sec. 3.3). This case also fits well with a relational account of calibration of trust in which the strength of the calibration is dependent on the epistemic character of the researcher performing the calibration, since, in this view, junior researchers are not in as strong a position to perform the calibration as senior researchers—even more so in a situation where part of their training has been deficient.

III.2. Calibrating Collaborators in a Shared, Collaborative Activity

While the relational account of calibration explains the difference between what may be expected from calibrations performed by junior and senior collaborators, there still needs to be an explanation of why there are different expectations for the calibrations performed by collaborators and other scientists outside the collaboration.

In the Schön case, the committee explicitly argued that co-authors "often have access to technical details that other parties, such as management, referees, editors and award committees do not have" and that, consequently, co-authors "represent the first line of defense against misconduct" (Lucent Technologies, 2002, p. 16). At the same time, the committee emphasized that collaborative research requires trust, but that "such trust must be balanced with a responsibility to ensure the veracity of all results" (Lucent Technologies, 2002, p. 17). Similarly, in the Stapel case, the investigative committee emphasized that "reciprocal trust is an essential part of all teamwork, also in scientific research" (University of Tilburg, 2011, sec. 3.2).

One way to understand the various reflections provided by the investigative committees of the misconduct cases described above is to view a group of collaborating scientists as participating in a shared cooperative activity of delivering—and being able to defend—a new and interesting result *p* which cannot be fully derived or justified by any of the scientists alone, but only by integrating knowledge from each of the collaborators' individual areas of expertise.[15] In this situation, the group members are not only mutually epistemically dependent on each other in establishing that *p*; in order for their work to succeed, they also need to have meshing subplans for collecting and integrating all the necessary pieces of knowledge, and in the iterative process of their collection and integration of knowledge they need to be mutually responsive in tracking the status of the joint activity until they have finally arrived at a result that can be justified by the group members together.

On such an interpretation, what is expressed by the committees above is that, given that it is the shared intention of collaborators to defend, for example, an extraordinary, unprecedented and spectacular result, then their meshing subplans must include preparing for serious critical inquiries, including mutual responsiveness that enables group members to assist each other. This interpretation explains some of the apparent confusion and uneasiness expressed by the investigation committee. For a shared, cooperative activity of producing a new research result, the mutual responsiveness is part of the *epistemic* process through which a group of epistemically dependent scientists ascertain that they can together justify their results. To consider possible critical objections to spectacular results is part of their working carefully and thoroughly, and to keep track of the justificatory status of the result is part of their epistemic self-assessment. In other words, mutual responsiveness is an important ingredient of scientists' *epistemic* character when they collaborate in groups. But, in contrast, it is not a *moral* category. Hence, in cases involving scientific misconduct only the scientist who has performed the falsification or fabrication can be blamed for his or her *moral* character, while it his or her collaborators' *epistemic* character that may be questioned, in exactly the same way as when dubious data arises from the use of a defective instrument rather than a defective collaborator.

Acknowledgements: I would like to thank the participants at the conference "Rethinking Science After the Practice Turn", and especially the commentator to my talk, Cyrille Imbert, for valuable comments and suggestions, and Claire Neesham for linguistic revision of the final manuscript. I would also like to thank members of the Philosophy of Contemporary Science in Practice group, as well as Douglas Allchin for important input as this paper developed. Finally, I would like to thank the Danish National Research Foundation, Humanities, for supporting the project "Philosophy of Contemporary Science in Practice".

NOTES

1. In legal terminology, a distinction is made between fraud that is understood as *intentional* deceit, and malpractice that is understood as *negligence* in which a professional fails to observe or follow generally accepted professional standards. Here, scientific malpractice is used in a broad sense that includes intentional, reckless, as well as negligent acts in the production of scientific knowledge.
2. Obviously, the case that A is completely incompetent within the domain in which it can be held that *p* and that A is therefore not at all in a position to understand B's reasons for believing *p* and why they are good reasons is the one extreme end of a spectrum. At the other end of the spectrum, A may also have to rely on B in cases in which A is fully competent within the area but where the relevant evidence is for some reason or other simply too extensive or too costly for A to gather.
3. See also Chang's discussion of actors' capabilities and limitations in Chapter 2 of this volume.
4. Further, being interested in scientific change and how epistemic conservatism can be overcome, Kitcher includes in his analysis prestige effects stemming from training or employment at privileged institutions as well as backscratching effects stemming from mutual adjustments between scientists in order to explain splits in alliances and distributed community decisions.
5. This is especially true for interdisciplinary research teams, see Wagenknecht (*forthcoming*).
6. See, for example, Heinz and Chubin (1988) and Steneck (1994), as well as U.S. Congress Committee on Science and Technology, Subcommittee on Investigations and Oversight (1981) and U.S. Congress Committee on Governmental Operations, Subcommittee on Investigations and Oversight (1988).
7. Case summaries on U.S. misconduct investigations that have led to administrative action against researchers can be found on the web pages of the Office of Research Integrity: http://ori.hhs.gov/case_summary
8. While there is some variation between the regulations in different countries, the U.S. regulations, which focus exclusively on fabrication, falsification, and plagiarism (FFP), will be used as the primary sample of regulations for the purposes of this discussion.
9. Prominent scientists, in particular, have argued that the whole issue of misconduct in science is "a storm in a teacup". For example, the president of the National Academy of Science expressed the view at the congressional hearings on fabrication and falsification of research results that "The matter of falsification of data, I contend, need not be a matter of general societal concern.

It is, rather a relatively small matter which is generated in and is normally effectively managed by that smaller segment of the larger society which is the scientific community. This occurs in a system that operates in a highly effective democratic, self-correcting mode—the . . . 'peer review system'". (quoted from Woolf, 1981, p. 10)

10. This difference is not just of abstract, philosophical interest. Several studies show that, for example, even retracted papers continue being cited quite extensively. Especially in the biomedical sciences, this is a cause for concern because treatments based on fraudulent research can put patients at risk (Steen, 2011). For a general discussion of the problem of dealing with fraudulent results once they have been published, see, for example, Couzin and Unger (2006).

11. The committee also pointed out that some fields associate authorship with credit for accumulated labor, others with particular, identifiable intellectual contributions, and others again with joint responsibility for the work in its entirety.

12. In accordance with this analysis, a survey performed among researchers funded by the U.S. NIH showed that people were more likely to take action if they were senior to the suspect (see Koocher & Keith-Spiegel, 2010).

13. Similarly, in the U.S. NIH survey previously mentioned, respondents indicated a higher willingness to, or higher success with, intervening when they "felt sure, or fairly sure, that the errors were unintentional" (Koocher & Keith-Spiegel, 2010, p. 439).

14. Discussions of the same sort can also be raised on historical cases; see, for example, Van Dongen's (2007) analysis of the presumably fraudulent research of the physicist Rupp and Rupp's collaboration with Einstein.

15. See Bratman (1992, 1999, 2009) for the notion of shared, cooperative activity.

The Identification and Prevention of Bad Practices and Malpractices in Science

Commentary on "Epistemic Dependence in Contemporary Science: Practices and Malpractices", by Hanne Andersen

Cyrille Imbert

According to Hanne Andersen, "an analysis of [malpractices] goes beyond research ethics and includes important epistemological aspects" (p. 161). Her purpose is to point at a new area for philosophy of science in practice, which she does by highlighting different epistemological issues about malpractices and showing how documenting them in a precise way is beneficial to their solution. She articulates in particular two questions, namely the issue of the identification of bad practices and malpractices, and the ways of preventing the latter from happening. I shall discuss how Andersen contributes to these issues and make additional suggestions.

Before going further, I first want to clarify a few points.

a) Scientists can fail to match scientific standards, as scientists, in various ways. In what follows, I shall focus upon ways in which scientists fail to follow scientific standards in their *research* practices, which leaves aside other important circumstances in which they may scientifically misbehave. For example, casting doubt on scientific issues by challenging scientific evidence on non-scientific grounds in fields in which one is not an expert, as Frederick Seitz or Fred Singer did regarding issues like global warming, is an example of using one's scientific reputation to smuggle in pseudo-expertise, a clear scientific misbehavior (Conway & Oreskes, 2010), but not a research malpractice.

b) Something that is presented as a good research practice can fail to be so for various reasons. To mention just a few: A practice may not contribute to a target result in the way her author claims that it does, it may have been carried out by its author without the scientific vigilance or care expected by her community, or it may not be a token of a scientific research practice at all—typically, forging data is not an example of a bad scientific practice, it is simply no scientific practice at all. I shall follow Andersen in describing as malpractices all such cases, when the failure occurs either deliberately or by negligence. The difference between the two is that, in the latter case, the author does not intentionally want to perform a bad practice, but consciously fails to follow an accepted standard of scientific rigor, while knowing it may adulterate her practice and results.

c) One may wonder whether the above working definitions are a sufficient basis for tackling issues about malpractices. From a logical point of view, analyzing intentional bad practices requires possessing a sound characterization of what good/bad practices are, and therefore having a sound definition of scientific practices themselves, a question that is still being investigated. Further, the dichotomy between *good* and *bad* may be too coarse. For example, a trichotomy of *bad/acceptable/good* may catch more precisely the epistemic stances of scientists towards actual practices. Typically, a reviewer may deem a practice acceptable for publication but may not wish to rely on the corresponding paper for her research. Finally, in a research community, it is unlikely that there is a consensus about where the boundaries between good, bad, and acceptable practices lie: While shared hypotheses and methods provide a common basis for scientific discussions, they do not determine every single aspect of practices, and this provides room for disagreement about the validity of this or that aspect of a practice. But do we really need to solve all these issues to start analyzing malpractices? Probably not. If we wanted to provide a principled analysis which would be sufficient for understanding what would happen if, for each scientific practice, we asked scientists to discuss whether it is a good, acceptable, bad, or non-scientific practice, then we would have to answer all the above problems. However, as Andersen's paper illustrates, it is possible to fuel a valuable discussion about malpractices by focusing on uncontroversial cases, and this is enough for the present purposes.

I. IDENTIFYING BAD PRACTICES AND MALPRACTICES

While one of the declared topics of Andersen's paper is the identification of scientific malpractices, her paper deals mainly with the "calibration of trust [and distrust]" between scientists. Typically, she refers to direct calibration (when a scientist directly assesses the result of another scientist because it belongs to a field that overlaps with her own field of research) as a way to calibrate scientists, not their practices. Of course, the calibration of practices is often not an available option for scientists, and trust has to come into play (Hardwig, 1991). However, the direct assessment of the value of scientific practices is an important part of the identification of bad practices, and it should be analyzed in detail, if only to understand how delicate this assessment is, if it is to be carried out properly, and why bad practices and malpractices are difficult to erase in science.

I.1. The Identification of Bad Practices and Malpractices, or the New Trickier Version of the Demarcation Problem

Why are bad practices not more easily identified in science? I shall first hint at some principled reasons why this is so, and explain why the picture of science that emerges from the practice turn gives insightful, if not explicit, clues regarding this issue.

This identification problem is germane to the problem of demarcation between science and pseudo-science, which was seen as central by logical empiricists.[1] According to this tradition, focusing on features of scientific statements (theories, axioms, observational statements, etc.) and their treatment by scientific reasoning is a sound and fruitful way to analyze scientific activity. Then, the question of the demarcation between scientific and non-scientific practices, or good and bad ones (resp. research programs, agents, particular inquiries, or any relevant scientific item), can still be raised, but it should be seen either as having a solution that can be derived from an analysis of more central (that is, linguistic) features of science, or as "merely" related to the pragmatics of science.

On the contrary, once one adopts a perspective in which scientific practices are seen as the core of science, and the right level of analysis to tackle traditional questions like the meaning of scientific terms or the nature and possibility of scientific progress (see, in particular, Kitcher, 1993), it becomes of crucial importance that one understands better how good scientific practices can be distinguished from bad or non-scientific ones and how demarcation issues are solved in practice. The worry, as I shall now argue, is that this version of the demarcation problem is much trickier to solve.

There is in general a gap between the possession of a sound, clear, and precise definition of an X and the ability to recognize particular instances of this X when they are met: Knowing what prime numbers or proofs are is one thing, identifying instances of them is another. In the case of science, the identification problem requires studying the gap between the predicates "being a scientific (resp. acceptable scientific, good scientific, bad scientific) item" and "being identifiable as a scientific (resp. acceptable scientific, good scientific, bad scientific) item". And there is of course another gap between "being identifiable as a scientific item" and "being identifiable as a scientific item with limited resources and on the basis of limited information". As I shall now argue, there are various reasons why these gaps should be wide in the case of scientific practices.

First, even if science and scientific activity could be fruitfully described by merely analyzing scientific statements and whether they can be proved within a theory T, the identification problem would remain. As is well known, whether a sentence belongs to a language can be an undecidable question or a computationally intractable one, even if there often exist tractable procedures to answer such questions. So even if the demarcation problem boiled down to questions like "Is this sentence a consequence of this theory T?" and theories were axiomatized, the question of the identification of *bona fide* scientific items would be difficult in practice.

Second, because scientific practices are much more complex and multi-dimensional entities than meaningful statements of theories or proofs, the difficulty to identify the latter is a lower bound on the difficulty to identify the former.[2]

Third, in many cases, there is often no way to give a complete access to practices, which may make it difficult to settle in a transparent, explicit, and uncontroversial way potential debates about their validity. While scientists do share elements of practices—hence the notion of "consensus scientific practices" developed by Kitcher (1993)—many of these are not explicit (see the introduction of this volume, Section V.5; see also Collins, 2010, and Soler, 2011), or cannot be presented by linguistic means. This may be the case for elements like typical experiments, ways of identifying authorities, scientific values, typical instruments and familiarity in their use—or elements like methodological principles or rules of thumbs, which are conveyed in natural languages and do not have a precise semantics. As a consequence, even what is shared cannot be made common knowledge (in the sense of epistemic logic) and it can never be completely uncontroversial that an individual practice is an adequate token of an unexplicitly agreed upon practice. And, of course, individual practices include, in addition, specific instruments, particular experiments, runs of simulations, large sets of data, and so forth and all these particular items cannot be completely presented in articles. As a consequence, judgments about the value of practices often need to be made on the basis of irreducibly incomplete information, and this is more room for misidentification of valid practices. Importantly, this problem is not simply one of moral integrity or deliberately partial reports made by scientists about their individual practices. When honest scientists unconsciously fail to carry out good scientific practices, their peers may not detect their failures because the reports do not and cannot include all the relevant scientific information about potential causes of failures, and experts may fail to ask about all aspects that can conceal hidden unexpected problems.

Finally, as pointed out above, there can be vagueness remaining in some aspects of a research program, as well as disagreement between sub-communities or individuals about which practices should be considered as good ones, all of which can be potential sources of trouble for the uncontroversial identification of bad practices. And, of course, the more complex an item is, the more it is likely that we have partial disagreement about its nature and potential trouble in consensually identifying accepted instances of it—and scientific practices are complex, multi-dimensional entities for sure.

Overall, the difficulty in solving the identification problem means that, in many cases, even experts like reviewers have no infallible procedure for identifying and discarding results issued from bad practices. In other words, there are some principled reasons explaining why the dream of a perfectly and infallibly checkable science should be seen as a utopia. And it is no surprise that, as Andersen points out, the identification of malpractices is a difficult issue and there remains a potentially large grey zone in science—that is, published results that are the product of poor research due to "sloppiness", "incompetence" (p. 161) or "so-called honest error" (p. 166), given that—by definition—what belongs to this grey zone is not clearly known.

The implications of the above general negative conclusion should be drawn with great care. It should not, for example, be seen as implying that the situation is always desperate and that scientists have no good ways of checking individual results. To use an analogy, a problem can be in general undecidable, but be composed in large part of decidable problems; similarly difficult problems can have easy subproblems or easy approximate solutions. In the same way, saying that the identification problem cannot always be exactly solved is not informational about how often it can be solved, how much approximate methods exist that are often successful, and so forth. From this perspective, a reasoned analysis explaining how difficult it is to identify bad practices, depending on the type of practices involved, is still to be made.

As things are, it is not surprising that the examples of malpractices described by Andersen are drawn from experimental parts of the natural sciences since experimental reports can be falsified. But how much should one extrapolate? While one may agree that the analysis of science in terms of practice is a general one and includes the formal sciences—Kitcher's seminal analysis (1983) was indeed about mathematics (see also Giardino, Moktefi, Mols, & Van Bendegem, 2012, and Mancosu, 2008), the philosophy of scientific practices has devoted much attention to experimental science and it remains to be seen how much the practice turn differently or similarly impacts our understanding of the various sciences regarding each question, and this one in particular: In other words, a comparative analysis of what it takes to identify bad practices and malpractices in the various sciences, and which aspects of practices are responsible for these differences, would be most welcome.

To wrap up, once it is acknowledged that the identification problem cannot *always* be solved with total reliability, there remains to be analyzed how the *direct* epistemic identification of individual bad practices does work—that is, when, how much, how, and with what reliability it is possible to directly assess the validity and quality of scientific practices by analyzing information and clues about their nature and content. This philosophical agenda is compatible with the claim that the indirect calibration of practices, which is made by using clues related to the external circumstances in which they were carried out (like the reputation of their authors or the institution in which they were developed) also plays an important role in science.

Because evaluation based on the description of practices is difficult, scientists are led to use indirect, coarser strategies such as calibrating the practitioners themselves, all the more since calibrating individuals on the basis of external indicators can be less costly than analyzing in detail their practices. From this point of view, the difficulty of solving the identification problem is one more reason to follow Hardwig when he says that trust is an essential ingredient of science—perhaps "even more basic than empirical data or logical arguments" (Hardwig, 1991, p. 694): Because external agents cannot check the validity of practices, we have in part to trust authors, both epistemically and morally, for science to be possible. But indirect strategies cannot be all there is to the evaluation of practices; they need to be fed somewhere in the process by direct, partly reliable, epistemic evaluations of practices. Further, as already pointed out, authors can be honest and competent but fail to develop fully satisfactory practices—and there is then the need for external evaluators to help these authors find out when something is wrong and pin down why—and for this reason, trust towards authors cannot be the sole answer. Reviewing is often described as a practice of selection. It surely in part is, at least in a first stage, but it is also a practice of melioration, and the finally accepted papers are usually significantly better than they would have been, had not the reviewers requested revisions by calibrating practices and results. So even if the problem of identification of good and bad practices has no general solution, there are ways to partly solve it, and the possibility of reviewing, as a melioration activity, is an evidence of this.

Overall, because of the difficulty of the identification problem both strategies, direct and indirect, are sometimes jointly used. There is clearly the need to analyze how both strategies work. The study of the direct evaluation of practices (by reviewers and peers in general) probably needs to be rooted in case studies and is work for the philosopher of practices. Another issue to analyze is which balance of direct and indirect strategies is acceptable, if not optimal, if science is to progress reasonably, and this seems to be work for social epistemogists.[3]

I.2. Calibrating Good/Bad/Mal Practices and Practitioners: What Relations?

Let us now turn to the calibration of scientific practices via the calibration of practitioners, and the assessment of when they should be deemed trustworthy, which is how Andersen tackles this issue. Believing results without having calibrated the corresponding practices means that one becomes epistemically dependent on its author, who is the warrant of their validity. Andersen follows Hardwig's analysis of trust, which says that it is based on beliefs about the epistemic and moral character of the author. Accordingly, if B is trusted by A, B will be believed to be knowledgeable and truthful (given that it is unlikely that someone is trustworthy and untruthful—that is, reliable by accident); however, an "untrustworthy scientist may either be untruthful or unknowledgeable (or both)" and, to calibrate untrustworthiness, there may be the need to distinguish "between the moral and the epistemic component of a scientist's trustworthiness" and "assess [her] moral and epistemic character separately" (p. 164).

While it is commonly accepted that trust, and trust in moral integrity, is involved in science (Rennie, Yank, & Emanuel, 1997), there is still work to be done to delineate how much and when it does play a role. I shall, in the remainder of this section, content myself with analyzing when the calibration of practitioners and their moral character in particular seems—or not—to be required.

First, as partly discussed at the end of the last section, even if one is not a reductionist, who believes that all legitimate trust in scientific agents is rooted in the evidence-based assessment of how knowledgeable (and moral) agents are, calibration of agents cannot be the whole story, and the calibration of the very practices is needed somewhere in the process, if only to give clues about the reliability of agents. A good description of how trust is built should disentangle the role of the various components that contribute to it. As an aside, even if the calibration of practitioners was exclusively grounded on the results of the calibrations of practices, it would still make sense to acknowledge that it does play an independent role in science. Indeed, the direct calibration of practices (what reviewers typically do) and the calibration of practitioners do not arise in the same epistemic circumstances (see the third point below); further, how exactly the results of the direct calibration of practices (typically, acceptance in this or that journal) should be used in order to build a picture of the reliability and trustworthiness of practitioners is an independent question.

Second, *as researchers,* scientists are first and foremost interested in reliability. If it is shown that they do not always need to, why should they always get engaged in the moral evaluation of beliefs? Indeed, belief in a degree of reliability of an agent does not commit to any particular belief about the morality of this agent. For example, if I believe that an author A has a 0.9 overall degree of reliability on the ground of indicators like her publication records (that is, ways that do not require from me the direct moral calibration of A), my belief is compatible with different beliefs about her moral and epistemic character (e.g., having a 0.9 degree of integrity and 1 degree of competence, or vice versa). Therefore, I do not need to be committed to any particular belief about her moral or epistemic character—even if, in the process of acceptance of papers, primary evaluators, like reviewers, may have entertained beliefs about her moral character (especially if data were reported)—and I need to rely on these unknown evaluators or indicators. So, trust in the results of an author A at best implies having an implicit indefinite belief about her moral and epistemic character that is compatible with one's precise degree of trust, and no precise commitment is required. In other words, Hardwig's analysis of the implication of trust in terms of beliefs about the truthfulness of authors need not always accurately describe the actual beliefs of scientists about the authors that they epistemically depend upon.

Third, if trust, on the one hand, and belief in moral and epistemic character, on the other, are related, then assessing the epistemic and moral character of A *can* be a way to assess the reliability of A (and vice versa). For epistemology in practice, the question is then "Which ways to reliability and trust are usually taken?" There clearly seems to be cases in which the way that goes through the calibration of moral character will not be taken, and no *explicit* belief about the morality of authors will be developed. For example, clues about reliability can derive from external indicators like the scientific records of the author, the reliability of the journals she has published in, the credit of her co-authors, and so on. All this may be sufficient evidence for believing it is rational to trust a result and its author—and no explicit moral trust, let alone direct calibration of moral character, are then required.

This does not, however, imply that moral calibration plays no role in the scientific processes that lead to trust by an agent towards a particular result. Potential reliability indicators (reputation of scientists, quality of a journal, etc.) also need to be trusted and calibrated by agents, and this may be done on different grounds (information about the reliability of past results, moral character of editors, trust in the judgment of esteemed peers, etc.). Primary procedures of evaluations of new results—typically reviewing—may also partly require trust in the moral character of authors (see below). Still, the existence of moral trust in the evaluation process differs in various ways from the existence and use of explicit moral trust towards authors by scientific users. First, even if beliefs in the moral character of authors play a role in the evaluation process, these need not propagate downstream. Indeed, belief does not seem to be a transitive relation (if I believe that a journal editor is reliable, the journal editor believes that the referee is reliable, and the referee believes that the author is morally honest, it does not follow that I believe that the author is honest). Second, even if reviewers have to morally trust authors, the dependence on moral trust may vanish later in the process, when the reliability of published results is further checked by the community, and the result becomes well-entrenched (or not). One may finally note that it may be safer to rely on beliefs about the moral character of editors or colleagues regarding a journal than on beliefs about the moral character of authors regarding their own results.

Fourth, it is probably the case that, in practice, calibration is performed differently in different scientific contexts and that calibration of bad and good practices (or authors) work differently (even if 0.8 trust is 0.2 distrust). Here are examples to illustrate this point and show how it could be developed.

i) **Scientific judgments between peers involved in research.** First, it is likely that only calibration of "good enough" practices will usually be completed in research contexts. Arguably, scientists are not after a detailed evaluation of the practices of their peers, but only want some good reasons for trusting a subset of very reliable results that pass a chosen threshold, and the distinction that matters is between results under or above this threshold. As soon as it becomes obvious that a result will not pass this threshold, it needs not be calibrated any further—precise assessments take time—since the reasons of this reliability failure do not matter for research—unless perhaps the result is of crucial importance and one is compelled to precisely assess its value on exclusively scientific grounds. To take one of Andersen's examples, once Nobel Laureate Peter Medawar was convinced that there was something fishy about Summerlin's work, he did not bother to push forward the investigation and determine whether this was a case of incompetence or dishonesty.

In other words, scientists are like diggers that try to find gold nuggets in a mine and procedures for finding these nuggets need not be similar to (never actually used) potential procedures for classifying all clearly-non-golden nuggets. This analysis is coherent with the fact that only a small fraction of the literature is cited and it is not unlikely that this fraction complies on average with the highest epistemic standards. So, the argument goes, as far as the advancement of research is concerned, no calibration of moral character is needed for the most dubious results, for which moral calibration would be most needed.

ii) **Moral trust without moral calibration as a default rule.** In some cases, typically in experimental research, having extremely high trust on purely epistemic evidence may not be possible and authors may have to be morally trusted regarding the uncheckable aspects of their work. This does not, however, imply that moral calibration is then performed. Blind moral trust for what cannot be checked may have to be the default rule, as acknowledged by some scientists (Rennie et al., 1997, p. 579). Similarly, when they accept a paper after a thorough epistemic scrutiny, reviewers are more epistemically vulnerable than scientists using the literature, since they cannot benefit from the expertise of other members of the community, which is present once a result has been published and discussed and the result has become entrenched. Reviewers are on their own, and, for this reason, they may have to trust to a greater extent the epistemic and moral character of authors. This does not, however, imply that reviewers develop a specific activity of assessing the morality of the author—again, how would they do this? It can be argued that they simply try to eliminate bad practices and select good ones on the basis of available epistemic evidence—and defectors that do not play the game honestly regarding uncheckable aspects of their work and are afterwards identified will be given a tit for a tat, by no longer being trusted.[4]

Actually, it is not even sure that reviewers are committed to believing that submitting authors, whose papers are accepted, are honest, since they may be simply described as doing *as if* authors were honest, given that honesty and trust are the condition of possibility of scientific activity, and are therefore the rules of the game. So they may be described as saying something like this: "To the extent that it can be checked, the content of the paper is worthwhile and the amount of requestable epistemic evidence that has been provided is proportionate to the importance and novelty of the paper—given that any evidence cannot be requested, in particular in the case of simulations or experiments; so I have the conditional belief that if the author has been honest and conformed to the ethos of scientists, this is a good result".

iii) **Fraud detection.** Once malpractices are publicly suspected for a result, it clearly becomes an important issue to determine whether an author should be convicted of malpractice. Inquiries aimed at assessing the morality of authors may then have to launched, with distinct—heavier, institutional, more collective—procedures than what happens in the usual processes of peer calibration, and the cases presented by Andersen nicely fit this description.

iv) **Collaboration.** As clearly highlighted by Andersen, collaborators are epistemically vulnerable to their co-authors. Because being engaged with fraudulent co-authors is risky, and co-authors are sometimes in a position to have additional clues about the honesty of their colleagues, calibration of malpractices is more likely to take place.

Overall, it is dubious that beliefs about the moral character of authors and moral calibration are explicitly involved or at play in all cases of epistemic dependence, even if potential beliefs in the moral character of authors may be an implicit consequence of an actual belief—but as is well-known, having the actual belief A needs not imply actually believing all the consequences of A. So, in which scientific circumstances moral beliefs and moral calibration are actually important would have to be investigated in more detail.

This being said, even if one subscribes to this mitigated skepticism about the importance of moral beliefs and calibration in scientific activity, and even if one believes that identifying malpractices is much more difficult than identifying bad practices, and that researchers do not frequently engage in such activity, there is room to agree with Andersen that malpractices are an important object of inquiry for philosophers of science and epistemologists. Indeed, even if bad practices and malpractices are hard to identify, understanding what they are, how they occur, why they occur, and so forth, cannot but be useful to make them less frequent. A parallel can be drawn with the detection of driving infractions. While identifying all cases, intentional or not, of dangerous driving infractions is hardly possible, understanding which mechanisms contribute to generate them can be instrumental in reducing their number. So it is clearly important that epistemologists investigate which scientific policies (like editorial policies about co-authorship) can be adopted to reduce malpractices.

II.   HOW TO FIGHT AGAINST BAD PRACTICES AND MALPRACTICES—AND WHEN?

Scientific bad practices may be a threat both for science and society, and their detection is by no means easy. What can be done, then, to make them less frequent in science? Because co-authors have information that reviewers and readers do not have, they are in a better position to be aware or suspicious of scientific faulty practices. Accordingly, one of Andersen's suggestions for improving detection is to make co-authors partly responsible for the fault of their colleagues and thereby compel them to be whistleblowers.

II.1.   Differentiating Questions About Efficiency

Before discussing Andersen's suggestion and its justification, let us analyze more sharply the general issue of fighting against malpractices. Bad practices are faulty actions, which can have detrimental consequences. When trying to eradicate them, one must be careful to assess all the effects of the policies that one may want to apply. Accordingly, it is important to distinguish among the following problems.

> P1. **Efficiency of prevention problem.** Which policies can be adopted to keep bad practices and malpractices as low as possible?
>
> P2. **Scientific efficiency problem.** Which policies are most scientifically efficient (and, in particular, are the policies that keep bad practices and malpractices low scientifically efficient, once all their epistemic and scientific effects are taken into account)?
>
> P3. **Social efficiency problem.** Which policies are socially efficient, once social and scientific advantages and drawbacks are taken into account?

Let us be more explicit. It is hardly controversial that, everything being equal, if there are ways to decrease the importance of bad practices and malpractices in science, they are welcome and it is important to identify such possible ways and answer P1. The distribution of responsibility among co-authors is such a potential repellant against malpractices. Still, in trying to fight against malpractices, one must take care not to make scientific activities, and collaborative practices in particular, too risky, and thereby hamper the development of science, which may crucially require collaborations. So it is important to analyze how beneficial such potential policies against bad practices are for science in general, which means answering P2. Finally, given that bad practices can also have dramatic consequences for society, we should not content ourselves with balancing scientific advantages and drawbacks only. Some policies may be globally detrimental for science, because they slow down its dynamics, but beneficial for society, because they filter out some bad practices that may have devastating social consequences. Answering P3 therefore requires an epistemological and social analysis of the effects of such policies; it requires weighing issues like how acceptable it is to tolerate minor scientific risks of not detecting bad scientific results if they imply major social risks for society, which can be the case when health or environmental issues are concerned.

II.2. Malpractices and the Distribution of Responsibility Within Collaborative Works

There are various ways of fighting against bad practices and malpractices. Some are targeted at individual authors and consist in the development of policies aimed at preventing and detecting bad practices or malpractices at the individual level. Others can be organized at the institutional level of scientific research, such as policies controlling the funding of research and its transparency or potential conflicts of interests. Andersen focuses on the individual level when she analyzes how, in the context of collective works, responsibility for individual malpractices should be shared by co-authors, based on what she calls a relational account of calibration, in which "the strength of the calibration is dependent on the epistemic character of the researcher performing the calibration" (p. 170). She gives as a potential justification for shared responsibility the fact that co-authors have access to technical details that other agents, such as referees, editors, or scientific users do not have: with greater knowledge and the possibility of making valuable calibrations comes greater responsibility. One may wonder, however, whether this responsibility should be seen as rooted in the *actual possession* of knowledge about a faulty behavior or *the possibility* of having easier access to evidence and a correlative epistemic duty of calibrating co-authors. This latter option seems more appropriate to describe the Woo Suk Hwang case and the position of Gerhard Schatten, the senior researcher who, given his role and knowledge, was in an epistemic position to be suspicious about the possibility for his junior co-author to have really carried out the research he pretended he had, and who could gather evidence to prove the fault. Andersen also points at that "collaborative research requires trust", which must "be balanced with a responsibility to ensure the veracity of all results" (p. 171). Here again, the claim is suggestive: Scientists who give their trust should be epistemically accountable for giving it. But then, are co-authors partly responsible for the malpractice or simply for their epistemic failure to detect it—a much more benign fault, given that, the lesser the fault, the lesser the punishment, and the less efficient the prevention?

Andersen finally grounds the potential responsibility of co-authors in an analysis of the epistemic duties of members of scientific groups, who, "participating in a shared cooperative activity of delivering—and being able to defend—a new and interesting result *p*" (p. 171), have the duty of being "epistemically responsive" and having "meshing subplans", which "must include preparing for serious critical inquiries" (p. 171). But here again, there is a difference between being responsible for failing to organize an inquiry—a benign epistemic fault—and the actual responsibility for malpractices to which co-authors did not actively contribute. Andersen concludes that the very author of a malpractice can be morally blamed, but that co-authors can be blamed only epistemically "in exactly the same way as when dubious data arises from the use of a defective instrument rather than a defective collaborator" (p. 171). While this is a suggestive distinction, it does not help one assess how much the responsibility should be shared, since there are indeed cases (like parents for children or ministers for their administration) in which it is possible to be responsible, and even liable, for a fault that one did not want.

How to attribute responsibility and credit is a difficult question. The directions indicated by Andersen seem to be valuable ones, but they raise significant conceptual and philosophical questions about the nature of agency (something that responsibility is often rooted in), the various responsibilities and accountabilities involved (moral, epistemic, scientific, legal), and how potential punishments should be tied to how responsibility distributes. Since scientific practices are actions having their specificities, it surely requires input from scholars studying science. My final suggestion is that the treatment and clarification of these questions may also benefit from, if not require, taking into account the existing rich debates in philosophy of law and action—in particular, issues like collective responsibility and cases in which there may be responsibility "with non contributory fault" (Feinberg, 1968, p. 681). Malpractices are, after all, faulty activities among others and they are sometimes embedded in larger criminal activities (e.g., when the malpractice is deliberately aimed at favoring some industrial interest) by making apparently legal an activity that would not be permitted, had the right results about the corresponding scientific questions been made public. As indicated by the ongoing debates about authorship, responsibility, and accountability in scientific journals (Rennie et al., 1997; Eggert, 2011), these are questions that are in present need of treatment for the development of a healthier science, and philosophers can contribute to shaping the forms that the authorial and editorial scientific practices should take in the future.

NOTES

1. Logical empiricists claim that they want to distinguish between metaphysics and science. Because they provide a solution based on a characterization of scientific statements, they de facto answer the more general problem of the demarcation between science and non-science (for more details, see Hansson, 2012).

2. If a subproblem A' of a problem A has difficulty K, the more general problem A cannot have difficulty less than K. Among other things, scientific practices include linguistic components, and judging the quality of a practice often requires, among other things, judging the quality or validity of a linguistic, mathematical, or computational item.

3. Given that there are various ways of using testimonies and external clues in general, there are many ways of indirectly calibrating authors and practices. Finding which ones are most efficient is another object of inquiry (see, in particular, Mayo-Wilson, *forthcoming*).

4. See (Blais, 1987) for applications of the tit for tat strategy to knowledge contexts.

6 Values in Engineering
From Object Worlds to Socio-Technical Systems

Louis Bucciarelli and Peter Kroes

Within an undergraduate engineering curriculum, whether Mechanical, Electrical, Civil, Chemical, and so on the Engineering Sciences constitute the core requirements for a degree. Courses in Thermodynamics, Electronics, Structural Mechanics, Fluid Mechanics, Controls, Heat Transfer, Materials Science, and the like are normally studied beginning the second year at university. Courses in mathematics, physics, chemistry, and now biology are generally required as prerequisites. This sequence has contributed to the notion that engineering science is but science applied.

But this is too general, an oversimplification. The engineering sciences develop, not fully independently of the classical sciences, but with sufficient gusto and on their own terms—constructing their own useful conceptual schemes, models, and methods, doing their own kinds of tests and data collection—that to see them as a straightforward application of mathematics and physics and chemistry and biology is a gross injustice.[1] Their central place in the thought and practice of engineers designing, testing, and bringing a product to market illustrates more clearly their ontological distance from the classical sciences. As such, they join with discipline specific ways of modeling a product's behavior, with special methods developed for problem solving and established notions about what constitutes a robust solution, with their own body of codes for use in design, with their own forms of prototypical hardware and suppliers' catalogues—all the resources an engineer has to call upon in practice—to constitute what we label an object world.

Those who fail to see all of this and focus only on the engineering sciences alone—presuming that engineering practice consists solely of the solution of practical problems via the straightforward application of science—are likely to conclude that engineering is value free. Note, too, that there is no single engineering science "discipline", but many. And these differ among themselves in significant ways, differences that have important bearing on engineering practice, especially engineering design and product and system development.

Engineering curricula, with their promotion of the engineering sciences alone as fundamental, are deficient because they do not offer students a realistic picture of engineering practice—in particular, with regard to the role of social features and social values. These values enter engineering practice because engineering work nowadays requires ongoing teamwork—a mode in which engineers with different disciplinary backgrounds, responsibilities, and interests (i.e., from different object worlds)—must work together. This gives engineering work a social dimension because negotiations between engineers then become an unavoidable aspect of their work.

Engineering curricula are also deficient because of their almost exclusive focus on object-world problems in which only instrumental engineering values play a role, whereas de facto engineers often have to deal with problems of socio-technical systems—which, by their very nature, are loaded with often conflicting social values. In other words, the kind of problems that engineering students are trained to solve does not exhaust and is not representative of the kind of problems they have to face in daily engineering life.

We claim that this can be fixed: Engineering curricula may be improved by taking into account the social nature of practice and the socio-technical nature of the products and systems they design and develop. We first discuss the nature of engineering object worlds (Section I) and how engineering education is geared to, adheres to, and is constrained by object-world values (Section II). We then briefly discuss how social values, alongside engineering values, become important when object-world systems are replaced by socio-technical systems as the focus of study (Section III). The paper concludes with the suggestion that the engineering educational system itself may be—must be—considered a socio-technical system if it is to meet the needs of contemporary engineering practice.

I.   ENGINEERING OBJECT WORLDS[2]

There is a cartoon familiar to most aerospace engineers that purports to depict the design of an airplane: It shows in some half dozen images, within a single frame, the different visions of the final product which accord with the different interests of those responsible for its design. The vision of the structural engineer includes massive I-beams which assure the craft does not fall apart; the vision of the design participant responsible for powering the craft shows very little structure other than that required to support the huge twin engines.

The aerodynamicist's representation is as sleek and slim as one might imagine; there is hardly room for the pilot. And so it goes; indeed, that is akin to the way it goes—engineering design is a process that engages different individuals, each with different ways of seeing the object of design. These different ways of seeing find their source in the different engineering sciences that sustain the different models and methods that the structural engineer, the heat engine specialist, the controls person, the aerodynamicist, and so forth bring to the task.

The differences in readings of the artifact-to-be, the object of design, are rooted in the different work experiences as well as educational histories of participants in a task, and they correlate with the different individual's responsibilities. Each participant sees and understands the object of design in accord with the standards of thought and practice within their domain of specialization. It's like they live in different worlds.

We might even claim the participants in a design task live in incommensurable worlds. The structural engineer has a certain way of dealing with, describing, and speaking about the world. He or she relies upon the theories and methods of structural analysis to figure how the structural elements of the object will behave when subject to external forces, how it will deform and displace or vibrate. Stress and strain, force and displacement, are the variables that matter most to the structural engineer. There are computer programs designed specifically for modeling complex structures of all shapes and sizes. But it's not just the engineering science of mechanics of solids and particular mathematical methods that define the world of the structural engineer. Certain machinery exists for testing materials and completed structures; instrumentation and sensors have been developed specifically for use in this regard. There are particular standards and regulations, and particular suppliers and consultants who can be called upon for support. The existence of standardized, "off the shelf" structural elements and fasteners makes life easier. And structural engineers have their own professional journals and professional societies.

The world of the electronics engineer is different. He or she relies upon theories and methods of circuit analysis, knows the characteristics of electronic components, and speaks of voltage and current, of power. A different infrastructure of standards and regulations both guides and constrains the electronic engineer in design; there are (many) different off-the-shelf devices to choose from and build with, different forms of mathematics to apply (e.g., Boolean), different ways of sketching and modeling—the block diagram figures largely here—and different computer tools for modeling their systems. Testing instruments and apparatus are different. Time even has a different quality. Dynamic response predominates. And of course their professional journals and societies differ from those of the structural engineer.

And so it is, down through the list of all those who have a significant say in a design process. Each inhabits a world of things particular and employs specialized modes of representation. A world with its own unique instruments, reference texts, prototypical bits of hardware, special tools, suppliers' catalogues, codes, regulations, and unwritten rules. There are exemplars, standard models of the way things work from the disciplinary perspective of the particular world, and particular metaphors which enlighten and enliven the efforts of inhabitants. There are specialized computational methods, specialized ways of graphically representing states and processes. And each participant works with a particular system of units, with variables of particular dimensions, with certain ranges of values, perhaps. Different participants work within different *object worlds*.

This has important implications for understanding, none the less managing, the design process—or any engineering task that requires the collaboration among different participants from different domains for that matter. It means that, in order for the ordinarily conflicting proposals and analyses of different individuals to be brought into harmony, participants must rise above their respective object worlds and reconcile their differences. While recourse may be made to mathematical methods for setting priorities, making choices, resolving trade-offs, and reconciling differences, these methods do not have the same degree of authenticity as those that participants rely upon within their respective object worlds. Instrumental rationality only goes so far. Negotiation becomes necessary; designing is a social process.

II.   ENGINEERING EDUCATION[3]

The full import of this claim is slowly being acknowledged by faculties of engineering. Over the past few decades, engineering faculty, the professional societies, and perhaps most importantly, ABET, the agency responsible for setting criteria for accreditation, have come to recognize that while mathematical and scientific knowledge and know-how is necessary to do engineering, such are not sufficient. Recognition that the social/political features of practice merit attention is reflected in new (circa 1998) criteria and statements of educational objectives: Now communication and teamwork and ethics matter; so, too, an understanding of the "impacts" of technology on society. But the emphasis here remains on the acquisition of certain "skills"; there is little recognition of the social realities of day-to-day engineering practice. Engineering science budges not.

The dominance of the culture of engineering science is reflected in the exercises assigned within the core courses of the student's major, whether mechanical, electrical, civil, chemical, and so on. Their purpose is to convey a well-established body of instrumental, disciplinary knowledge from faculty to the student. The abilities stressed are problem solving within the discipline's paradigm using its concepts and principles alone. Here, for example, is an excerpt from a well-known textbook in engineering mechanics.

> The main objective of a basic mechanics course should be to develop in the engineering student the ability to analyze a given problem in a simple and logical manner and to apply to its solution a few fundamental and well-understood principles.[4]

The mechanics problem is *given*—not to be formulated by the student; it demands a *simple and logical analysis*—not a conjectural, inferential thinking up and about; and is to be solved using a *few fundamental and well-understood principles*—not by trying several, alternative, perhaps conflicting, approaches and perspectives. The work-life of an engineering student, hence graduate, from this perspective is neat, well posed, deductive, and principled.

Solving well-posed, single-answer problems is the dominant learning experience of the undergraduate. It is a solitary activity; one is engaged in competition with one's peers. It is an essential activity in engineering practice, but it is not all. Solving problems is at the core of object-world work. It is necessary work, but it does not suffice. The (over)emphasis on solving well-posed, single-answer problems, with its reductionist, deterministic ideology, works against taking seriously the social/political features of tasks that require rising above object-world work.

This rising above object-world work may take two forms, each of which lays bare the social/political nature of engineering practice. In most cases engineering work is teamwork; as noted above, this requires cooperation between inhabitants of different object words that brings all kinds of social processes into play. So, engineers will have to leave the comfort zone of their own object worlds in order to arrive at an engineering solution for a given problem. But engineers will have to rise above object-world thinking in an even more fundamental way. Real-life engineering problems do not fall from heaven, like manna, in the form of well-posed, single-answer textbook problems. On the contrary, they are the outcome of complicated social/political processes between different stakeholders, among which are engineers. Real-life engineering problems are socially constructed, and once a problem has been defined as an engineering problem, it may still take a lot of engineering work to transform that problem into one that, from an object-world perspective, may be considered a well-posed, single-answer engineering problem. So, a lot of engineering work transcends object-world thinking in the sense that it is part and parcel of engineering work to define engineering problems in such a way that object-world thinking can do its work.

The conception of engineering as solving well-posed, single answer problems severely constrains any attempts at curriculum reform that take seriously the social nature of engineering practice. Any such attempts to include subjects of study that do not yield to mathematical modeling or use of a spreadsheet on an equal footing with the engineering sciences are immediately challenged. How to fit all of this new material into an already overbearing curriculum? And what do we leave out?

Something is fundamentally wrong with this way of framing the challenge of reform. The source of error lies in our ways of thinking as engineering teachers and researchers—best characterized as thinking that is dominated by instrumental rationality, that is, a form of rationality that is strictly confined to choosing the most effective and efficient means (instruments) for achieving a given end, but which is mute when it comes to choosing the ends themselves. This way of thinking is evident in the ways we describe a "problem" to be "solved" and what is deemed a legitimate "solution".

Beer and Johnston's definition of the main objective of a mechanics course provides an example. And it's this way of thinking that leads us to write and speak of knowledge as if it were some kind of material stuff: We *gain* knowledge, *store it away* somewhere in our head, *transfer* it to our students; students claim that my course is "like taking a drink from a fire hydrant". Our research contributes to the *body* of knowledge and we measure this in large part by the *number* of our publications. We know *more* now than before.

The "knowledge as stuff" metaphor leads us astray—down the path of curriculum reform that constrains our discussion to what material we must *cover*, what we must *leave out*, what we should *keep in*. Instrumental rationality is essential to engineering, but it fails when taken too far, when presumed a basis all of our thoughts, for dealing with events and features, phenomena and people, beliefs and values, that cannot—ought not—be reduced to quantitative measure.

The exercise of instrumental rationality requires abstraction and simplification. This is key to methods for problem solving in all fields of engineering. Abstraction requires:

> Simplification of a complex problem by breaking it down into manageable components. Specifically modeling in *quantitative terms* critical aspects of the physical and human world, and necessarily simplifying or *eliminating* [my emphasis] less important elements for the sake of problem analysis and design.[5]

For a problem to be treated as an engineering problem, it must be expressed in quantitative terms. Only factors, aspects, and features of the "real" world that can be construed as measurable and quantified matter may count in a literal sense. Numerical measures of inputs, outputs, parameters, variables, behavior and performance, costs and benefits are the essential ingredients of a problem. One might wonder what criteria are used in eliminating (or deforming) more qualitative elements for the sake of problem analysis and design. Is it perhaps the case that only those "elements" that can be quantified are considered at all? Anything that can't be measured is, ipso facto, irrelevant, not of interest or significance?

Some, constrained to think in this way, can only express their desire for improving engineering education in terms like ". . . our 'solution spaces' must be extended".[6]

However, in our opinion, broadening the solution spaces will not do; that is not sufficient to bring into view the essentially social nature of engineering. For that it will be necessary to give up the idea that engineering work starts with a "given quantitative problem" and to step back in order to include the problem definition as an integral part of engineering work; therefore, in whatever way the "solution spaces" may be extended, it will have to be complemented with a "problem-spaces" approach, since the engineering problem to be solved is the outcome of a social process.

Instrumental rationality is but one mode of thinking in engineering; we must allow that much more goes on in a classroom than the learning how to solve well-posed, single-answer problems. Normally never explicitly brought to the fore, there are particular ways of seeing the world, ways of identifying and constructing a problem to be solved, and particular ways of *not seeing* what cannot be so easily included in a problem to be solved, that reside in between the lines so to speak. Attitudes, perspectives, and values—other than instrumental—pervade our engagement with students. Most of all, we implicitly or explicitly instill in our students the value that only instrumental rationality matters in engineering practice, a value that itself is taken for granted and that itself lacks, ironically, any grounding in instrumental rationality!

The attitudes and values we convey to our students—in teaching engineering design as well as the engineering sciences—are the attitudes and values appropriate to object-world work. But narrow, disciplined approaches, while necessary, do not suffice in practice. Not just in design, but throughout all kinds of engineering tasks, participants' proposals, as well as analyses, conflict; they must be brought into harmony with one another if a task is to be brought to completion. For this, constructive critique and clarity of explanation, negotiation, and exchange is necessary. In this—what's best described as a social process—traditions of the firm and of the nation matter. Norms and beliefs—about what is a "robust" design, about the capabilities of the user—of citizens matter. Ethics matters. Culture matters. The "context in which engineering is practiced" is value laden. Our current undergraduate programs in engineering miss all of this; we do very well at preparing the object-world worker—but pay little attention to the rich and varied and social/political contexts of engineering practice.

* * *

The reader ought not conclude from this that thought and practice within object worlds is mundane, done mechanically (looking up in tables. . .), routine, or uninteresting. That's the case for some tasks, but not true in general. Quite the contrary: The challenges engineers face within these worlds are never so neatly defined as problems to be solved (they first must be constructed), nor bounded so narrowly (defining interfaces requires more than a look-up table), nor devoid of opportunity for creativity (even in the smallest item) as the general public might presume.

Engineers derive great pleasure and satisfaction from getting things to work right in accord with their conjectured solutions, their proposals, their designs. Finding an elegant solution to a problem, or going from ideas, words on paper, a statement of specifications, to a device that actually does what the boss or a client says it should do is quite an amazing achievement—and it is sensed that way. For there is no rule book for doing such. Object-world work is immensely satisfying, albeit constrained and instrumental, quantitative, and material.

The narrowness of the domain, the instrumental nature of object-world work, frees the engineer from social concerns. Working like a scientist in this regard, uncontaminated by human foibles, varying opinion, subjective judgment—or this is the way it seems—one can dream of reinventing and saving the world, through the miracle of modern technology—oblivious to what goes on in the world outside.

This kind of thinking is reflected in the tripartite model of engineering responsibility according to which engineers can only be held responsible for the correct solution of strictly technical problems—that is, for the correct solution of object-world problems.[7]

This fascination with technology in and of itself alone is characteristic of the exciting part of object-world work. And it is what sustains the energy and engagement of faculty in their teaching of undergraduate as well as graduate students. In this fascination lie the roots of the value system fundamental to engineering education.

III. FROM OBJECT WORLDS TO SOCIO-TECHNICAL SYSTEMS

But the educational system is deficient: It is deficient because not only does it ignore context—the context of practice, the context of use, the context of the individual psyche (barely acknowledged in the teaching of engineering)—but also because the object of study is ill-chosen. The system implies that solving single-answer problems or finding optimum designs, uncontaminated by the legitimate interests of others who see the world in other ways than we do, is what engineers do all of the time. It rarely explores or shows how social and political interests contribute in important ways to the forms of technologies we produce. It assumes that engineering knowledge and know-how is universally accessible and understood by all in the same way—free of cultural variety or individual expression.

With blinders on, what is seen is only the "hard" stuff; what is discussed in earnest is limited to how to get from a well-posed problem statement to the unique solution, from a list of functional requirements in design to specifications of the product. The educational system harbors a value system that glorifies the material to the extent that the system will not allow any serious discussion of values and visions other than those co-opted within itself. Object-world work might be perceived to be value free, but that is but one part of engineering competence that we need to develop in our students.

***

Picture, then, the arena of object-world work as a closed space with quite impermeable boundaries that limit thought and discourse. These boundaries determine what belongs to the object world and what to its context.[8] To test these limits, to explore what these walls are made of, we can observe a problem in the making or solving and ask: What's in, what's out? What questions are legitimate? Which are made a laughing matter? If, within a problematic situation in which an appeal to technology and engineering is made to "solve a problem", we were to confront students and ask them: Who has what problem and why? Or who has the power to "frame" the situation as problematic? Or why is the problematic situation "framed" in such a manner that technology is supposed to be necessary to solve it?—then students are likely to feel uneasy, if not mystified, by this turn away from solving the problem at hand.

Consider, as an example of a "problematic situation", the U.S. government's position on genetically modified organisms (GMO): The FDA has deemed that there is no credible, scientific evidence that GMO foodstuffs are hazardous to the consumer's health. But is human health the only issue to be taken into account? If one frames the "situation" to include the effects of corporate dominance through the distribution of "terminator" seeds and control of the cost of seed stock, a different kind of question emerges—one not answered by quantitative analysis of the results of consumption, but one that requires the qualitative analysis of social forces—that is, the motivations and interests of different "stakeholders". Strictly instrumental scientific/technical analysis and tests will not suffice if one takes seriously these concerns. Social/political interests, including the justification and value to society, enter and cannot be ignored.

Or we might ask: "Should food products containing GMO ingredients be labeled as such?" Although an analysis of costs and benefits might be included in response, staying within bounds of an economic object world would most likely fail to address the justifiable concerns of consumers that cannot be captured in some (probabilistic) quantitative measure. Consideration of the value to society of a public who can distinguish and make judgments given appropriate information is even less likely to be taken seriously, not the least because of the difficulties in answering questions such as: What should be the form of that information? Who is the audience? How to convey scientific information in a "usable" form? None of these kinds of questions would be considered appropriate if one confines one's analysis to the quantification of costs and benefits staying within the bounds of economic analysis.

One way to de-mystify (or to introduce thoroughly) questions of values—to make sure that our field of view includes more than a problem to be solved through instrumental methods—is to engage students in a discussion of the distinctions between an *engineering system's* and a *socio-technical system's* approach to a problematic situation and bring them to understand that these differ, not in the nature of the technical ingredients of a situation, but in the range of perspectives, priorities, methods, and competencies of those who define the scope of the system and direct the work. Examples of socio-technical systems are civic air transport systems, electric power supply systems, or car road transport systems. Engineering (technical) systems are integral parts of these systems, and the functioning of these systems depends crucially on the functioning of these engineering systems. When the technical infrastructure of a socio-technical system breaks down, the whole system breaks down, but the same goes for the social infrastructure. For instance, without traffic laws and the institutions to enforce compliance with these laws, road transport systems as we know them would be impossible. The functioning of socio-technical systems depends on the functioning of the technical hardware and the "social software" and the way these two subsystems are attuned to each other. Socio-technical systems are hybrid systems; they combine technical and social elements and, because of this, an object-world approach will not do. When dealing with problems within socio-technical systems, other values besides the ones that reign supreme in object worlds will have to be taken into account.

A *socio-technical system approach* can be defined as one where the *social/political features* are clearly primary and the *technical features,* while essential, are secondary, while an *engineering system's* approach would remain within bounds of an object world and give priority to the *quantitative analysis* of the interplay of features which can be represented as *variables and parameters in a model.* Another mark of distinction is the kind of methods used in addressing the system: Scholars of *engineering systems* rely primarily on *instrumental* methods of the kind taught in departments of operations research and schools of economics or business, as well as engineering. Those who work on *socio-technical systems* rely as much on *qualitative* description and argument of the kind taught in departments of anthropology, social psychology, or STS. "Actor-network theory" serves as a relevant example.

Going from engineering systems to socio-technical systems will involve the analysis and description of things usually unseen—namely, all kinds of social/institutional aspects that may be of far-reaching importance when implementing (new) technologies. But more is involved in diagnosing socio-technical systems. It is not only a descriptive matter, but also a normative one. It involves the evaluation of whether such systems are performing their societal function well or not, and whether it is desirable to keep the socio-technical system as it is or to change it. It is clear that if one wants to change/intervene into a socio-technical system, then normative matters cannot be avoided because any change/intervention will be related to realizing certain ends and one may dispute whether the ends chosen are (morally) appropriate.

When it comes to analyzing socio-technical systems, it is very difficult to separate descriptive from normative issues. For a simple technical artifact— say, a pump—it may be possible to analyze descriptively whether it functions well or not in the sense of whether the pump meets the list of specs. However, for socio-technical systems, things become much more complicated because there is no unambiguous list of specs. So, if someone is claiming that such a system is not functioning well, that cannot be simply an empirical statement; it also involves normative/evaluative elements. That is why, in those cases, the question "Who says so?" becomes so important.

Students therefore need to learn how to come to a decision about how to proceed in a problematic situation: whom to involve in those decisions, what aspects or problems to solve by technology, what to solve by social/institutional measures, how to match technical and social measures/actions, and so on.

IV. CONCLUSION

We have described how in the rush to solve "the problem", certain kinds of questions—in particular, those that require consideration of values—are out of bounds if they don't yield to instrumental modeling and quantitative response. But ought we be satisfied with this explanation and leave it there, accepting the autonomy of the engineer to define "what's in, what's out"? After all, all of this problem solving has contributed mightily to the welfare of society in many respects.

Evidently not. There are too many other voices, and indeed some within the profession itself, claiming that today's graduates are sorely lacking the ability to deal with extra-object-world dimensions of a problematic situation. Many, in response, seek a remedy through additions or alterations to the traditional engineering curriculum—for example, adding a course in ethics, or sociology, or STS.

We suggest that this will not suffice; curriculum reform of this sort will not, by itself alone, lead to the kind of cultural change we see as necessary if the walls around object-world work are to be leveled and questions of value taken seriously. Prerequisite to change, we must consider *engineering education* itself as a *socio-technical system*. The engineering education system is a socio-technical system in the broad sense of the notion "technical"; it is a system that makes use of educational means and techniques (including hardware means, such as computers, and software means, such as educational tools). But it is also a social practice with a social infrastructure that comprises, for instance, rules for awarding Bachelor's, Master's, and PhD degrees; membership of professional societies; and accreditation rules and laws about academic freedom of teaching and research. In particular, we can ask: What is the social function of this socio-technical *engineering educational system*, and is it, in its present form, performing its social function well or not?

In this, we ought not rest satisfied with an engineering system's analysis; if the purpose is to bring about change that would respond to the need for graduates to have developed an awareness of the importance of context (broadly conceived as social and political), then we need to see that the context of engineering education itself stretches well beyond the walls of the classroom, the problems at the end of the textbook, and the research interests and expertise of the faculty. We must broaden our field of view to include the whole cultural mix of engineering and its institutions—including its professional societies, the organization of the firm or government agency within which the graduate labors, the accrediting bodies, and even society's lure for technology and the media's hype of the latest innovation. The analysis of this socio-technical system, we invite others to engage in.

NOTES

1. See Vincenti (1990).
2. This section appeared previously in Bucciarelli (2002) and in the second chapter of Bucciarelli (2004).
3. Much of this section has been taken from Bucciarelli (2012).
4. Beer, Johnston, and DeWolf (2006, p. xiii).
5. Massachusetts Institute of Technology (2005, p. 4)
6. Grasso and Burkins (2010).
7. See Van de Poel and Royakkers (2011, pp. 21–22).
8. For a discussion of the notion of (social) context of technology, see Kroes and van de Poel (2009).

Practice Turn as Paradigm Shift

Commentary on "Values in Engineering: From Object Worlds to Socio-Technical Systems", by Louis Bucciarelli and Peter Kroes

Sjoerd Zwart

I. THE TURN TO ENGINEERING PRACTICES

Four decades ago, scholars, inspired by the emerging social study of science (SSS), turned to examine the practices of engineers with the aim to come to a more realistic picture of engineering and technology than the one provided by philosophy of technology predominant at that time.[1] In the present chapter, Larry Bucciarelli and Peter Kroes focus on the lessons to be learned from this turn toward engineering and design practices. Before considering their insights, I will first sketch some highlights of the literature on these practices as far as relevant for this chapter. After that, in Section II, I will argue that the turn to practice is most appropriately interpreted as a shift of paradigm in science and engineering studies. Section III is dedicated to the mismatch or discrepancy Bucciarelli and Kroes observe between engineering curricula and the practices in engineering design. And finally, in Section IV, I turn to the advantages of interpreting this mismatch as originating in the shift of paradigm of Section II, rather than as just a lack of societal norms and values.

Let us consider now some highlights in the history of the turn towards engineering practices. I discern three origins of practice-based studies in engineering science: the history of technology, design methodology, and, most importantly, the social studies of science. I will end this section with the quest of Kroes and Anthonie Meijers for empirically better-informed analytical philosophy of technology. Long before the 1980s, historians of technology had already been occupied with detailed historic case studies of technology that could be viewed as investigations of technological practices, but their work had virtually no, or only restricted, impact on the philosophy of technology. This situation changed drastically when Walter G. Vincenti, an accomplished practitioner and educator in aeronautics, started in 1979 to publish his detailed historical investigations into the developments of aeronautics, taking into account theoretical and practical aspects mostly neglected at that time. What prompted Vincenti to launch his meticulous studies of engineering was, among other things, a question asked by Stanford economist Nathan Rosenberg somewhere in the early 1970s over lunch—a question that can retrospectively be recognized as typical of the turn to practice, viz., "What is it you engineers really do?" (Vincenti, 1990, p. 3). Vincenti's endeavors culminated in his generally acknowledged 1990 book on engineering knowledge.

Vincenti's aim in this book was to characterize the nature of engineering knowledge and how this knowledge came about. Based on fine-grained case studies, he described engineering practices in aeronautics in the first half of the 20th century and portrayed the distinct character of engineering knowledge in comparison with scientific knowledge. Against the (at the time) widespread view of engineering as applied science, Vincenti convincingly argued that engineering knowledge comprises much more than only the application of scientific knowledge. The relevance of his empirically based studies was acknowledged; they fundamentally changed the epistemology of the engineering sciences. His book became an incentive for any serious study into technology to no longer ignore detailed historical case studies of engineering practices.[2]

The second line in the development towards the turn to engineering practices originates within the work of practitioners of engineering *design.* Typically, these practitioners were teachers of engineering design, primarily concerned with standardization and corresponding improvements in engineers' education. In the 1980s, these practitioners started to claim that the dominant positivist model of the engineering design method—that is, roughly put, a linear model mainly based on instrumental rationality—hopelessly failed to account for what actually happened in design practice. Accordingly, they developed alternative, more empirically adequate, characterizations of the methods actually applied when engineers solved design problems in practice. Donald Schön, for instance, started with Argyris to publish on practices in this spirit in 1974 (Argyris & Schön, 1974), and in his later work (1983) Schön developed a new reflective epistemology of practice to replace the backward-looking paradigm of the positivists (Schön, 1992). In their endeavors to understand and teach engineering design, design epistemologists turned to ethnographical methods, participatory observation, and protocol analysis.[3] Unlike the social constructivists, however, the design epistemologists did not focus on the social aspects of design practices, which are so central to the work of Bucciarelli (as we will see below)—although they acknowledged that industrial design was often carried out in groups.

Besides the historical and design trends towards engineering practices, the most important impetus to study them came from within SSS. The landmark in these studies is, of course, *Laboratory Life* (1979), in which Bruno Latour, together with Steve Woolgar, report on Latour's visit to the Salk Institute in San Diego. Latour entered this molecular biology laboratory as an anthropologist gathering empirical material about the daily activities of its scientist community. Latour and Woolgar anticipated that this material would help them to "obtain an interesting analytical handle on [their] understanding of scientific practice" (Latour & Woolgar, 1979, p. 34) Besides collecting empirical data about scientific practices, the goal of their study was also to find out how the "idiosyncratic, local, heterogeneous, contextual, and multifaceted" (1979, p. 153) scientific practices ended up in well-organized, systematized, and tidied-up research reports (1979, p. 29). In addition, they came to conclude how scientists, in their daily scientific practices, constructed scientific facts. Which conclusions are to be drawn from Latour and Woolgar's constructivist endeavors is still subject of debate.[4] In any case, *Laboratory Life* has been a role model for many other scholars carrying out ethnographic research of scientific and engineering practices.

Latour and other SSS scholars refused to distinguish between science and technology, a reluctance characteristically enshrined in the notion of "techno-science". One of the first seminal works in science studies that did explicitly distinguish between the two, and concentrated on a social constructivist analysis of *technology,* is the edited volume of Wiebe Bijker, Thomas Hughes, and Trevor Pinch, published in 1987. This volume reprints the Pinch and Bijker (1984/1987) paper in which the authors develop their theses about technological artifacts as social constructs, and in which Bijker introduces his "Social Construction of Technology" (SCOT) model of the developmental processes of a technological artifact. In contrast to the standard linear model of technology development, it concerns a multidirectional model.[5] In the same volume, Edward Constant II investigates "The Social Locus of Technological Practice". He claims that researchers up to that time had considered three ways in which technological practices were rooted in society, viz., via technological communities; via complex, usually corporate, organizations; and via the technological systems that found their way to society (Constant, 1987, p. 224).

Probably inspired by the work of Latour and his co-workers, in the 1980s, scholars started to apply ethnographical methodologies, in particular, to engineering practices. These methodologies are characterized by carrying out empirical field studies in which people's actions and accounts are investigated in their everyday settings. According to Martyn Hammersley and Paul Atkinson (2007), ethnographical research usually concentrates on a few cases in which participant observation and informal conversations are the most prominent method of data collection. The interpretation of the data in terms of meaning, function, and intention is usually a qualitative affair, without using statistics. Bucciarelli (1988), being one of the first to apply ethnographic methods in engineering, reported on his participatory observation in two engineering firms (one making solar panels and the other x-ray equipment).[6] Taking the frame of an ethnographer, he concludes that for engineers *designing is a social process.* Bucciarelli characterizes the various designers as living in their own, rather isolated, "object worlds", which are "worlds of technical specializations, with their own dialects, systems of symbols, metaphors and models, instruments and craft sensitivities" (Bucciarelli, 1988, p. 162). Bucciarelli concludes that mechanical, electrical, and manager engineers, who live in their own worlds of specialization, produce designs (as blueprints of the future artifact) in a social process of mutual dependence and collaboration.[7]

No short sketch of research into engineering practices could go without mentioning recent attempts to come to more interdisciplinary empirical accounts of engineering practices that *integrate* the social studies of science and cognitive science. The work of Nancy Nersessian (2005, 2012; see also Nersessian & Patton, 2009) provides a typical example. It reports on ethnographical studies into the model-building behavior of engineers in biotechnological laboratories. In her work, Nersessian opposes strong reductionist tendencies apparent in both SSS and cognitive science. Instead of focusing on the engineering actions alone, she maintains that if we are to come to grips with engineering practices in laboratories, we should concentrate on the interplay between these actions and the resulting *products* (i.e., the models that come out of these actions). We should take these models as lynchpins interlinking the cognitive and the sociocultural aspects of the practices of solving engineering problems.

Finally, let us turn to analytically oriented philosophy of technology that values being empirically well informed, and consider the volume *Empirical Turn in the Philosophy of Technology* (2000), edited by Kroes and Meijers. The editors of that volume want philosophy of technology to become better informed by detailed case studies of engineering, and they aspire to analytical philosophy of technology becoming more empirically based. Moreover, parallel to maxims in SSS, they maintain that philosophy of technology should not consider technology as a monolithic whole. It has to evade claims about technology in general, as general claims about science also threaten to make philosophy of science too abstract an undertaking. Neither do they want the study of technology to become a strictly descriptive affair, as preferred by some sociologists of technology. According to Kroes and Meijers, philosophy of technology may sometimes be strictly conceptual, and it should take a normative stance whenever necessary. In sum, Kroes and Meijers (2000) put the content of their work under the heading of analytic philosophy of technology that is conceptual, empirically well-informed, and possibly normative. They acknowledge the detailed differences in the various fields of technology and would like to avoid treating technology as a monolithic whole.

II. TURN TO PRACTICE AS CHANGE OF PARADIGMS

In this section, I argue that the turn to practice in the science studies is most appropriately interpreted as a paradigm shift. Then I show that Bucciarelli's frame belongs to the practice-turn paradigm, and that he is well aware of taking the perspective of an ethnographer in his field studies. The paradigm-shift perspective will turn out to be crucial for my diagnosis of the mismatch that Bucciarelli and Kroes observe between engineering education and the practices of engineering design discussed in Section III.

Practices in general started to become subject of study when the foundations of sociological methodology were discussed by towering figures such as the American ethno-methodologist Harold Garfinkel, the French empirical sociologist Pierre Bourdieu, and the English theoretical sociologist Anthony Giddens. These discussions matched well some aspects of the work of the late Ludwig Wittgenstein and of Thomas Kuhn. Regarding the definitions or at least delineations of the meaning of "practice", Schatzki et al. (2001) claim that according to most theorists, "practices are arrays of human activity" (p. 11). On the same page, they write: "A central core . . . of practice theorists conceives of practices as embodied, materially mediated arrays of human activity centrally organized around shared practical understanding". Leaving aside the apparent circularity of defining practice in terms practical understanding, I want to draw the reader's attention to the all-embracing conception of theory that prevails in the studies of practices. A "theory", according to Schatzki, is "simply, [a] general, abstract account" (2001, p. 12).[8] This notion of theory is so radically different from the received view notion of scientific theories that it suggests the turn to practice to be a shift of paradigm.

Schatzki et al.'s definition also illustrates how the turn to *scientific* practices has been the change of the focus on the *products* of scientists and engineers towards their *activities*.[9] Whereas traditional philosophers of science were mainly occupied with the systematic analysis of scientific products such as theories, laws, models, phenomena, measurement results, empirical data, and so on, the practice scholars were much more involved in studying the actual processes of knowledge production. They endeavored to frame the formulation and acceptance of scientific and engineering claims in terms of actions of scientists in the context of social structures. Or, in the words of, for instance, Pinch and Bijker: "explanations for the genesis, acceptance, and rejection of knowledge claims are sought in the domain of the social world rather than in the natural world" (Pinch & Bijker, 1984/1987, p. 18).

The change of focus from product to process went along with a strong dissatisfaction with the backward-looking character of the more traditional, philosophical studies into science and technology. For instance, after having reviewed the literature, Pinch and Bijker conclude, "In any case it is clear that a historical account founded on the retrospective success of the artifact leaves much untold" (1984/1987, p. 24). They propose to apply Bloor's symmetry thesis to the realms of technology studies and to treat technological success and failures on a par: "[it is] our intention of building a sociology of technology that treats technological knowledge in the same symmetric, impartial manner that scientific facts are treated within the sociology of scientific knowledge" (1984/1987, p. 24). They did so because "preference for successful innovations seems to lead scholars to assume that the success of an artifact is an explanation of its subsequent development" (1984/1987, p. 24). Thus taking successful artifacts as explanans, one arrives retrospectively at too simplistic and erroneous linear models of technological innovation processes. By studying successful and unsuccessful technological innovations in a symmetrical fashion, and taking the successes as explanandum rather than as explanans, Pinch and Bijker arrive at, according to their own account, a much more interesting, non-linear, multidirectional model of the innovation processes.

In what follows, I will use the term *forward-looking* or *prospective* studies of science and engineering for those studies in which the researcher puts herself beside the investigators who carry out the scientific research. From within the researchers' context, quasi-participating in the course of the problem solving process, she looks sideward and forward and describes how the actors behave during the development of the research process.[10] Most importantly, the analyst is not allowed to use any knowledge about the outcome of the research process if such is available. Thus, history of science takes a forward-looking perspective when it refrains from using scientific results arrived at after the period of investigation. The Introduction of this volume calls the prospective view the *view from within science.* Additionally, I will use the adjective *backward-looking* or *retrospective,* in the broad sense of the term, for those studies of science that take the present-day, stabilized, well-established, and generally accepted scientific knowledge as point of departure for their research into historical scientific developments.[11] Retrospective studies into science in this broad sense often, but need not, go hand in hand with the following additional features. They are easily combined with the belief in cumulative progress of science according to which the present state of established scientific knowledge is an improvement on previous states (realists may claim they are closer to the truth and antirealists maintain they may be empirically more adequate). Moreover, they typically concentrate on scientific products such as theories and laws and scientific facts, rather than on the (daily) activities of scientists. And finally, backward-looking analyses of science are apt to carry out rational reconstructions of these present-day stabilized and accepted scientific products. When retrospective studies of science are accompanied by these three features, as for example in much of the work of the logical empiricists, one could call them retrospective or backward-looking in the strong sense. In what follows, I will use "retrospective" and "backward-looking" in this last sense.

The change from retrospective towards prospective studies of science comes with a comprehensible critique and abandonment of many distinctions proposed by scholars carrying out retrospective studies of science and technology. The distinctions between technology and science, for instance, or between being internal or external to science and technology, or between the contexts of discovery and justification, turned out to be hard to maintain when studying science and technology prospectively.[12] They were replaced by overarching concepts such as, for instance, the seamless web of society and technology.

Taking the observations of the last four paragraphs into account, I come to the following conclusion. The turn to the study of scientific and engineering practices starting within SSS in the late 1970s, and gaining momentum in the 1980s, is well-suited to be considered as a *shift of paradigm*, where paradigm is understood as a *disciplinary matrix*. Kuhn characterized the latter by shared conceptual generalizations, metaphysical beliefs, shared values and exemplars (1970, postscript). According to Kuhn "the switch of gestalt . . . is a useful elementary prototype for what occurs in full-scale paradigm shift" (1970, p. 85). The shift from equating science with its successful or stabilized products towards equating it with all activities of scientists provides a clear example of such a gestalt switch. Below I will refer to this gestalt switch regarding science as *taking a different frame on science*. Also, the change from using retrospective to prospective research methodologies is a clear indication of the shift of paradigm, and, finally, the abandonment of distinctions crucial for the traditional philosophy of science comes close to loosening "stereotypes", which according to Kuhn is an indication of a paradigm shift (1970, p. 89) as well. All in all, the turn to practice is adequately characterized as a Kuhnian shift of paradigm, which implies a different frame on science.

Bucciarelli's ethnographic study of engineering design, then, forms a typical example of science studies after the turn to practice. He writes, "my frame is that of the ethnographer; my first premise, based upon what I have observed, is that designing is a social process" (1988, p. 160). This statement discloses that, using "participant-observation techniques" (1988, p. 160), Bucciarelli carries out a forward-looking study of engineering practices. He puts himself in a prospective manner next to the actors that he is studying without knowing the outcomes of the developments. Within the turn-to-practice paradigm, Bucciarelli focuses on the designing *process*, rather than on the product or hardware that comes out of it. He defines designing to be "all that goes on within the subculture of the [engineering] firm" (1988, p. 160), and not only the actions of some individual designer. He denies explicitly that the item "represented in formal drawings, detailed lists of performance specifications, lists of materials, subcontractor orders, prototypical hardware" exhaust a design (1988, p. 161). Moreover, Bucciarelli explicitly blurs the internal versus external to technology distinction when he writes: "[a]ll members of the firm and those they call upon outside the firm are potential contributors [to the design process]: individuals in marketing and production, purchasing and finance, as well as in engineering along with subcontractors, suppliers, and the customer", (1988, p. 161). All these citations indicate that Bucciarelli uses prospective methodologies when he carries out a study in engineering practices.

As Bucciarelli acknowledges the prospective character of his work, he is well aware of alternative ways to study engineering design. Adopting the *frame* of an ethnographer, he recognizes *different ways of framing* the object of study. For instance, he states: "I want to frame my study so that technology is more integral, more ideational, more fully social than is allowed by a framing that sees it only as a material element of culture"(1994, p. 50). Moreover, he uses the word "frames" to refer to the object worlds being "worlds of technical specializations"; he also describes the shift of attention from the object worlds towards design-as-a-social-process as a change of frame. Eschewing therefore the approach of the metaphysician who would search for the "essence" of design, Bucciarelli is well aware of, and remains carefully faithful to, his much more subtle framing methodology. He remains aware of the fact that the outcomes of his observations are relative to his forward-looking perspective and does not naively believe that he is revealing "the truth" about engineering design.

III. MISMATCH BETWEEN ENGINEERING EDUCATION AND PRACTICE

Let us now turn to the specifics of Bucciarelli and Kroes' contribution. The authors start with a description of the object-world notion, which Bucciarelli used in his 1988 and 1994 publications. Subsequently, they show how current engineering curricula are almost exclusively geared to, and focused on, the isolated object worlds of mechanical, electrical, civil, chemical, and so on, engineers. Considering the outcomes of Bucciarelli's ethnographical studies, he and Kroes conclude that engineering curricula are deficient for two reasons: They do not offer a realistic picture of engineering practice, and they almost exclusively focus on object-world problems without considering social values that are important in practice. They maintain that to repair engineering curricula, they should at least take into account the social-technical nature of engineering practices, products, and processes. The authors finally suggest that if the engineering educational systems are to meet the demands of contemporary engineering practice, it helps to consider them to be socio-technical systems (st-systems). In this section, I describe the mismatch that Bucciarelli and Kroes observe between the engineering curricula and design practices. In the final section, I discuss possible ways to cope with it.

What is exactly the discrepancy between engineering education and practice as conceived by Bucciarelli and Kroes? As they perceive it, engineering *students* are educated as inhabitants of the object worlds of the mechanical, thermo-dynamical, civil, electrical, chemical, and so forth, engineering textbooks. The authors characterize these worlds as "real", hard, technical stuff, being materialistic and strongly related to empirical science and mathematics. These worlds are quantitative, closed, and internally focused, while a design is taken to be a product and not the process that brings about this product. In those worlds, the students need only solve well-posed engineering problems with only one legitimate solution. These problems are given beforehand and have a definite solution space, characterized by instrumental rationality: The student should only consider the question how to achieve a given end in the most efficient way, without deliberating the reasonableness of this end. In these educational settings, knowledge is mainly knowing-*that,* and it can be stored and transferred "as stuff".

As professional *engineers,* however, those students will *end up* in the real world as practitioners participating in heterogeneous engineering teams solving real-world, ill-structured, and intractable problems. In this real and open world, the internal-external distinction is almost impossible to draw. It is strongly social in character, qualitative, and concerns mainly soft stuff; moreover, design is considered to be a process rather than a product. In this context, the design engineer herself has to frame and formulate "the" design problem and should identify the problem owner. In real life, engineering problems and their solutions are socially constructed, and a design is the outcome of social and political processes. According to Bucciarelli and Kroes, students should therefore learn to identify, isolate and formulate the problem at stake, and should not only learn to search problem spaces, but should also get acquainted with creating them. And parallel to the latter, the knowledge-as-stuff metaphor prevalent in the "object worlds" should be replaced by knowledge as skill, as knowing-*how.*

Bucciarelli and Kroes go on to maintain that the object worlds as sustained in the engineering curricula are considered to be descriptive, value-free, and, as much as possible, separated from society. Applying a few fundamental and well-understood principles, its inhabitants are trained to use mainly simple and deductive methods to solve the neat, principled problems. Bucciarelli and Kroes ascribe a reductionist and deterministic ideology to the object worlds of abstraction and simplification in which logical analysis prevails. Subsequently, they again emphasize the enormous difference from the real world of engineering practice. Here, the inhabitants have to accept the complexity and the context specificity of design problems, and should try several, alternative, perhaps conflicting, approaches and perspectives to identify and describe the problem. In addition, the method to solve the problem eventually defined typically depends on conjectural and inferential thinking, as well as on the traditions of the firm that employs the engineer. Moreover, the outcomes of the design process are profoundly normative and society related, and the problem solvers should be aware of the social and political features of their tasks.

In addition to the dramatic substantial differences between the object worlds and the worlds of engineering practice, students and engineers also act in very different settings. Problem solving in the object worlds is mainly a solitary activity predominated by instrumental rationality. Solving a design problem, the inhabitants of the different object worlds seldom collaborate with subjects from other object worlds. In engineering practice, however, designing is a group activity and teamwork. It is a social process and requires cooperation, negotiation, and efficient mutual exchange of views. To accomplish these facets, design engineers need to be social, open to alternative solutions, and constructively critical. They should be able to communicate their ideas and explanations clearly and convincingly. Many of these social designing skills are rarely, if ever, a subject of education in the traditional engineering curricula.

Considering the differences between the object-world education and engineering design practice, we may agree with Bucciarelli and Kroes that the two do not match.[13] Below I will refer to the problem explained as the "mismatch" or "discrepancy", rather than the more normative qualification "deficiency".

IV.   THE DIAGNOSIS AND ITS ADVANTAGES

So far, I have argued that the practice turn is a paradigm shift in the studies of science, and we have seen that engineering education does not match engineering practices. In this section, I first contend that the mismatch probably originates in the fact that authors of traditional engineering textbooks and curricula frame science in accordance to the standard logical empiricist model of science, and equate science typically with its products. After this "genealogical" point, I come to the observation that the turn to practice helped us to become aware of this mismatch, and emphasizes the inadequacy or at least the one-sidedness of the frame on science embedded in the engineering textbooks and curricula. Next, the methodological issue is discussed of how we should cope with the discrepancy. Finally, it appears that Bucciarelli and Kroes' proposal to put st-systems at the center of engineering curricula fits my change-of-frame diagnosis rather well.

An important observation regarding engineering curricula after WWII is that most of them are grafted on the "standard" logical empiricist model of science. The logical empiricists' paradigm is backward-looking in the strong sense. It frames science as a collection of scientific products, such as theories, models, laws, and explanations; logical empiricists were mainly concerned with the rational reconstructions of these products. Many authors of engineering textbooks and designers of engineering curricula took over this, at the time, predominant frame of the (engineering) sciences. Moreover, an ancestor of the logical empiricist paradigm, positivism, matched well the post-war needs of rebuilding society and the belief that most societal problems could be solved through the application of science.[14] Much of the object-world construction that Bucciarelli and Kroes consider therefore fits perfectly the logical empiricist paradigm of science. Object-world features such as hard, abstract, value-free, quantitative, deductive, and instrumental rationality are all closely related to the backward-looking view on science in the strong sense. The predominance of the object worlds in many engineering curricula, therefore, seems to a considerable extent inspired by, and based on, the logical empiricist frame on science.

A somewhat more trivial but nevertheless equally important remark concerns the observation about how the mismatch of Section III came to be recognized and acknowledged. Accepting the practice-turn paradigm, and taking on the frame of an anthropologist, Bucciarelli studied engineers in their daily practices. His studies revealed engineering practices being heavily constrained by social values and group phenomena. Regarding the social context of the design practices, he observed only superficial similarities between the daily activities of engineers and engineering curricula. Retrospectively, the divergence was not surprising at all because the practice of teaching student engineers in university was typically based upon the products of science and engineering and on the retrospective frame on science. We may therefore safely conclude that Bucciarelli's observation of the mismatch has been brought about mainly as a result of the practice turn described in Section II.

Even if we accept the genealogical and the awareness issues unquestioningly, they are of little help if we want to find out how to cope with the observed mismatch of Section III appropriately. To that end, I claim, it is important for at least two separate reasons to put the two issues at the backdrop of the *paradigm shift* from the logical empiricists' toward the practice-oriented studies of science. In the first place, the shift clearly harbors the origins of both issues just mentioned. It implicates the change in the frame of science—from its products toward its practices—and it shows how Bucciarelli came to the observation of the mismatch between education and practice. But in the second place, and even more importantly, putting this paradigm shift at the backdrop of the mismatch deepens our understanding of the problem at hand and of Bucciarelli and Kroes' proposals how to cope with it.

To see how the paradigm-shift interpretation deepens our understanding of the mismatch, I start with observing that, at first sight, Bucciarelli and Kroes' discontent with the present engineering curricula is somewhat surprising. Many engineering educational practices have changed since the publication of Bucciarelli's ethnographical work. Numerous project-based learning initiatives have been implemented at MIT, the home base of Bucciarelli, and at the Delft University of Technology, the employer of Kroes. Despite these initiatives, both authors still consider the changes to be insufficient. How are we to understand this discontent? Is it the mere number of learning projects? I tend to doubt it. After all, Bucciarelli and Kroes suggest that a solution to the problem of mismatch has to be found in the "[a]ttitudes, perspectives, and values—other than instrumental—[that] pervade our engagement with students", and in this engagement, "the attitudes and values appropriate to object-world work" should not prevail, as is presently the case (Chapter 6, this volume, pp. 193–4).

If we take Bucciarelli and Kroes' formulations one step further and paraphrase them in terms of a change of paradigm—as disciplinary matrix—their uneasiness becomes more understandable. Changing the discourse from "adding skills to the learning aims" to that of "adding social and ethical values" or even to "adding attitudes and habits" is too shallow. It only fights the symptoms, and does not touch upon the foundation of the question, viz., the clash between the "standard" and the "practice-turn" paradigms of engineering sciences and the different frames on engineering (science) that are implicated. Just *adding* skills, values, or attitudes is insufficient since the fundamental attitude towards all elements within the current curricula should be changed profoundly. Engineering departments should train their students almost from the start with the social ramifications of their work.[15] Students should be provided with an alternative *perspective of their professional engineering activity*[16]—engineering should be conceived of as solving societal problems for which one needs hardcore scientific knowledge, rather than as hardcore knowledge that can be applied to achieve a solution. Instead of being only a lack of interest in societal norms, the mismatch concerns, rather, a substantially clash of paradigms. The paradigm-shift interpretation of the mismatch therefore deepens our understanding of Bucciarelli and Kroes' discontent with current curricula.

If we turn to the solution of the mismatch, the clash-of-paradigms view on the deficiency also has the advantage of helping to gear actual curricula toward the practice of engineering design *in a smooth way*. It changes the competing "either-or" reorganization of curriculum into an "and-and" way to deal with engineering practices. Practices therefore need not be taught to the detriment of the object-world theory. Much of the old curricula content remains valuable, of course, or even indispensable. It should, however, be taught in completely different education process. Teaching theoretical knowledge, for instance, should start with, and be embedded in, explanations about why and how engineers use that form of knowledge in practice. Moreover, students should become acquainted with forward- and backward-looking perspectives on the development of knowledge and design, including the fact that design and research are carried out in teams.

Let us turn to the proposals of Bucciarelli and Kroes on how to handle the problem of the mismatch between engineering curricula and practices. An important ingredient of these proposals is to put *socio-technical (st-)systems* much more prominently on the student's menu than before. St-systems are systems that comprise human actors and their social relations, which are inextricably intertwined with artifacts and their technical infrastructure. In these overarching systems, the one cannot function without the other because of their relation of mutual dependence. Energy grids of a country, a city's traffic infrastructure, or international air traffic control systems provide telling examples. Emphasis on st-systems in engineering curricula helps to close the gap between engineering education and practices in two important ways. In the first place, they are closely related to society and its norms and values. Nowhere along the line of their development can their social or political dimensions be bracketed away because this development proceeds in immensely complicated networks. These networks consist of diverging stakeholders, such as different users, regulators, entrepreneurs, servicing people, politicians, action groups, and so on, who all have a relation of mutual interdependence with the design engineers. Moreover, the outcomes of social interventions are much less predictable than interventions in the physical world. Consequently, the evaluation and assessment of the social functioning of st-systems includes complicated dynamic social processes and is established in a design environment comprising many feedback loops between designers, users, and regulators, who have diverging "perspectives, priorities, methods, and competencies" (Chapter 6, this volume, p. 196). Consequently, st-systems are inextricably mixed up with societal, juridical, ethical, and technological norms that cannot be left out.

In the second place, st-system development helps to close the gap between education and practice because their design and development epitomizes the practical complications engineering projects typically encounter in daily practice. Theoretically—and academically—"normal" engineering design starts from scratch, with the definition of the design brief, on the basis of which final design will be assessed. The development of st-systems—and design projects in practice—deviates from this scheme in two important ways. First, this development never starts from scratch; new improvements are always built on technical and juridical remnants of previous systems. And second, it typically starts with an incomplete and open design brief, which develops "on the fly". Or, in the words of Bucciarelli and Kroes, "for socio-technical systems things become much more complicated since there is no unambiguous list of specs" (Chapter 6, this volume, p. 197). It is this observation that renders instrumental rationality unsuitable as a model of choice for the development and introduction of st-systems or social systems in general. Both aspects, the inextricable society-relatedness and the co-developing assessment, render st-systems efficient means for letting current engineering curricula engage with the practice turn. Moreover, as these two aspects of st-systems concern changing the focus of attention and of method, they illustrate the productive fit between Bucciarelli and Kroes' solution to the problem and its paradigm-shift interpretation.

Bucciarelli and Kroes are well aware of the two specific aspects of st-systems and use them to strengthen their own position. At the end of the paper, they apply their arguments to their own case and propose to consider *engineering curricula themselves* to be st-systems. By making this cunning argumentative move, they guard their ideas in advance against instrumental rationalist objections claiming that their proposals will not achieve the ends they pursue. Perhaps, the argument would go, these proposals will turn out to be counterproductive and produce engineers who are less technically capable than before. Consequently, their proposal to restructure engineering curricula is putting the cart before the horse. By considering educational curricula to be st-systems, however, Bucciarelli and Kroes maintain that these objections miss an important point, viz., that at the start of the process of curricula reorganization not all objectives are readily given beforehand, but are likely to be filled in during the process of implementation.

Finally, let us consider the relevance of Bucciarelli and Kroes' contribution for the research agenda of science studies. To make engineering curricula pay due attention to the retro- and prospective frames of science raises the question about their mutual relation. Unfortunately, a full-blown picture of science and technology taking both frames seriously into account does not exist yet. But we cannot do without. As long as scientists in their day-to-day work use, formulate, and produce what they consider to be products of science and engineering, such as theories, models, explanations, experimental results, (un)successful designs, and so forth we cannot escape the consequence that these products play a crucial role in engineering practices, and should be dealt with accordingly. By emphasizing the importance of st-systems for engineering curricula, Bucciarelli and Kroes emphatically put these systems back on the science-and-engineering research agenda. At the backdrop of the observed paradigm change, the development of such systems provides a challenging example of the interplay between the retro- and prospective frames on engineering knowledge, with st-systems assuming the typical role of a lynchpin, which connects the two frames.

We may conclude that the turn to engineering practices has rightfully drawn our attention to an important discrepancy between engineering education and practices, whereas the paradigm-clash interpretation of the practice turn has deepened our understanding of this discrepancy and the possible ways out. Moreover, this interpretation renders st-systems crucial subjects of research in an endeavor to come to an overarching account of science and engineering that takes into account seriously the retro- and prospective frames on science and engineering.

NOTES

1. At that time, philosophy of technology was well established (see Carl Mitcham, 1994). Although Mitcham maintains "it is good for philosophy, if it wishes to reflect on technology, to engage engineering practice" (p. 271), the term *practice* does not appear in the index.
2. For instance, David Bloor (2011), in his impressive recent study on airfoil research in the start of the 20th century, extensively acknowledges Vincenti's "firsthand experience of aerodynamic research". He writes that Vincenti's comments "have been invaluable to me in learning to find my way in this new field" (2011, p. xiii).
3. See, for instance, Frankenberger and Badke-Schaub (1996), Dorst (1997), and Dorst and van Overveld (2009).
4. In (2004), even Latour himself muses: "What were we really after when we were so intent on showing the social construction of scientific facts?" (p. 227).

5. For a nice description of the linear model of technology development, see Godin (2006).
6. In addition to these two cases, Bucciarelli (1994) reports on his findings in a large firm producing copying machines.
7. Other interesting SSS works focusing on the turn to engineering practices are Downey (1988, 1998), Vinck (1992, 2003), and, more recently, Trevelyan (2007, 2010). For the general role of social science in engineering, the reader may consult Sørensen (2009).
8. He adds: "Systems of generalizations (or universal statements) that back explanations, predictions, and research strategies are theories. But so, too, for example, are typologies of social phenomena; models of social affairs; accounts of what social things (e.g., practices, institutions) are; conceptual frameworks developed expressly for depicting sociality; and descriptions of social life—so long as they are couched in general, abstract terms" (Schatzki, 2001, p. 12).
9. See also the Introduction to this volume.
10. Kuhn's characterization of a particular paradigm would be an example of studying science prospectively, as would Pickering's reconstruction of particle physics in the years 1960–1980.
11. Backward-looking constitutional history has been called "Whig-history" (Butterfield, 1931); whether retrospective history of science should be renounced as "Whiggish" is still subject of discussion (Harrison, 1987; Jardine, 2003).
12. For an elaborated analysis of the difficulties regarding the internal-external distinction spelled out with regard to empirical case study in sewage water treatment technology, see Zwart and Kroes (to appear).
13. Very similar sentiments have recently been expressed in Trevelyan (2009) and Williams and Figueiredo (2011).
14. Bush (1945/1995).
15. For the ways that societal values intervene in engineering design and its modeling, see Zwart, Jacobs, and Van de Poel (2013).
16. See, e.g., the paragraph in Section II composed of the sentence: "Some, constrained to think in this way, can only express their desire for improving engineering education in terms like '. . . our "solution spaces" must be extended'"; and the two paragraphs that follow (Chapter 6, this volume, p. 193).

7 The Impact of the Philosophy of Mathematical Practice on the Philosophy of Mathematics[1]

Jean Paul Van Bendegem

I. INTRODUCTION

There are always several ways to read a paper on any topic. This paper, for instance, can be read as a two-part text, the first major part dealing with the history of the study of mathematical practice, and the second, somewhat shorter, part with a proposal as to how the "traditional" philosophy of mathematics and the philosophy of mathematical practice can fruitfully interact. Or it can be read in a different way. Anybody who is interested in and who believes that mathematics, through its practices, is also subject to external influences, say from the other sciences or from society/culture at large, will probably know the famous quote of Jean Dieudonné:

> To the person who will explain to me why the social setting of the small German courts of the 18th century wherein Gauss lived forced him inevitably to occupy himself with the construction of a 17-sided regular polygon, well, to him I will give a chocolate medal. (Dieudonné, 1982, p. 23)[2]

The conclusion of this paper will show what the appropriate answer to Dieudonné's question could be, an answer that is inspired by the ideas presented in this paper.

II. A SHORT AND INCOMPLETE HISTORICAL OUTLINE

II.1. Lakatos as the Starting Point

Taking into account that almost any historical outline is partially a reconstruction, one wants nevertheless to find a historical starting or turning point of a sufficiently symbolic nature. Or, if you like, if the philosophy of mathematics underwent a similar change as the one that "shook" the philosophy of science (at least, according to some), then it seems fair to take Imre Lakatos' 1976 seminal *Proofs and Refutations* (1976), a work that was in part inspired by Georg Pólya's *How to Solve It* (1945), which

discusses heuristics and problem-solving techniques in an educational setting.[3] Its focus on mathematical practice was clear by the mere fact that
it boldly presented nothing less than a "logic" of mathematical discovery.
What did the book offer? Nothing less than the *history* of a mathematical
statement and its proofs. That fact alone should amaze everybody, for is not
mathematics supposed to be timeless, eternal even, unchangeable for sure,
and is mathematical proof not supposed to be absolutely certain, undoubtable, secure? This implies that, if a proof is found for a statement A, then
A has been proven and that is that. To which is usually added: and a wrong
proof is not a proof; it is wrong. Lakatos disagrees and takes as an example
the statement that for polyhedra (in three-dimensional Euclidean space),
V(ertices) – E(dges) + F(aces) = 2. Take, for example, a cube. There are
8 vertices, 12 edges, and 6 faces, and indeed 8 – 12 + 6 = 2. Euler himself
had found an ingenious proof, or so he believed. As soon as the proof was
around, counterexamples appeared, making it necessary to (sometimes seriously) modify the proof and, in that sense, the proof has a history. Even
more important is that Lakatos found patterns in this game of proof and
refutation, explaining thereby the title of the book. In this book, writing in
a dialogue form, stressing the dynamics of the search for a "final" proof, a
totally different picture was drawn of what mathematics is and what it is
about.

An important warning should be made at this point. Selecting a symbolic
starting point is not without its dangers. The most prominent one is the
exclusion of important contributions not related to that particular starting
point. One such exclusion—and this would require a thorough study of
its own—is the French historico-philosophical approach to mathematics,
including such thinkers as Léon Brunschvicg, (see Brunschvicg, 1912) and
his pupils Albert Lautman (who incidentally brings us quite close to the
Bourbaki group and hence to the already-mentioned Jean Dieudonné; see
Lautman, 1977) and Jean Cavaillès (see Cavaillès, 1962). In its turn, this
would invite us to consider the work of François Le Lionnais and Raymond
Queneau and, inevitably, the OuLiPo (*Ouvroir de Littérature Potentielle*;
see Van Kerkhove & Van Bendegem, to appear) for a more detailed presentation of this latter development. Exclusion in this case is to be understood
as "to be included at a later stage".

II.2.   Kitcher as the Next Step

There was not an immediate follow-up, as far as I can see, on Lakatos'
ideas. We had to wait until Philip Kitcher proposed in his book *The Nature
of Mathematical Knowledge* (1983) a more or less formal model of how
mathematics as an activity can be described, which was clearly inspired by
the developments in the philosophy of science, where attempts to develop
formal models have always been present. It is sufficient to think of the
logico-mathematical work of Joseph Sneed, formalizing a Kuhnian outlook,

as a prime example (see Sneed, 1971). I will describe the model itself in some more detail in the second part of this paper, but let me stress here that Kitcher's approach is definitely not a continuation of what Lakatos had initiated. For one thing, Kitcher discusses philosophical problems that clearly belong to the traditional philosophy of mathematics (and I will return to that topic as well in the second part), such as questions of realism, of the existence of mathematical objects, of our capability (or lack of it) to get to know these objects, and so forth. In short, in Kitcher's approach we get a mixture of the more radical Lakatosian view and more traditional philosophy of mathematics, using a Kuhnian framework to accommodate both.

The Kitcherian outlook proved to be more successful, as several authors embraced this trend, as is shown in subsequent volumes that made connections between philosophy and history of mathematics, as did Thomas Tymoczko's *New Directions in the Philosophy of Mathematics* (1985) and, later and more explicitly, Aspray and Kitcher's *History and Philosophy of Modern Mathematics* (1988). It must be explicitly stated here that historians of mathematics have always paid attention, in varying degrees to be sure, to practices and have produced significant studies with often implicit, but also sometimes explicit, claims about their philosophical relevance (see, e.g., the recent volumes of Ferreiros & Gray [2006] and Gray [2008], in which history and philosophy clearly interact).

II.3. A Tension Is Introduced to Stay

This initial tension between Lakatos and Kitcher has never left (up to now) the study of mathematics through its practices. The nature of that tension is anything but new. We have known it in the philosophy of science in the form of the context of discovery versus context of justification divide. A justification is preferably seen as something independent from the discovery process. In other words, if I need to justify something, I need not wonder about how it is has been found. In the case of mathematics, if such an independence holds, then I only need to look at the finished proof, the final version, and wonder whether or not I can justify that this text that claims to be a proof, indeed is a proof. The processes that led to the proof are of no importance. Lakatos' claim is exactly the opposite: To make sense of how mathematics develops, an understanding of these discovery processes is essential. Kitcher is not prepared to defend the *total* independence view but is willing to allow such elements from the practices that are necessary to understand the final results—in most cases, the proofs. All this means that, right from the start, two approaches were being initiated and developed. Has the situation changed much since then? To be honest, not that much.

In the introduction to *The Philosophy of Mathematical Practice* (2008, p. 3), Paolo Mancosu identifies two main traditions within this new research "paradigm", the study of mathematical practices. The first is what he calls the "maverick" tradition, which remains close(r) to the Lakatosian approach,[4] while the second one settles itself within the modern analytical tradition, thus remaining closer to Kitcher than to Lakatos, and focuses, among other things, on the naturalizing program that started with Willard V.O. Quine, and where, for example, Penelope Maddy (1997, 2007) has played and still plays an important role. That being said, there is not a particularly impressive unity in this analytical approach, as the following quote, taken from the entry on naturalizing epistemology in the *Stanford Encyclopedia of Philosophy*, makes immediately clear:

> Naturalized epistemology is best seen as a cluster of views according to which epistemology is closely connected to natural science. Some advocates of naturalized epistemology emphasize methodological issues, arguing that epistemologists must make use of results from the sciences that study human reasoning in pursuing epistemological questions. The most extreme view along these lines recommends replacing traditional epistemology with the psychological study of how we reason. A more modest view recommends that philosophers make use of results from sciences studying cognition to resolve epistemological issues. A rather different form of naturalized epistemology is about the content of paradigmatically epistemological statements. Advocates of this kind of naturalized epistemology propose accounts of these statements entirely in terms of scientifically respectable objects and properties. (Feldman, 2012, introduction)

Similar oppositions to the ones sketched in this quote exist in the analytical tradition in the philosophy of mathematical practice. One example: Some, such as Maddy, claim that philosophers do not have to interfere with what mathematicians are doing and, in fact, have to listen to what they have to say about, for example, the concepts they use—there is a certain Wittgensteinian flavor present here—but some do believe that mathematicians have to listen to philosophers when, for example, the latter show to the former that some aspect of their practice does not deliver what they believe it should. Roughly speaking, this too is an old tension, namely the tension between a more descriptive outlook versus a more normative outlook. Let me illustrate this tension with an example: proofs that rely partially on computers. We now have proofs, such as the proof of the four-color theorem[5] or Kepler's conjecture,[6] that consist of two parts. A first part reduces the statement to be proven to a finite set of instances that need to be checked one by one, and then in the second part, a computer program is presented that does the actual checking of these cases, involving extremely clever computational methods to speed up the process as—and that is the problem—the set of instances is not a small set; indeed, it is huge, and hence, the computer power needed. The philosophical question to be asked is whether or not this counts as a genuine proof. At least two attitudes are possible: either (a) ask the mathematicians and if they tell you that they consider it to be a proof, that is what should be reported, together with the reasons and arguments why they believe so, or (b) make a study of the nature of such proofs, how mathematicians handle them and establish whether they can be considered to be genuine proofs and, if not, report back to the mathematicians that they have a problem. Many more illustrations can be provided, ranging from specific questions, such as what proofs mathematicians consider to be beautiful, or what their explanatory power can be, to the question of whether revolutions (in the Kuhnian sense) occur in mathematics (see Gillies, 1992).

So, at the end of this section, we have already identified three major approaches in the philosophy of mathematical practice:

(a) the Lakatosian approach, also called the "maverick" tradition;
(b) the descriptive analytical naturalizing approach; and
(c) the normative analytical naturalizing approach.

This, however, is absolutely not the end of the story.

II.4. Enter the Sociologists, Educationalists, and Ethnomathematicians

One might be amazed by the fact that I did not explicitly include historians in this section's title. The obvious answer, I believe, is that they do not need to enter, for they are already inside (and for quite some time for that matter). In other words, there are well-established connections between mathematics, the history of mathematics, and the philosophy of mathematics. For sure, Lakatos could not have written his seminal work without the sources made available by historians. There is, however, a story to be told: namely, how these relations changed over time, for there is no reason to believe that these are in any sense stable, especially between history and philosophy of mathematics. But that requires not so much a separate paper as a separate research program. In summary, the focus in this section will be on the "newcomers" in the field.

Independently of these developments in the philosophy of mathematics, but also partially inspired by Lakatos, as well as Thomas Kuhn's *The Structure of Scientific Revolutions* (1970), some researchers developed a sociology of mathematics, where one of the major focuses was mathematical practice as a group or community phenomenon. Two works should be mentioned in this line: David Bloor's *Knowledge and Social Imagery* (1976/1991) and Sal Restivo's *The Social Relations of Physics, Mysticism, and Mathematics* (1985; but see also Restivo, 1992). In contrast with history and philosophy of mathematics, the sociological approach did not merge easily with the mentioned traditions, although some "brave" attempts in this direction should be noticed (Löwe & Müller, 2010; Restivo, Van Bendegem, & Fischer, 1993; Rosental, 2008). Plenty of reasons can be listed, but surely one of the corner elements is the internal-external debate, yet another tension to be introduced in the framework: Does mathematics develop according to "laws" or patterns that are internal to mathematics (and, as a consequence, independent, or at most marginally influenced by external "laws" or patterns) or, on the contrary, do external elements contribute and, if so, in what way(s)?

Let me briefly sketch a classic example. Consider the story we get presented in almost every mainstream handbook on the history of mathematics about the flooding of the Nile, the need to redivide the fertile land, the need for a right angle to be able to measure, and the rope with the twelve knots[7] as a device to solve this problem. It is not difficult to describe this episode solely in terms of an externally given problem (divide the land), which is then transformed into a mathematical problem (how to measure a right angle) and this problem receives a mathematical solution (take a rope with twelve knots). But, it is sufficient to ask the simple question: why a right angle? What was the need to divide the land in pieces that, apparently, had a rectangular shape? The answer is that the surface of a rectangle is easy to calculate, and this is interesting from the perspective of taxes. It is therefore not difficult either to see the whole process as primarily an economic-political problem, wherein a mathematical problem is posed, and that needs to be solved to satisfy an economic-political need.

One outcome of these discussions has been to draw the attention to other, so far neglected, areas where mathematics is involved—prominently, mathematics education and ethnomathematics. To avoid all misunderstandings, I do not mean, of course, that these domains did not exist before the practice turn in the philosophy of mathematics: Witness, on the one hand, the pioneering work[8] of Pólya, which has already been mentioned, and, on the other hand, the equally pioneering work of Ubiratan D'Ambrosio.[9] Quite the opposite, but they were not perceived as being *relevant* to the philosophy of mathematics. In the standard view, mathematics educators are interested in how pupils can learn to grasp the concept of a mathematical proof, understand the certainty involved in a geometric construction, or develop the ability to translate a verbal problem into a mathematical problem. That, of course, had little or nothing to do with questions such as what are reliable foundations for the whole of mathematics and is set theory a better or worse candidate than category theory (to name the two, rather unequal, rivals at the present moment). In that very same standard view, ethnomathematics is a branch of anthropology and, as such, not relevant to a study of a high-level abstract mathematical problem in western mathematics.

From the practice point of view, however, the links are evident. Firstly, practices are "carried" by people and people have to be educated; that forges the link with education. Secondly, practices are socially embedded and thus culturally situated; that forges the link with ethnomathematics. Or, if you like, education concerns the *diachronic* dimension of how mathematical knowledge is situated in time, whereas ethnomathematics concerns the *synchronic* dimension of how mathematical knowledge is situated in space.

Finally, it must be added that mathematics education and ethnomathematics have an extensive, non-empty intersection.

And still the picture is not complete.

II.5.   Brain and Cognition Complete the Picture

Other sciences play a part as well, and I just mention the two most important ones: namely, (evolutionary) biology and (cognitive) psychology. In the first field, the challenge is to determine how much mathematical knowledge is biologically, perhaps genetically, encoded in the human body and how it affects our mathematical abilities. The work of, for example, Stanislas Dehaene (see Dehaene & Brannon, 2011, for an excellent overview) is the best illustration of this type of work. It involves, as well, comparisons between humans and other animals, leading, by the way, to quite intriguing results, such as the fact that basic mathematical capabilities are definitely not exclusively human. In the second field, one of the main topics is the study of human thinking and, quite trivially, since mathematical thinking is a perhaps highly particular and extraordinary form of thinking, it should and does attract their attention. What immediately comes to mind is the well-known study of George Lakoff and Rafael Núñez (2000). And, to further increase the complexity, these studies can range from brain activity studies of mathematical tasks human subjects have to perform to interviews with mathematicians on the way they (think they) solve mathematical problems.

We did already identify three major approaches in the philosophy of mathematical practice whereto five approaches need to be added:

(a) the Lakatosian approach, also called the "maverick" tradition;
(b) the descriptive analytical naturalizing approach;
(c) the normative analytical naturalizing approach;
(d) the sociology of mathematics approach;
(e) the mathematics educationalist approach;
(f) the ethnomathematical approach;
(g) the evolutionary biology of mathematics; and
(h) the cognitive psychology of mathematics.

A complex picture indeed!

II.6.   Working in Different "Registers"

One of the major consequences of the complexity of this situation as sketched above, is that, as a philosopher-mathematician myself, I find that I have to work in different "registers" when reading different texts in the field (or that I, at least, consider to be relevant). There is a world of difference reading papers (a) on representations of numbers in the brain, including all the nicely colored pictures of brain scans, or (b) on a cognitive-analytical approach to the use of diagrams in mathematical reasoning, where the thought processes are described in detail, or (c) on a sociological analysis of the mathematical concepts of space and time, where anthropological methodologies are of crucial importance to be aware of possible cultural biases, or, finally, (d) a formal approach to what explanatory strength of proofs could be.

Let there be no mistake: One might think that perhaps we are dealing here with a division of labor of a vast field to explore, but such a division suggests that all the parts can be put together again to form a minimally coherent whole. And that is (at present) definitely not the case. There are, as I have indicated, fundamental oppositions and tensions at play. Therefore, it follows, I believe, that the two major tasks for the future are, first, to develop a greater coherence in the field and, two, to keep the conversation going with the other philosophers of mathematics (that so far have not yet been mentioned). These are difficult tasks, no doubt, but without any such attempts, in the former case, incoherence can prove to be fatal for the survival of this new and emerging domain, and, in the latter case, one should not forget that the mainstream in the philosophy of mathematical practice is itself not mainstream at all in the larger field of the philosophy of mathematics where, for example, foundational studies still form a major part. Setting up the dialogue can, quite frankly, again be a matter of survival.[10]

The next section presents some ideas and suggestions as to how bridges can be constructed between mathematical practice and the "traditional" philosophy of mathematics.

III. OF PRACTICES AND FOUNDATIONS

Although one might think that, due to its century-long tradition, the philosophy of mathematics is a well-defined and well-outlined domain, nothing is less true. If I were forced to write a similar short and incomplete historical outline, it would have to encompass more than twenty centuries and the most obvious, plain trivial observation to make is that mathematics itself has dramatically changed over these centuries. One cannot seriously claim that Plato's (original) insights would have been so formidable as to be applicable to 17th or 18th century, let alone present-day, mathematics.[11] In a gross historical exaggeration, one could claim that the end of the 19th and beginning of the 20th century had a profound influence on the philosophy of mathematics by the attempts to reduce the agenda to foundational studies.

At first sight, nothing seems further away from the study of mathematical practice than these foundational studies. Just check an issue of *The Journal of Symbolic Logic* and it is obvious that a different language is spoken, a different outlook is shared by its practitioners, and it would seem normal that both have little to say to one another. But is this necessarily so? I think not, for the sketch in the first part of the paper shows that the analytical tradition, whether naturalizing or not, is present in both. More specifically, a major part of that tradition relies on formal tools and modeling techniques to explore philosophical questions. These have the power to perform a bridge function between the study of mathematical practice and "traditional" philosophy of mathematics.

I see at least two roads to explore. Either one can try to build local bridges or to sketch outlines of global bridges.

Examples of "local bridges" are, for example, (the already briefly mentioned) diagrammatic reasoning and (formal models in) argumentation theory. Let me say a few words about these two topics. At the present moment, diagrammatic reasoning has been approached from different angles. On the one hand, there are studies closer to mathematical practice (such as, e.g., Carter, 2010), and, on the other hand, there are logical studies to understand the correctness of diagrammatic reasoning (such as, e.g., Shin, 1995). Recently, the work of John Mumma (to appear) brings the two together, so the first bridge has been built. That being said, a lot of "bottom-up" work will still be necessary, not only to better understand the process in practice, but also to refine the formal models. In argumentation theory there have always been, quite early, attempts to formalize arguments in some sense. One can think of the pioneering work by Jaakko Hintikka (1985), and Else Barth and Erik Krabbe (1982). In recent times, a new impetus has been given to the field from studies in Artificial Intelligence—namely, the study of so-called argumentation networks and problem-solving networks (see Dung, 1995) for a pioneering study. And in even more recent times, groundbreaking work has been done by Andrew Aberdein (see, e.g., Pease & Aberdein, 2011). What these studies show clearly is that the incorporation of insights from argumentation theory is a quite natural move, and the study of formal models that will allow us to represent and study the interactions between arguments and counterarguments will make the bridging function possible.

The more daring task, of course, is to construct "global bridges". Here, too, several options are available. One can think of models where mathematical "agents" exchange information with one another, where they argue together, where they practice a division of labor, where they convince one another of the relevance of a bit of mathematics, and so on. The outlines of such models exist at the present moment (see, e.g., Van Benthem, 2011, as one of the major contributors), although rarely specified for mathematics. But all indications point in the same direction: It can be done. And it should be done, for it will allow the practitioners of the study of mathematical practice to tap into these rich sources.

Another option is not to forget our heritage, so to speak, and to return to earlier attempts. I refer here to the early and already-mentioned work of Philip Kitcher. The formal model he proposed is of an entirely different nature, but it also shows plenty of promises. In "The Unreasonable Richness of Mathematics" (Van Kerkhove & Van Bendegem, 2004), we suggested a generalization of Kitcher's model along the following lines. In the original version, the model consisted of a quintuple of components $<L, M, Q, R, S>$, containing

- a *language L,*
- a set of accepted *statements S,*
- a set of accepted *reasonings R,*
- a set of important *questions Q,* and
- a set of philosophical or *metamathematical views M.*

Although this model could do a lot of work, an extension was needed and the generalization we arrived is that, instead of a quintuple, we now have a seventuple <*M, P, F, PM, C, AM, PS*>, containing:

- a mathematical *community* M of individual *mathematicians* m_1, $m_2, \ldots, m_i$,
- a *research program* P within the framework of which specific problems $p_1, p_2, \ldots, p_j, \ldots$ can be formulated,
- a *formal language F,* including axioms, definitions, and a body of *formal proofs* $f_1, f_2, \ldots, f_k, \ldots$, as typical answers to the above problems; note that here too metamathematical considerations can play their part,
- a set *PM* of *proof methods* $pm_1, pm_2, \ldots, pm_k, \ldots$
- a set C of *concepts* $c_1, c_2, \ldots, c_n, \ldots$
- a set *AM* of *argumentative methods* $am_1, am_2, \ldots, am_s, \ldots$
- a set *PS* of *proof strategies* $ps_1, ps_2, \ldots, ps_t, \ldots$

We will not go into details here, but instead just mention two things. The first one is the most obvious one: This Kitcher-like extension is certainly not at odds with the formal-logical approach. Quite the contrary, it is easy to see how both could (and should) interact. The second one is a heuristically important remark. The framework, briefly outlined here, allows for the following technique. Take the seventuple and keep six of the seven components (fairly) constant and then investigate what happens to the seventh, remaining, element. Thus, for example, we could keep everything "constant" except the set C of concepts and that would entail that we now describe the history of a particular concept, assuming a kind of *ceteris paribus* rule by fixing the other elements. Now a history of a mathematical concept sounds much closer to "traditional" philosophy of mathematics than to the study of mathematical practice, and that looks promising. Even more interesting is to keep five elements fixed and let the two remaining elements vary, leading to, in the best of cases, correlations between the elements in the seventuple.

If all of this looks rather sketchy, it is important to realize, as had been said before in this paper, that we are looking, in comparison with developments in the philosophy of science, at a very young discipline. Nevertheless, I do think it is important, right from the start, to look for collaborations and not exclusions. Exclusions are only to be accepted when everything else fails, and this is definitely not the case at the present moment.

IV. CONCLUSION

To conclude the paper, let me return briefly, as promised, to the Dieudonné quote. We are challenged to show how life at the small German courts of the 18th century forced Gauss to occupy himself with the construction of a 17-sided regular polygon. Of course, if any possible answer is to show a direct connection, the task is outright silly. But, if we are allowed to have a broader outlook, then we could think of the following.[12] What were the political-social circumstances that gave rise to a system such as the small German courts of the 18th century? This question is answerable and it will also shed light on the status of the sciences, including mathematics, during that process. We will understand why mathematicians were given the opportunity to continue their work and we will probably also understand why particular concepts were deemed more important than others—for example, because of theological connections (I am thinking here about infinity). This in its turn invites us to see how this relative isolation of the mathematicians at these courts led to an internal dynamic wherein certain topics were favored over others, where the research community of 18th-century mathematicians decided what problems were interesting and what problems not, to a certain extent co-determined by societal elements. Now we can focus on the internal, more local development and understand the importance of the study of polygons and, if we focus on the most detailed level, the importance of the 17-sided regular polygon. Then we see Gauss drawing such a polygon and understand the social act that he has now performed.

NOTES

1. The starting point of this paper is based on my contribution to the introduction of the special issue of *Philosophia Scientiae* (2012), edited by Giardino et al. In a sense this paper is an elaboration of the first-order sketch in that introduction.
2. Our translation of:

   Celui qui m'expliquera pourquoi le milieu social des petites cours allemandes du XVIIIᵉ siècle où vivait Gauss devait inévitablement le conduire à s'occuper de la construction du polygone régulier à 17 côtes, eh bien, je lui donnerai une médaille en chocolat.

3. This, in itself, is a quite interesting phenomenon. One of the main sources of inspiration for Lakatos came from an educational setting and not solely from a reflection on academic mathematics. I will come back to this connection as it presents a separate difficulty in the philosophy of mathematical practice. And then, I do not even mention the other source that was just as important, but just as often forgotten, due to the difficulty, I assume, of giving it its proper place, namely the influence of Georg Wilhelm Friedrich Hegel. See Larvor (1998) for a thorough discussion of this relation.
4. The author of this paper is supposed to belong to this tradition and I do most certainly not object!

5. This is the statement that says that, given a planar map, divided up in regions, this map can be colored using four colors only, satisfying the condition that neighboring (i.e., sharing a border) countries get different colors.
6. This statement is about sphere packing. In classical three-dimensional space, one can think of spheres that have to be arranged in such a way that the space between the spheres is minimized. The regular packing, familiar from supermarkets where spheres are oranges, turns out to be the most efficient.
7. 12 is equal to 3 + 4 + 5. If a triangle is made with sides 3, 4, and 5, then, because these numbers satisfy Pythagoras' theorem, $3^2 + 4^2 = 5^2$, one knows that the angle has to be a right one.
8. Today one of the core figures in the field is Alan Bishop (see Bishop, 1988). For a "local" contribution, see François and Van Bendegem (2007).
9. Although the first time the concept is mentioned in a paper is 1985 (see D'Ambrosio, 1985), the term was circulating informally much earlier; even in the 1985 paper, it is clear that the Lakatos approach did not play a role (although Thomas Kuhn is mentioned). See, also, D'Ambrosio (1990, 2007.)
10. A historical note: In 2002, a conference was organized in Brussels, Belgium, where the organizers, Jean Paul Van Bendegem and Bart Van Kerkhove, tried to realize their ambition in bringing together representatives of some of the disciplines mentioned. This conference has led to the book or proceedings titled *Perspectives on Mathematical Practices,* with the overambitious subtitle *Bringing Together Philosophy of Mathematics, Sociology of Mathematics, and Mathematics Education* (Van Kerkhove & Van Bendegem, 2007). There was a follow-up conference in 2007, whereof the proceedings were published in two volumes, Van Bendegem, De Vuyst, and Van Kerkhove (2010) and Van Kerkhove (2009), with our ambitions slightly moderated, but another important outcome from all these developments, overall, has been the confirmation of the rather heterogeneous character of this field known as the "philosophy of mathematical practice", as can be seen in the recently founded *Association for the Philosophy of Mathematical Practice (APMP;* see the website of the association at http://institucional.us.es/apmp/), wherein both of Mancosu's traditions are clearly present and at times happily interacting.
11. In a certain sense, one could invert the Dieudonné quote and challenge anyone to explain, relying only on the vocabulary used by Plato himself, why there seems to be a "natural" closure of number fields, from reals to complex numbers to quaternions to, finally, octonions. I will happily return the chocolate medal.
12. A strong inspiration for this answer is to be found in Restivo:

> The sociological way is first to look to both "external" and "internal" contexts, networks, and organizations. Dieudonné's error is to imagine that only "external" milieux hold social influences. Second, the sociological task is to unpack the social histories and social worlds embodied in objects such as theorems. Mathematical objects must be treated as things that are produced by, manufactured by, social beings through social means in social settings. There is no reason why an object such as a theorem should be treated any differently than a sculpture, a teapot, or a skyscraper. (2011, p. 47)

Conceptions of Mathematical Practices: Some Remarks

Commentary on "The Impact of the Philosophy of Mathematical Practice on the Philosophy of Mathematics", by Jean Paul Van Bendegem

Caroline Jullien and Léna Soler

I. INTRODUCTION

Jean Paul Van Bendegem proposes a mapping of the "major approaches" that constitute the relatively new "philosophy of mathematical practice", gives insights about the relation between this new trend and "traditional philosophy of mathematics" and makes suggestions to build "bridges" between the two. A mapping of the different constitutive trends of practice-based studies of mathematics is especially welcome, since even a quick glance at the works susceptible to be viewed as instances of practice-based studies of mathematics is enough to convince that the corresponding studies have diversified and heterogeneous objects, aims, and methods. In our commentary, we focus on this diversity. We start with an overview of Van Bendegem's cartography considered as a whole (Section II). Next we attempt to specify further what it means to study mathematical practices, following first the *philosophical* approaches listed by Van Bendegem (Section III), and then, more briefly, the *other* approaches of his repertory (Section IV). Along the path, we compare some aspects of the philosophy of *mathematical* practice and the practice-based philosophy of *natural* sciences.

II. DIFFERENT WAYS TO TAKE MATHEMATICS AS AN OBJECT OF STUDY

Mathematics can be taken as an object of study in multiple perspectives. Van Bendegem is primarily concerned with the *philosophy* of mathematics. The philosophy of mathematics, however, can itself be differently conceived, and can be practiced in more or less close interaction with other, strictly speaking non-philosophical approaches, such as, for example, the history of mathematics, or the sociology of mathematics, or any other taken-as-relevant field. Van Bendegem first divides the territory of the philosophy of mathematics into two areas, which correspond, respectively, to the "traditional" and the practice-based perspectives on mathematics. Then he maps the second area into eight sub-domains: "(a) the Lakatosian approach, also called the 'maverick' tradition"; "(b) the descriptive analytical naturalizing approach"; "(c) the normative analytical naturalizing approach"; "(d) the sociology of mathematics approach"; "(e) the mathematics educationalist approach"; "(f) the ethnomathematical approach"; "(g) the evolutionary biology of mathematics"; "(h) the cognitive psychology of mathematics" (p. 221).

About this cartography considered as a whole, it is worth stressing, first, that the five latter approaches are not, strictly speaking, approaches "in the philosophy of mathematical practice" (p. 221). They are, rather, five *non*-philosophical perspectives on mathematical practice that are *used by* philosophers of mathematical practice or, more prudently, on which *some* philosophers of mathematical practice *can find relevant* to rely. The introduction of these five non-philosophical approaches into the picture can perhaps be viewed as an indication that the philosophy of mathematical practice is more open to interdisciplinary work than traditional philosophy of mathematics. In any case, there are substantial reasons why a philosophy *of mathematical practice* may find such non-philosophical approaches relevant. These reasons are related to the fact that "mathematical practice" is a wide-scope term that can refer to any dimension of the activity of doing mathematics. People who practice activities akin to mathematics practice them *in a particular society,* and according to the society under scrutiny, mathematics may be differently conceived, organized, and transmitted; may occupy a different place; may be differently valued; may be differently related to other activities; and so on. Hence it becomes possibly relevant, for a philosophy *of mathematical practice,* to take into account the sociology of mathematics, mathematical education, and ethnomathematics. Moreover, mathematics are practiced *by human beings,* that is, by a certain kind of living species, characterized by some determined biological equipment (body and brain) and by some determined cognitive abilities. Hence it becomes possibly relevant, for a philosophy *of mathematical practice,* to take into account evolutionary biology and cognitive psychology.

Second, approaches (a) to (c) are, in contrast with (d) to (h), indeed philosophical approaches. Van Bendegem provides historical landmarks about their origin, putting forward two seminal works. Imre Lakatos, with his 1976 *Proofs and Refutations,* is presented as the first origin, and the father of a rather radical "maverick tradition" (an expression borrowed to Mancosu). This tradition had some, but few, descendents, among which is Van Bendegem himself.[1] Kitcher is viewed as a second, later, less radical starting point, initiator of an analytic naturalizing trend in the philosophy of mathematical practice. This analytic naturalizing trend has inspired more scholars than the maverick tradition. Van Bendegem distinguishes two branches within the analytic naturalizing tradition. The first one is descriptive: It intends to build philosophical characterizations based on records of what mathematicians actually do, without interfering with their work. The second one is normative: It intends to provide mathematicians with philosophical analyses of mathematical concepts and methods (for example, analyses of what a genuine proof should be) that might help mathematicians to become aware of problems in their ways of practicing mathematics (for instance, if they use computers to establish mathematical propositions and if it appears that this use departs from genuine demonstration as defined by philosophical analyses). We can note that within the practice turn, such a way of being normative, according to which philosophers are supposed to know better than scientists how to proceed, has often been considered as typical of *traditional* philosophy of science and has been strongly criticized.[2]

Third, one may be astonished not to find the history of mathematics within Van Bendegem's repertory of major approaches. Van Bendegem motivates this absence by a restriction to "newcomers" (p. 219) in the field. He stresses that "historians of mathematics have always paid attention, in varying degrees to be sure, to practices and have produced significant studies with often implicit, but also sometimes explicit, claims about their philosophical relevance" (p. 217). He adds that historians of mathematics "do not need to enter" into his picture of practice-based studies of mathematics, because "they are already inside (and for quite some time, for that matter)" (p. 219). Thus, clearly, the history of science, or at least *some* histories of science, should be added to the list. However, the sense in which historians of science have studied mathematical practices, and the nature of the relations between the history of mathematics and practice-based philosophy of mathematics, should be further specified to complete and clarify the picture. This would be all the more important because Van Bendegem suggests that one essential reason to identify both Lakatos' and Kitcher's works with pioneering philosophies *of mathematical practice* is, precisely, that these works were among the first to take seriously into account the *historical* dimension of mathematics.[3]

Fourth, relying on Van Bendegem picture, the unity of the philosophy of mathematics is highly problematic. This is the case, unsurprisingly, between traditional and practice-based philosophy of mathematics; but perhaps more surprisingly, this is also the case inside of the philosophy *of mathematical practice* itself. Van Bendegem insists that his eight approaches do not correspond to "a division of labor of a vast field to explore" (p. 222). It is not the case that "all the parts can be put together again to form a minimally coherent whole" (p. 222). To the contrary, "fundamental oppositions and tensions [are] at play" (p. 222)—and Van Bendegem considers as a crucial task for the future to attempt to reduce them. As for the contrast between traditional philosophy of mathematics and the philosophy of mathematical practice, it is manifested at multiple levels, including aims, languages, and methods. First of all, the questions of interest are neatly different. This is suggested by Van Bendegem, but since no systematic contrast is provided at this level in his contribution to the present book, let us say a little bit more about this point. Traditional philosophy of mathematics is primarily concerned with "foundational studies" and with ontological and conceptual abstract issues: with questions such as "What are reliable foundations for the whole of mathematics?"; "Is set theory a better or worse candidate than category theory?" (p. 220); "What is the ontological nature of the number 2?"; "What is a geometric concept?"[4]; and so on. In contrast, the philosophy of mathematical practice is interested in the concrete ways through which mathematics has developed historically and is conducted nowadays. This introduced new issues with respect to traditional ones, among which are the following: "How does mathematics develop?" (In particular, are there Kuhnian revolutions in mathematics?); "How do diagrams, drawings, and other supporting materials contribute to the notion of proof and the activity of proving?"; "How do aesthetic considerations contribute?"; "Are explanatory proofs more interesting than non-explanatory ones? And do we know what we mean when we say that a proof is explanatory?"; "Why do mathematicians consider some problems interesting and others not?"

Fifth and finally, traditional philosophy of mathematics is, according to Van Bendegem, in a dominant position with respect to the philosophy of mathematical practice. The latter is presented as a "new and emerging domain" (p. 222), still in its infancy, and marginal in the philosophy of mathematics considered as a whole. Moreover, Van Bendegem depicts the relations between the two trends as a *competing* relation. Even stronger, he suggests, using a vocabulary reminiscent of Darwinian evolutionism, that the very existence of a philosophy of mathematical practice is precarious and might be threatened. Developing Van Bendegem's metaphor, we could say that the philosophy of mathematical practice has to struggle for life and that in order to survive, it should not only avoid strengthening competition, which might be fatal to the weakest; not even just try to find an ecological niche that would ensure the pacific cohabitation of the two coexisting species; but, instead, try to institute, if not a complete symbiosis, at least some relations of mutual support—that is, collaborations and bridges around shared questions or methods. Although at first sight, "a different language is spoken, a different outlook is shared by its practitioners, and it would seem normal that both have little to say to one another" (p. 222), Van Bendegem nevertheless identifies common points that might provide the missing links. Namely, the "analytic tradition", "present in both" traditional philosophy of science and the philosophy of mathematical practice, and more specifically, "formal tools and modeling techniques to explore philosophical questions" (p. 223). As a personal contribution along these lines, presented as a "global bridges" (p. 223) between traditional and practice-based philosophy of mathematics, Van Bendegem sketches a formalized model of mathematical practice, developed elsewhere in collaboration with Van Kerkhove, and understood as a "generalization" (p. 223) of the model of practice elaborated by Kitcher (see

Section III of this commentary for more about Kitcher's model). Considering such type of contribution, Van Bendegem might be viewed as a descendent of the analytical naturalizing approach as well as of the Lakatosian one—yet an unlikely hybridization given his emphasis on the discontinuity between the two approaches.

III. PHILOSOPHICAL APPROACHES OF MATHEMATICAL PRACTICE: LAKATOS AND KITCHER

In what sense is Lakatos' project a philosophy *of mathematical practice?* The more direct explicit clue given by Van Bendegem is the following sentence: Lakatos' "focus on mathematical practice was clear by the mere fact that it boldly presented nothing less than a 'logic' of mathematical discovery" (p. 216). More precisely, Lakatos (i) provides a *"history* of a mathematical statement and its proofs" (p. 216), and (ii) "found patterns in this game of proof and refutation" (p. 216). As we understand Van Bendegem's view, Lakatos' work is classified as a study of mathematical practice because it is interested in the *process* of mathematical *"discovery"*—that is, on the genetic trajectory through which some mathematical conjecture has been historically formulated, explored, associated with evolving elements of justifications and counterexamples, modified and refined through this process, and finally proved—rather than being restrictively focused on mathematical *products*—that is, accepted mathematical propositions and their "final proofs" as they typically appear in specialized publications. Two distinctions are intertwined here, that are often involved in discussions of the novelty introduced by the practice turn: the classical opposition between the so-called "context of discovery" versus "context of justification" and the distinction between scientific processes and scientific products.[5] The latter distinction conveys a very broad negative sense of scientific practices, namely, what scientific products are not. Scientific "discovery" does not exhaust scientific practices so understood—in particular, nothing prevents in principle to consider scientific *justification* also *from the standpoint of practice;* but since in facts, discovery was claimed to be irrelevant to epistemology by most traditional philosophers of science before Lakatos, and since justification was correlatively mostly treated in a logical, largely anhistorical perspective, the turn to discovery processes has often been perceived as emblematic of a turn *to scientific practice.*

The assimilation of Lakatos' project with a study *of mathematical practice* could, however, be questioned. This is because in the case of Lakatos, the scientific processes of discovery are characterized through the frame of *rational reconstructions* (see Lakatos, 1970a). Lakatos' account is, explicitly and self-consciously, not viewed as a faithful description of what actually happened in the history of science. Lakatos sometimes adds notes to his rational reconstructions, which are supposed to connect them to the "real history" (1970a, p. 5), but his characterization of "mathematical practice", and the "patterns in this game of proof and refutation" he claimed to find, are rational reconstructions, that is, an *idealized, retrospective, present-centered* picture, which made *active selections* in the history of science, *taking present-mathematics as the reference.* Obviously, this is not what is commonly targeted, within the practice turn, under the term "scientific practice". To the contrary, it was precisely *against* rational reconstructions that the early supporters of a turn to practice in philosophy of the natural sciences elaborated their positions.[6] About the fact that Lakatos' picture was intended to depart from history of science, Van Bendegem says nothing. He describes Lakatos' import as an attention to the history of mathematics: "What did the book [*Proofs and Refutations*] offer? Nothing less but the *history* of a mathematical statement and its proofs" (p. 216, italics in the original). At the beginning of his paper, among qualifications about his choice of Lakatos as the first origin of philosophy of mathematical practice, he stresses, about his own account, that "almost any historical outline is partially a reconstruction" (p. 215). However, if we can concede that no history of science is a "pure replication" of the actual temporal unfolding it targets, it remains that there is a sharp difference between the aim of a rational reconstruction and the aim of a descriptively adequate picture. Different projects are involved here. Taking that into account, even if it is perhaps the case that "Lakatos could not have written his seminal work without the sources made available by historians" (p. 219), what Lakatos himself developed does not seem appropriately named a philosophy *of mathematical practice* in the *usual present understanding* of "practice". Alternatively, we could concede that Lakatos' philosophy is a characterization of mathematical practice *in a highly idealized sense.*

It remains, however, that even if Lakatos' account does not have the status of an historical account of the actual practices of past mathematicians, one important moral of his work was, as Kitcher writes in his 1976 review of Lakatos' book, and in agreement with Van Bendegem's main message, the following: "Philosophers of mathematics should not continue to ignore the fact that mathematics has a rich and exciting history" (1976, p. 783). More precisely, Lakatos indeed contributed, in the history of philosophy of mathematics, to draw attention to mathematics *as a dynamic process,* to institute the process of mathematical *"discovery"* as an object of interest, and to show that mathematical proofs can have *heuristic functions* in addition to justificatory ones. Such dynamic, genetic, and heuristic aspects were largely ignored from the pre-practice turn philosophy of science (including mathematics), and the simple fact to take them as an object of study has often worked as a sufficient reason to classify an author as an actor of the practice turn.

Let us now turn to the work of Kitcher: In what sense is Kitcher's project a philosophy *of mathematical practice,* and in what sense is this project "definitely not a continuation of what Lakatos had initiated" (p. 217)?

Kitcher, contrary to Lakatos, explicitly formulates the object of his interest in terms *of mathematical practice,* and moreover offers a precise definition of a mathematical practice: "I suggest that we focus on the development of mathematical practice, and that we view a mathematical practice as consisting of five components: a language, a set of accepted statements, a set of accepted reasonings, a set of questions selected as important, and a set of metamathematical views (including standards for proof and definition and claims about the scope and structure of mathematics)" (1983, p. 163). Kitcher's approach, as Lakatos' one, is an approach of mathematical practice, first of all because it is interested in mathematical *processes* rather than just on mathematical finished products. In contrast to Lakatos, however, Kitcher is primarily interested in processes *of justification* rather than of discovery. Van Bendegem describes this difference as an "initial tension" between Lakatos and Kitcher, and as the old "divide" between the so-called "context of discovery" and "context of justification" (p. 217). He adds, moreover, that this initial tension "has never left (up to now) the study of mathematics through its practices" (p. 217). Concerning this latter point, we can note that the situation differs in the practice-based philosophy of the *natural* sciences. In the latter, the "divide" between the two contexts has been extensively and strongly criticized, to the point that most scholars nowadays claim that the distinction is inapplicable when one considers the actual flow of scientific practices.[7] To come back to Kitcher and Lakatos, whether there is a tension or not, we can at least concede that they are not interested in the same object (justification versus discovery). Since contrary to discovery, justification was part of the central topics of traditional philosophy of mathematics, we can ask why, compared with traditional accounts of justification in mathematics, Kitcher's perspective can be categorized as a philosophy of mathematical *practices* of justification.

Several reasons can be proposed. First, Kitcher characterizes mathematical processes of justification not in terms of a universal logic, but through a set of collective fundamental resources involved in mathematicians' work at a given time in a given community. His "more or less formal model" (p. 216) of mathematical practice, inspired by the Kuhnian idea of a disciplinary matrix but developed in a more analytic naturalized style, provides a structural characterization of a given type of mathematical collective activity, which identifies pivotal shared elements of the activity of justifying. Practice, here, refers to normal ways of doing mathematics in a given stage of mathematical development—where "normal" has to be understood, as in Kuhn's phrase "normal science", in the double sense of "often" and "standard". Second, Kitcher's definition of mathematical practice includes components that were not part of traditional accounts of justification, namely, taken as interesting questions, and meta-mathematical considerations. To include these components into the framework is to offer a less idealized characterization of processes of justification than traditional ones, and a characterization which immediately suggests that standards of mathematical justification might evolve through time—since interesting questions and meta-mathematical views are clearly not invariant, trans-historical and universal elements. Third, Kitcher (1983) explicitly discusses the historical evolution of his five components of mathematical practice, thereby substituting a dynamic picture of justification processes to the logical, a-temporal, and absolute, traditional one. Fourth, Kitcher's dynamic picture is less idealized and "more practical" than traditional pictures—and also than Lakatos' one—in the sense that it put forward, as the main motor of mathematical growth, the "pragmatic concerns of mathematicians" (1983, p. 271). Most of the time, Kitcher argues, "Foundational study is motivated by the need to fashion tools for continuing mathematical research" (1983, p. 271), often in relation to "problems which have physical significance" (1983, p. 236), rather than "because of apriorist epistemological ideas" (1983, p. 246), as philosophers of mathematics have misguidedly suggested.

IV.   SOME ADDITIONAL REMARKS TAKING INTO ACCOUNT NON-PHILOSOPHICAL APPROACHES OF MATHEMATICAL PRACTICE

Considering the two main philosophical approaches to mathematical practice identified by Van Bendegem, we can say that to study mathematical practices means, first, to study mathematical processes (rather than just mathematical products) and, second, to characterize these processes in an empirically based, historically informed perspective (rather than through a priori, idealized logical schemes). This general idea of what it means to study mathematical practices could certainly be accepted by, or at least does not seem to be in contradiction with, the five others, non-philosophical approaches that Van Bendegem put forward as possibly relevant to the philosophy of mathematical practice. Beyond this possibly shared general idea, however, differences and tensions between conceptions of mathematical practices also exist, that Van Bendegem's five non-philosophical approaches help to make salient. Let us stress some of them.

First, the idea of mathematical practices as historically situated processes by opposition to products is very broad. Mathematical practice, so understood, might encompass anything that has been involved in the historical elaboration and stabilization of some mathematical product. Inevitably, various analysts of mathematics will assess differently what is relevant and important and what is not. The corresponding differences are sometimes due to the fact that analysts ask different questions and have different interests (for instance, are interested in different types/aspects of practices, such as in practices of exploration of new mathematical objects, or in practices of justification of mathematical propositions, or in practices of development of new symbolisms, or in aesthetical and rhetorical aspects of mathematical activities, etc.). But often, the differences are due to the fact that analysts don't assess what is determinant and what is anecdotic in the same way. The so-called "social factors", however understood in detail, are a case in point. What place and weight should we give to them in a philosophy of mathematical practice? Can we characterize mathematical practices without including them into the picture, or by assigning them only a marginal role? The answer engages the relation between the sociology and the philosophy of mathematical practice. And Van Bendegem—certainly reflecting a widespread position in the philosophy of mathematics—locates the central tension between the sociological and the philosophical approaches of mathematical practice in divergent answers to these questions. After having mentioned David Bloor and Sal Restivo, he writes that "the sociological approach did not merge easily with the mentioned traditions", adding that "surely one of the corner elements is the internal-external debate . . . : Does mathematics develop according to 'law' or patterns that are internal to mathematics (and, as a consequence, independent or at least marginally influenced at most by external 'law' or patterns), or, on the contrary, do external elements contribute and, if so, in what way(s)"? (p. 220). Beyond the hard problem—not discussed by Van Bendegem in this paper but often stressed by advocates of the practice turn—of drawing the frontier between social-external and cognitive-internal components of mathematical practices, the point, here, is that even in cases in which we can concede that different analysts of science operate with a similar general conception of practice, very disparate, conflicting characterizations can nevertheless result in the end, because of disparate judgments about what is actually determinant in mathematical practices, and hence about the reasons why mathematical proposals have acquired the status of taken-for-granted mathematical products.

Second, we can note that taking into account Van Bendegem's five non-philosophical approaches of mathematical practice, it becomes clear that the question "the practices *of whom?*" is susceptible to a plurality of answers. Studies of mathematical practices are not restricted to the activity of specialists and professional mathematicians with which philosophers of mathematics are most of the time primarily concerned. They include as well the activity of any people performing tasks susceptible to be associated with mathematics: of "lay people" asked to solve some specified mathematical problems (in the cognitive psychology of mathematics); of pupils and students involved in learning processes (in the mathematics educationalist approach and sometimes in the cognitive psychology of mathematics); of "nonliterate" people in "traditional or small-scale cultures"[8] (in ethnomathematics); or even of non-human animals for the sake of comparison (in the biological approach). As a first corollary, the practices under scrutiny can be more or less complex, technical, and esoteric—from basic operations and notions mastered by most members of the society under scrutiny, if not by any intelligent beings, to strategies and techniques developed by a very small sub-set of competent persons in order to cope with innovative esoteric problems—and their identification as *mathematical* activities can even be problematic.⁹ As a second corollary, the study of mathematical practice can be conducted at different scales—from the particular practice of one singular mathematician, to the mathematical practices of more or less wide groups—and the study can aspire to a more or less large domain of validity (sometimes even a universal one, through claims about any "mathematical being").

Third, the characterization of mathematical practice can be framed in different types of categories, which can *more or less depart* from those used by the actors under study if asked to describe what they are doing. The categories of the analyst are sometimes radically different from, if not incommensurable to, the ones of the actors. This is most strikingly the case for the evolutionary biology of mathematics. We could actually even question whether accounts of "how much mathematical knowledge is biologically, perhaps genetically, encoded in the human body and how it affects our mathematical abilities" (p.221) are accounts *of mathematical practices*. Such accounts would perhaps be more adequately described as *materials* possibly useful to achieve a better understanding of certain aspects of mathematical practices (for instance, to help separate inner/natural and acquired/cultural features of these practices).

Fourth, the methods through which mathematical practices can be studied prove to be diversified: historical inquiries, which can be more or less attentive to the details of the particular contexts under scrutiny; ethnographic methodologies, including participant observation, interviews, and so forth (especially in social studies of mathematics and in ethnomathematics); experimentation and creation of artificial controlled situations (in biological, cognitive, and educational studies of mathematical tasks); and of course conceptual analysis, in different possible associations with one or several of the previous methods.

V.   CONCLUSION

As a conclusion, we would like to indicate what we identify as one fundamental but neglected epistemological stake of a philosophy of mathematical practices: How much are taken-for-granted mathematical products (propositions and proofs) essentially dependent, or on the contrary 'detachable', from the details of diachronically evolving and synchronically diverse mathematical practices? Following the practice turn, the idea of a genuine path-dependency, and hence of a radical contingency, of taken-for-granted achievements in the *natural* sciences, has acquired some plausibility, or at least has been seriously discussed by some scholars (Soler, 2008a, 2008b; Soler, Trizio, & Pickering, in progress). In contrast, very few philosophical works have investigated this issue in relation to mathematics. Perhaps because the traditional conception of mathematics that Van Bendegem credits Lakatos (1976) to have shaken—mathematics as "timeless, eternal even, unchangeable for sure" and "mathematical proof [as] . . . absolutely certain, undoubtable, secure" (p. 216)—has finally not been so deeply shaken?

NOTES

1. According to Mancosu, but Van Bendegem writes that he does "not object" (p. 225), and in private communication, he spontaneously presented his work as "basically an elaboration of" "the seminal work of Imre Lakatos".
2. See the introduction of this volume, Section V.3. For ways to be normative after the practice turn, see Sections VI.3, VI.4, and VI.5 of the introduction, and Chapters 3, 4, and 5, of this volume.
3. In relation to Lakatos, this is more than a suggestion: It is an explicit claim (see the quotation reproduced below, Section III of this commentary). In relation to Kitcher, the point is less explicit, but it is strongly suggested by the following passage: "The Kitcherian outlook proved to be more successful, as several authors embraced this trend, *as is shown in subsequent volumes that made connections between philosophy and history of mathematics*" (p. 217, italics added).
4. All the quotations for which no page is indicated have been provided in private communication.
5. See the introduction of this volume, Section V.7.
6. See the introduction of this volume, Section V.3.
7. See the introduction of this volume, Section V.6.
8. "Ethnomathematics is the study of mathematical ideas of nonliterate peoples" (Ascher & Ascher, 1986, p. 125); "Our focus . . . is elaborating the mathematical ideas people in these lesser known cultures, that is, the ideas of peoples in traditional or small-scale cultures" (Ascher, 2002, p. 3).
9. Ethnomathematicians, in particular, need to specify what they are ready to identify as mathematical practices in the foreign culture they study. For example, Ascher and Ascher (1986, p. 125) write: "We recognize as mathematical thought those notions that in some way correspond to that label in our culture. For example, all humans, literate or not, impose arbitrary order on space".

8 Observing Mathematical Practices as a Key to Mining Our Sources and Conducting Conceptual History

Division in Ancient China as a Case Study[1]

Karine Chemla

The present chapter can be read as a reflection arising from the working conditions of the historian of ancient mathematics. In contrast to colleagues working on the early modern or modern age, the historian of the ancient world generally has very few documents to rely on. In addition, these sources were often produced decades, or even centuries, apart. Owing to complex historical processes usually difficult to investigate, these sources, unlike many other documents from the past dealing with similar topics, happen to have survived.[2] The historian has to work with them even though at best, if at all, they can only be placed in a rarified historical context. These conditions demand that historians devote maximal attention to methods allowing them to derive the greatest possible amount of information from these rare documentary resources. Whereas the issue of how to make sources speak is of interest for historians generally, in relation to ancient history it is a matter of vital importance.

The difficulty is compounded when we want to address questions our sources evoke only tangentially or indirectly—we shall encounter specific examples below. Similar issues have been debated for decades in the field of general history, particularly when historians like Carlo Ginzburg attempted to derive information about actors who left no written records by means of documents produced by others.[3] Ginzburg (1989, 2012) has addressed this issue theoretically, using the notions of "clue" or "trace", and sketching the history of the "evidential paradigm".[4]

As far as I know, the implications of these debates for conceptual history, let alone conceptual history of mathematics in the ancient world, have not yet been addressed. How can we account for concepts and bodies of knowledge for which our sources have only left clues? The present chapter is an attempt to deal with this question. The thesis I shall propound is that reconstructing mathematical practices involved in producing our sources yields key resources for inquiring into issues of conceptual history that our sources do not tackle in detail.[5] I would argue that the description of practices allows us to understand elements of actors' knowledge for which our sources provide indirect evidence, but that would otherwise remain out of range. In other words, the description of practices helps with interpreting clues. Moreover, it allows us to perceive changes in the knowledge possessed by actors, and thus to discover questions that actors addressed, even without direct evidence for this. In brief, this chapter aims to demonstrate *how* the description of practices can be essential for doing conceptual history by providing the resources to interpret clues meaningfully.

To illustrate more clearly the issues addressed and the kind of evidence one can use to support my argument, I shall develop the argument in the context of a case study. This case study will be introduced in the first section of the chapter, and is devoted to part of the mathematical work done on arithmetical operations—more specifically, on division—in ancient China. In the second section, I shall outline aspects of the state of knowledge on this topic as we can ascertain it on the basis of a book compiled in China in the 1st century CE, and from related evidence. The argument here will aggregate evidence on this state of knowledge and information we have acquired on mathematical practices in that period. The third section will contrast the results obtained with features of the state of knowledge on the same topic as indicated by other writings produced in China in the preceding centuries. The comparison highlights that the state of knowledge evidenced in the 1st century must have been the result of a work for which we have no other written evidence as yet. The conclusion will derive some general observations from this specific case study.

I.   THE DIFFICULTY OF RESEARCHING ARITHMETICAL OPERATIONS IN ANCIENT CHINA: A CASE STUDY

In various parts of the ancient world, arithmetical operations were a topic of theoretical reflection. However, historians have not explored these reflections systematically. The main obstacle was probably the lack of direct evidence. Identifying traces of these reflections and interpreting them are delicate undertakings. This is precisely the topic I shall address in this chapter. Let me explain why these enterprises are complicated in the case of ancient China.

Mathematical documents that have survived from early imperial China, that is, documents produced between the 3rd century BCE and the 1st century CE, show that the practitioners of mathematics focused their attention on "procedures"—in present-day terminology, "algorithms", in Chinese "*shu* 術". Accordingly, we perceive that operations were key objects of study.

Firstly, arithmetical operations were the building blocks of procedures. One can illustrate this, for example, with the procedure for computing the volume of a half-parallelepiped. It consists of a sequence of two multiplications followed by a division by 2, and its aim is to carry out a task formulated by a mathematical problem. This rough description shows that building blocks—in this case, multiplication and division by 2—were created to fulfill this function. These building blocks are what we call operations. When in procedures we encounter operations like "dividing in return *baochu* 報除", we understand that we cannot take the shaping of building blocks for granted, and this demands closer examination than it is usually given.

Secondly, operations were also executed by means of procedures. Square root extraction, for instance, is an operation for which algorithms are provided in some ancient Chinese mathematical sources. However, as regards most operations in the time period under consideration, we usually have only indirect knowledge of procedures for executing them. These examples suffice to show that the relationship between operations and procedures is far from self-evident.

In fact, we can perceive *how* operations were objects of study, and the results of such an inquiry, by means of sources consisting of texts mainly composed of mathematical problems, algorithms that solve them, and numerical tables. These documents involve operations: names for operations occur in texts of procedures. Moreover, these sources contain evidence on ways of working with operations beyond the text. For instance, they provide fragmentary evidence on the computing instruments used to execute operations, or on diagrams associated with them. However, they evidence ideas about operations only indirectly. In other words, these sources probably referred to actions, such as the execution of operations. They belong to a practice of mathematics for which they provide clues. However, they are certainly not treatises about operations. In fact, no discursive treatment of operations survived from early imperial China, if any ever existed.

Consequently, the state of the extant documentation raises a problem for the historian who wants to inquire into the theoretical work done on operations. Looking at the subject matter, which methods can we use to discover the knowledge obtained in ancient China about operations? How can we ascertain ideas that were formed about operations, or the goals assigned to inquiries on this topic? In addition, with respect to the practice, how can we reconstruct the processes designed to work with operations? More generally, how can we write a history of the bodies of knowledge formed about operations when we are relying on sources that document all these aspects indirectly?

This is a complicated issue. However, we cannot leave these questions unanswered, since something essential is at stake. Clearly, significant efforts were devoted to operations, and their results represent an important, perhaps even an essential, part of the theoretical work on mathematics carried out in ancient China. As already mentioned, I shall assume the conclusions of the articles in which I have addressed the question of reconstructing practices, and shall focus here on *how* one can reveal the conceptual work done on operations.

II. WEAVING CLUES TOGETHER: ASPECTS OF ONE OF THE STATES OF KNOWLEDGE ON DIVISION AND RELATED OPERATIONS

The main document I shall rely on in this section is the book that was probably composed in the 1st century CE and that became a classic soon afterwards, *The Nine Chapters on Mathematical Procedures (Jiuzhang suanshu* 九章算術). Although in what follows I refer to the book as *The Nine Chapters,* it is interesting that the original title shows the importance the book's compilers gave to the topic of procedures. Like all classics, this was the subject of various commentaries, two of which survived in the selection process of the written tradition and were handed down systematically together with *The Nine Chapters.* These are, first, the commentary that Liu Hui completed in 263 and, second, that supervised by Li Chunfeng and presented to the throne in 656. For the purpose of our argument, we will need to refer to these texts below.

This section proposes to show that *The Nine Chapters* testifies to actors at that time being in command of fairly complex and elaborate knowledge about a set of operations in which division played the central part.[6] Yet, as we shall see, we have very little evidence on the operation of division. A complex argument is required to establish our proposition. Let us see, step by step, how we can weave clues together to reach this conclusion.

The texts of procedures in *The Nine Chapters* illuminate features of the computing instrument used in relation to the book. First, they often refer to the action of "placing 置 *zhi*" values, which testifies to the existence of an "instrument" beyond the text, probably a simple surface on which a representation of values was placed and with which one computed. Second, *The Nine Chapters* also indicates that "counting rods" were used to represent numbers on that base. In fact, the operations prescribed by its procedures were performed on numbers represented by counting rods on this surface, and apparently not on any writing material. What was the number system that recorded numbers using rods? *The Nine Chapters* does not give direct information on this, but as we shall see below, we can glean indirect evidence. Third, the procedures of *The Nine Chapters* show clearly that in the context of a computation, values could be inserted in different positions (above, below, in the middle, etc.).[7] Here are examples of clues that a text indirectly gives about a feature of mathematical practice—in this case, the practice of computing. These clues are evidence of an object which is outside the text, and which the text does not treat discursively.[8] Contrary to what most historians have claimed, the three features mentioned above do not suffice to identify the computing instrument and to assert that the instrument remained unchanged from the 3rd century BCE up to the 14th century. If this conclusion holds true, the case illustrates in general how we can find clues in writings that provide indirect evidence about mathematical practice, and how the interpretation of these clues involves a very delicate operation.

Indeed, all the manuscripts discovered recently, to which I shall return in the next section, also contain these three clues. However, I have reached the conclusion that they do not refer to the same instrument as *The Nine Chapters* does, unless they refer to a different way of using this instrument.[9] We can identify key differences between the practice of computing indicated in *The Nine Chapters,* and that to which the recently discovered manuscripts testify, if we observe the basic operations on the representation of values with rods to which their texts of procedures refer. For instance, procedures in *The Nine Chapters* regularly use the fact that rods, or representations of numbers, can be *moved* forward or backward on the surface. The utilization of these basic moves implies that the number system used in relation to *The Nine Chapters* must have had features that made these operations meaningful. By contrast, to my knowledge there is no reference to any basic operation of that kind in the recently discovered manuscripts. This contrast is essential to keep in mind. Indeed, these additional clues, derived from how operations echo the material properties of the number systems to which they are applied, seem to indicate that we should be careful not to assume that all these writings were composed with reference to the same number system, even though in every case the number system was written down with counting rods on the surface.

For the moment, let us limit our discussion to *The Nine Chapters.* We will subsequently encounter further evidence that the book provides about the practice with the computing instrument. Indeed, since operations are executed on the basis of a computing instrument and a given number system, it is not surprising that their execution reveals features of the system on which they operate. Conversely, the knowledge about the instrument and number system we can derive from texts of procedures helps us interpret these texts. We shall return to this interplay at the end of the chapter.

II.1. *Division in* The Nine Chapters

What evidence does *The Nine Chapters* offer with respect to division?

To begin with, the texts of procedures in *The Nine Chapters* abound in prescriptions of division. Several terms are used for this, and it will prove useful later to discuss them briefly now. Division can be prescribed by a verb, "divide 除 *chu*". However, the operation is also designated by one of a set of related expressions: "(quantity) then one 而 *er yi*", which indicates that the dividend is divided by the "quantity", with each part of the dividend equal to the "quantity" becoming 1;[10] "like (quantity) then one 如 . . . 而一 *ru* (quantity) *er yi*"; and finally, "when the dividend is like the divisor, then one" or "then it yields one", or else "then one (name of a measurement unit)", "then it yields one (name of a measurement unit)"—in Chinese: *shi ru fa er yi* 實如法而一, *shi ru fa de yi* 實如法得一, *shi ru fa er yi* (name of a measurement unit) 實如法而一(name of a measurement unit), *shi ru fa de yi* (name of a measurement unit) 實如法得一(name of a measurement unit).[11] In *The Nine Chapters*, prescriptions by other means than verbs exist only for the operation of division.[12]

Viewed from the way it is prescribed in the text of procedures, division differs from all the other arithmetic operations in another respect, as well. It is the only operation for which technical terms were introduced to designate the operands, "dividend *shi* 實", "divisor *fa* 法". This feature will also prove important for our argument later. When a multiplication is prescribed, for example, its operands are signified by values or expressions referring to the magnitudes whose values are to be multiplied. By contrast, the prescription of a division often describes how the dividend and divisor are obtained, before introducing one of the technical expressions referring to the operation to be performed on the basis of the operands.[13] Let us emphasize that in this case, technical terms evidence an understanding of an operation as, first, having operands and, second, being connected with a procedure that will be performed on these operands.

The Nine Chapters does not contain any text of procedure enabling the reader to do a division. The same applies to addition, subtraction, and multiplication. Only in later texts—for example, in the *Mathematical Classic by Master Sun* (*Sunzi suanjing* 孫子算經), completed in 400 CE—do we find explicit algorithms for multiplication and division (the latter is then called *chu*).[14] Is this procedure for division the same as that meant by the authors of *The Nine Chapters*? We shall demonstrate below how we can find indirect evidence in *The Nine Chapters*, based on our knowledge of mathematical practice at the time, to establish this. In addition, the *Mathematical Classic by Master Sun* makes clear how the algorithm is deployed on the surface that served as a computing instrument and how numbers are represented with rods. This evidence shows that the number system used to write down numbers with rods was a place-value decimal system. The same question arises: Can we establish that a similar system was used in the environment that produced *The Nine Chapters*? Here again, as we shall see below, the same type of indirect evidence will enable us to reply positively.

II.2. *Root Extraction in* The Nine Chapters

The examination of root extraction in *The Nine Chapters* will play a pivotal role in my argumentation. I shall not repeat what I have written elsewhere on this subject. Here I intend only to clarify the kind of knowledge about operations to which *The Nine Chapters* testifies and to show how historians need to rely on the description of mathematical practices to reach these conclusions. The type of clue the classic provides on root extraction is different from those found for division.

In texts of algorithms, *The Nine Chapters* refers to operations of square root extraction, and also cube root extraction, though the latter occurs less frequently. Two families of expressions are used to prescribe them. Sometimes root extraction, regardless whether it is square root or cube root extraction, is referred to by the verb "*kai* 開", for which I suggest the translation "extract the root of". This term is used only in the context of discussion on algorithms to perform root extraction. This explains why, even though a specific kind of extraction is meant in each case, the verb used can be non-specific. The context makes the actual meaning of the term clear.[15] The two operations are far more frequently prescribed by expressions of utmost importance for us. I translate them in a way that makes explicit the structure of the terminology in classical Chinese: "Divide this by extraction of the square root *kai fang chu zhi* 開方除之"—literally, "divide (*chu*) this (*zhi*) by opening (*kai*) the square (*fang*)"—and, "divide this by extraction of the cube root *kai lifang chu zhi* 開立方除之"—literally, "divide (*chu*) this (*zhi*) by opening (*kai*) the cube (*lifang*)". Let us comment on some features of these terms.

First, the anaphora "this *zhi*" designates the value to which the operations are to be applied. As in the case of division, and in contrast to the other arithmetical operations, a technical term is introduced to name the operand of root extractions. Interestingly enough, in both cases, this term is the one used in the context of division to refer to the "dividend", namely, *shi* 實. The fact and the term both connect root extractions and division. Note, incidentally that, at this stage, extractions appear as operations that bear on a single operand.

Second, in contrast to the previous terminology discussed ("*kai* 開 extract the root"), the latter expressions used to prescribe root extractions show a link between the execution of these operations and the operation of division. Note a detail that will prove useful later: Among the many expressions that could have been used to designate division in this context, it is the prescription by the verb *chu* that is chosen for this purpose. Each of the types of root extraction is prescribed by a formulation that qualifies the term *chu*, thereby stating which kind of division is meant. The qualifications introduced, that is, "by extraction of the square root" and "by extraction of the cube root", have parallel structures. As a whole, the terminology shows that division is the most fundamental operation, and that the two types of root extraction derive from it in similar ways. An examination of the ways of shaping terminologies in the context of *The Nine Chapters* seems to support the hypothesis that the choice of terms is a way of stating this fact about the relationship between the three operations.[16] Here we encounter an initial example in which knowledge about a practice can be used to gain insight into actors' knowledge, in this case about operations.

The terms prescribing root extractions and the term used to designate their operand both indicate that actors established a relationship between the three operations under discussion. We can confirm this conclusion by looking at the texts of the algorithms contained in *The Nine Chapters* that explain how to carry out the operations of root extraction (Li Jimin 李繼閔, 1990, pp. 91–105).[17] In fact, these texts enable us to gain a much deeper insight into how actors of that time understood the relations between the operations.

The texts describing algorithms for the execution of, respectively, square root extraction and cube root extraction are written in close parallel. They correspond mutually sentence by sentence, using the same terms. What the texts show mirrors what is stated by the expressions prescribing the operations. This method of writing texts for algorithms that display a relationship between the operations performed in the working of the algorithms is part of a practice that is recorded until at least the 13th century.[18] The fact that the practice presents a certain degree of stability does not imply, however, that the meanings thus expressed are the same. In fact, we can perceive differences in the relationships between the same operations which are expressed in this way, and consequently differences in how the relationship is understood.

The identification of this practice is essential for drawing information from *The Nine Chapters* in several ways. First, it enables us to assert that the authors of the algorithms shape and state knowledge about the relationship between the operations of root extraction by means of writing down texts prescribing how the operations can be performed. This means we can not only analyze this relationship, but also examine its reformulation in later texts.

Second, these texts for root extractions in *The Nine Chapters* are clearly written with reference to an algorithm for division. The terminology used includes: "dividend", "quotient", and "divisor", as well as several technical terms for operations similar to those found in the text of the *Mathematical Classic by Master Sun* mentioned above. This reference to division again echoes what was expressed by the mode of prescribing the operations examined above. We can deduce, from our knowledge of the practice of writing down texts for algorithms, that these texts also provide a statement of knowledge about the relationship between root extractions and division. Note that later texts similarly expressed relationships between the same operations that were partly similar and partly different from a theoretical viewpoint (Chemla, 1994b). This conclusion confirms that the practice of stating knowledge yielded texts that were read in this particular way and served as a basis for further reworking.

Third, if we consider the implication of the previous conclusion for *The Nine Chapters,* we can see that the texts describing algorithms to execute root extraction in the classic *indirectly* provide us with information about the procedure for division to which they refer. Although *The Nine Chapters* does not contain any algorithm to carry out division, from our knowledge of a mathematical practice we can nevertheless derive some understanding of the algorithm used at the time. In fact, the algorithm for division in relation to which these texts for root extraction are written appears to be the same as that described in the *Mathematical Classic by Master Sun*. We now identify the various types of insight that the description of practices yields with respect to actors' knowledge, in a case where the only documents we have are texts for algorithms. We also begin to understand that our information on their knowledge of operations would be greatly lacking if we only assumed actors' knowledge of certain algorithms—that is, merely what appears on the surface of the sources. But, as we shall see, there is yet more to be gained from this investigation.

Let us return to an earlier remark about the operand of root extraction. As I emphasized, root extractions are operations with a single operand, a "dividend". However, I also pointed out that the texts of algorithms given to execute these operations introduced a "divisor". How are these two facts consistent? The point is that in shaping root extractions as divisions in the process of their execution, the algorithms constitute, progressively in the course of the computation, a "divisor", or more precisely a "fixed divisor", which plays the part of the related operand in the process of performing a division. In the context of root extraction, the "divisor" is thus not an operand: It is a technical component of the process of execution of root extraction, which appears only when we describe the execution. This choice of terminology reveals an interesting fact about operations. Technical terms associated with them are not limited to the operands, or the name of the operation itself. They can also disclose actors' identification of specific entities needed for the process of computation that performs the operation. These elements show another dimension of the work actors have carried out in the analysis of an operation. The importance of this will be shown below.

The texts for the algorithms that perform root extraction do not "describe" how this is carried out on the surface that was the computing instrument in that period. They only provide incidental clues to the actual computation. What can be deduced from these clues? Clearly, the texts reveal that the sequences of actions used to perform the square root and cube root extractions presented strong correlations. Note that these correlations established connections between the dynamical processes of execution on the surface. For instance, the texts reveal that the processes for square root and cube root extractions make use of the same ways of moving the entities placed in corresponding positions on the surface. Likewise, the executions appear to rely on a similar opposition between key positions and auxiliary positions. The description of the practice with the computing instrument evidenced by *The Nine Chapters* and the *Mathematical Classic by Master Sun,* among other sources, provides a backdrop for interpreting these clues. I have shown elsewhere that establishing material relationships in this way between the processes for executing operations and the sequences of events occurring in each of the positions of the process on the surface was another method actors developed to express, and handle, the mathematical relationships between operations.[19] This brings us to yet another practice, related to the ones examined above, which could be used to inquire into operations. Here, again, the practice is characterized by great stability. It is recorded as late as the 13th century; yet the meanings expressed in different epochs— that is, the understanding of the relationship between operations—present both similarities and differences.

The Nine Chapters provides evidence showing how this practice was used for square and cube root extractions. The *Mathematical Classic by Master Sun* testifies to how the same practice expressed the relationship between multiplication, division, and root extraction. We have already seen that *The Nine Chapters* indirectly refers to an algorithm for division similar to that described in the *Mathematical Classic by Master Sun*. On this basis we can also note that the texts of the algorithms in *The Nine Chapters* show that the same type of material relationship was established on the computing instrument between the process executing a division and that executing a square (or cube) root extraction. This observation yields a key resource to supplement the information given by the texts of *The Nine Chapters* and to reconstruct completely the process of computation on the surface, not only for division, but also for root extractions. To recapitulate with respect to the topic of this chapter, the key practices that enable us to understand actors' knowledge with respect to the three operations, knowledge that is indirectly indicated in *The Nine Chapters,* included practices of naming, practices of writing down texts of procedures, and practices of computing.

With the means of expression offered by the computing instrument, in *The Nine Chapters* division appears once again as the fundamental operation from which root extraction derived, through the use of auxiliary positions. In conclusion, the structure of the set of operations referred to by the terminology prescribing the operations is parallel to the structure that is expressed, on the one hand, by the relationship between the texts of algorithms given to perform the operations and, on the other hand, by the relationship between the processes executing the operations on the computing instrument. We see that the knowledge actors displayed in this context includes not only knowledge of the relationships between the operations, but also knowledge of the inner structuring of each of the processes of computation performing the operations. It is clear what insight we can gain into actors' knowledge by means of a description of their practices. Moreover, as we shall see, this insight will allow us to suggest how this knowledge appears to have served as a basis for further developments.

With regard to the computing instrument, some of the features of the texts of the algorithms in *The Nine Chapters* provide clues to the number system on which the calculations were executed. Through some of the actions prescribed (such as jumping over columns, and moving values forward and backward), and through the iterative structure of the texts, we know that the algorithms operated on a place-value decimal system in a context in which numbers were written down with counting rods. This case again shows how our sources indirectly shed light on elements of actors' knowledge—and practice—that are not treated by explicit discourse, at least not in the remaining set of documents. To sum up, it appears that not only is the indirectly manifested algorithm for division the same as that described by the *Mathematical Classic by Master Sun,* but the number system tacitly assumed by *The Nine Chapters* is also the same as that explicitly outlined by the other classic composed several centuries later. In addition, the two classic volumes share practices of naming and practices of writing down texts for algorithms, as well as practices of inscribing processes of computation on the computing instrument. This should not surprise us: A number system, the algorithms based on it, and practices related to them, form a coherent system, and its coherence has consequences for how the system is transmitted.

II.3.   *Quadratic Equation as An Operation in* The Nine Chapters

Problem 19 of Chapter 9 in *The Nine Chapters* is solved by a procedure that is concluded by an operation.[20] The statement of the last operation of the procedure is characterized by the fact that its operands are designated by technical terms: "dividend" and "joined divisor". Again, we find the correlated situation that operands are associated with technical terms and that we are working within the framework of division. The latter conclusion is confirmed by the fact that the operation is prescribed by the verb *chu,* "divide", prefixed by a qualification. However, the identity of the operation is not immediately clear. Indeed, its prescription tells us to "divide this by extraction of the square root *kai fang chu zhi* 開方除之". The formulation that usually refers to a square root extraction as an operation deriving from division is thus used here. However, surprisingly, the operation is *not* a mere square root extraction: instead of having a single operand as expected, in the context of the procedure the operation designated by the same technical term is now applied to two operands, a "dividend", and a "joined divisor". In what follows, we shall simplify by calling them *a* and *b* respectively. The procedure solving problem 19 thus describes how to compute the value of the two operands *a* and *b,* before concluding with the prescription, "divide this by extraction of the square root *kai fang chu zhi* 開方除之". Apparently, the set of technical terms attached to the operation indicates that a new operation is introduced. What kind of operation is it?

There is something we should note before we interpret the nature of this operation and the reasons why the prescription takes this form. To formulate this, let us begin by outlining the conditions under which we interpret the text. *The Nine Chapters* contains only two textual passages that can help us to determine the nature of the operation. The first is the text of an algorithm prescribing how to execute a square root extraction. The other is problem 19 of Chapter 9 and the related procedure introducing the operation. The classic offers nothing more. Naturally, we also have the commentaries on *The Nine Chapters* and other writings that use the same or similar operations. Let us leave them aside for the moment. Where does the difficulty in the interpretation of the operation originate? In my view, it derives from the fact that we have an expression prescribing the operation as if it were simultaneously a square root extraction and not such an operation. Suppose the terms of the operation were called xxx and yyy. Suppose the operation were prescribed by the term zzz. We would have automatically done what we do for addition or multiplication. We know the required operation that should be applied to xxx and yyy to reach the solution of the problem at this point. Therefore, we would conclude that zzz must refer to this operation, no matter how the operation was executed. However, this was not the terminological choice to which *The Nine Chapters* testifies. The authors chose a completely different set of terms deriving from their practice of technical terminology. Their terms not only *refer* to an operation, but here, as above, they also *formulate* a state of understanding regarding the relationships between various operations. In our case, they assert an intimate connection between our operation and, respectively, division and square root extraction.

If we were confusing the two parts played by the terminology in this context, this could create a problem for the interpretation of the text. Interpretation would indeed become problematic, if, instead of reading the prescription, "divide this by extraction of the square root *kai fang chu zhi* 開方除之" as *stating* a relationship between the intended operation and root extraction, we were expecting the expression merely to refer to an operation. This would lead us to ask why an operation that is *not* a square root extraction can be prescribed as if it were, or whether the operation prescribed is actually a separate operation. In my view, problems of interpretation of that kind occur precisely when we neglect actors' practices in the context observed—here, their practice of terminology—and when, instead, we implicitly assume their practices to be identical to ours. The problem vanishes if, by contrast, we base our treatment of the sources on a description of the practices in the context in which they were produced. As we shall explain, this description prevents us from overlooking meanings expressed by the choice of terms. These two aspects mirror the two different parts played by the specific use of technical terms.

In the case under examination, the difficulty is compounded by two factors. As we shall see, the operation under consideration is not our idea of an ordinary arithmetical operation. Further, for us the mathematical entity intended does not have the identity of an operation. I would argue that these are the obstacles we have to surmount to get the interpretation of the text I am proposing.

Finally, before we turn to identifying the intended operation, let me stress yet another important point for my argument in this chapter. The case examined here not only shows once again why the description of practices is essential for interpreting documents, but also illustrates how such a description equips us to perceive the work actors carried out in that context. What is at stake here is not merely interpreting the intended operation or understanding, as in the case above, the relation actors established between that operation and other mathematical entities. Another important goal is to highlight an aspect of the effort actors devoted to the study of operations and the knowledge they gained in that respect. The reason these three features should not be separated is simple. The understanding reflected by actors' terminology derives from their examination of operations. One of the results of this work, that is, a conception of the structure of a set of operations, can be grasped in *The Nine Chapters only* by analyzing the terminology. In fact, it is precisely *because* work was done on the set of operations and *because* the results of this work are reflected in the terminology that we, as present-day exegetes, have difficulty interpreting the operation. We can also look at the situation from another angle. If we wish to account for actors' knowledge about the operations, we are not only interested in the operations they "knew"—the answer to the question of the interpretation—but also in how they explored and understood them.

From the previous analysis, it appears that we need to carry out a two-step analysis of the operation concluding the solution of problem 19 of Chapter 9. We must first identify the operation referred to as if it were named zzz. We must then interpret the results that can be gleaned from the structure of the terminology.

Let us thus begin with identification of the operation. Examining the problem from a modern viewpoint, we can see that the operation corresponds to our modern terms for a quadratic equation. That is to say, the execution of the operation is equivalent to solving the equation

$$a = x^2 + bx.$$

What we see as an equation thus presents itself in *The Nine Chapters* as a numerical operation such as division or root extraction.[21] I shall call this entity an "operation-equation". This may be surprising, but we must be prepared to realize that for entities that appear familiar to us, actors of the past shaped concepts that differ from ours. Here, the interpretation may seem even stranger because the operation-equation is described as having only two terms (corresponding to the coefficient in x and the constant term), and not three (including the coefficient of x^2), as we would expect. Records exist of such an understanding of the operation-equation at least until the 11th century in China, when it was replaced by a new concept with three terms for the operation-equation. By that time, in addition to the two terms a and b mentioned above, the term we know as the coefficient of x^2 had been identified. This interpretation of the operation in *The Nine Chapters* is confirmed by Liu Hui's commentary, in which the commentator establishes the correctness of the procedure. In this case, he shows how a square of side x, the unknown, and a rectangle of area bx combine to form the area a. His commentary ends at that point. He has thus shown that the "operation" was proved to conclude the procedure correctly, as it amounted to stating the relationship mentioned. This covers the identity of the operation. What can we say about its execution?

Liu Hui adds no information on how the operation-equation should be executed. Nor do the authors of *The Nine Chapters*. They seemed to think the prescription of a square root extraction was sufficient for the user of the procedure associated with problem 9.19 to know how to proceed. Observing some features of the practices to which the text of this algorithm adheres, and the execution of the extraction on the computing instrument, enables us to propose a hypothesis about this facet of the operation-equation. Suppose we apply the algorithm for extracting square roots described in *The Nine Chapters* to the number A. It determines the root digit by digit, beginning with the digit corresponding to the highest order of magnitude of base 10. Let us represent this part of the root by the modern expression $p.10^n$. The first steps of the algorithm in *The Nine Chapters* create an array on the computing instrument that we can represent in modern terms as follows:

Quotient	$p.10^n$
Dividend	A
Divisor	$p.10^{2n}$

In the left column I have indicated the technical terms borrowed from division that the text of the algorithm of root extraction uses to refer to the entities. We recognize, in the lower row, the technical term introduced by the text describing the execution of the operation. This term does not refer to any operand of the root extraction. On such configurations, the operation of "division *chu*" multiplies the digit of the "quotient", p, by the "divisor", and subtracts the product from the "dividend". In the case under consideration, the operation yields the following array:

Quotient	$p.10^n$
Dividend	$A - p^2.10^{2n}$
Divisor	$p.10^{2n}$

In order to prepare the determination of the next digit, the algorithm prescribes transformations to be applied to the "divisor", yielding

Quotient	$p.10^n$
Dividend	$A - p^2.10^{2n}$
Fixed divisor	$2p.10^{2n-1}$

The important point here is that if we omit the first phase of the algorithm of root extraction—that described above—the operation whose execution starts at this point solves the following quadratic equation[22]

$$x^2 + 2p.10^{2n-1}. x = A - p^2.10^{2n}$$

which is written on the computing instrument as follows:

Quotient	
Dividend	$A - p^2.10^{2n}$
Fixed Divisor	$2p.10^{2n-1}$

We can see that what remains on the surface are precisely the two terms that are the operands of the operation-equation, which concludes the procedure of problem 19 in Chapter 9. We can also observe that the technical terms designating the operands of this operation-equation in the procedure are correlated to those designating the values in the rows represented above.[23] Let us draw some conclusions from this.

* * * * * * * * * * * *

First, it appears that a subprocess of the execution of a root extraction was detached from this context and given the identity of an operation. This conclusion about the origin of this type of "quadratic equation" explains why the operation-equation was perceived as having only two operands. It also explains why the prescription could be made using the expression "divide this by extraction of the square root". Last, it explains why *The Nine Chapters* provides no new algorithm prescribing how the operation-equation should be executed. Seen from the viewpoint of the execution of the operation, the operation "quadratic equation" depends on the square root extraction in that it is performed by a subprocedure. This type of link between operations differs from the link shown above between root extractions, or between the latter operations and division. "Surgical operations" of that kind can be shown to fit the practice with the computing instrument of the period. Again, the description of the practice helps us inquire into features of the knowledge on which the sources remain silent.[24]

Second, we have emphasized above that an entity that was *not* an operand of root extraction, the "divisor", was nevertheless identified as a technical component of the process of execution, and even given a name. We now see that this correlates with the creation of a new operation by derivation from root extraction.

Let us now consider the knowledge reflected in the choice of terminology, and more generally recapitulate the facts so far. As we have seen, *The Nine Chapters* mentions four operations that are all interrelated at several distinct levels. The foundation of the set is constituted by the operation of division. Its name, *chu,* occurs in expressions prescribing all the other operations. Its process of execution is the basis for writing down and performing the executions of root extractions on the surface. In turn, the execution of root extraction provides source material for the derivation of a new operation, the operation-equation, or "quadratic equation". The type of relationship established between the operations varies depending on the case. However, the operations form a group that *The Nine Chapters* presents as having a fairly precise and complex structure. In fact, the terminology prescribing these operations has a structure that is transparent with regard to the specific relationships that the texts of algorithms or their execution determine and state.

II.4. *The Work Done With Operations as Reflected in* The Nine Chapters

So far, we have relied on clues about operations found in *The Nine Chapters* and related texts such as the *Mathematical Classic by Master Sun*. The clues included names used to refer to operations or operands, and texts for algorithms that carried out the operations. What we have shown is that these clues are evidence of work conducted on the operations and the relationships between them. They testify to knowledge about operations that the authors of the procedures recorded in *The Nine Chapters* evidently possessed. One result of this work, which is one of the components of the knowledge identified, is a conception of the structure of the set of four operations.

Illuminating the existence of this kind of work and its results is essential for history of mathematics. Indeed, this suggests that operations were not only tools in the ancient world, but also objects of study. Our analysis allows us to perceive the results of theoretical inquiry into operations.

Since our sources did not treat the pieces of knowledge acquired as discursive developments, we were only able to derive our results from the clues in the sources by relying on observation of various aspects of mathematical practice (practices of naming, of writing down texts for algorithms and shaping dynamic inscriptions on the computing instrument). Our understanding of the mathematical practice in relation to which *The Nine Chapters* was composed in ancient China was essential in helping us perceive the knowledge that underpinned our sources. As I suggested at the beginning of this chapter, it is in this sense that the description of mathematical practice can provide tools to do conceptual history in a new way.

The body of knowledge revealed by this process shows strong correlations with material in later documents from China. Clearly, it represents a stage in the understanding of operations that is reflected in *The Nine Chapters* and served as a basis for subsequent developments. As already mentioned, algorithms for the execution of root extraction remained a topic of work for centuries to come. This work was carried out by means similar to those examined above, and regularly reshaped the relationships between division (*chu*) and root extractions formulated by *The Nine Chapters*. Moreover, knowledge about algebraic equations was obtained in a similar conceptual framework in later centuries. The division *chu* appears to have played a central role in theoretical work done on operations in subsequent periods. Together with the opposite operation of multiplication (*cheng* 乘), it was a key factor in practices of proof as well as in the inquiry recorded in several texts aimed at uncovering the most fundamental operations (Guo Shuchun

郭書春, 1992, pp. 301–320; Chemla, 2010). These observations retrospectively support our interpretation of the knowledge possessed by the authors of *The Nine Chapters,* although they did not formulate it discursively. They also help to illuminate the essential role played in the history by the division *chu,* which will emerge as a key factor in what follows. The same observations show from another angle how conceptual history is enriched when we deal not only with explicit knowledge recorded in our mathematical sources, but also with the clues they contain.

However, the results stated so far also leave some questions unanswered. The first question to be addressed if we want to ensure the clues we interpret are not chance occurrences, is: What kind of history leads to the complex of operations described—that is, the body of knowledge outlined above? How can we grasp, from a historical viewpoint, the mathematical work whose results we observe from our reading of *The Nine Chapters?* Can we rediscover which guiding questions or goals on operations were pursued? Another aim may be to understand how the ways of working on operations took shape—that is, to inquire into the history of practices with operations. In line with our focus in this chapter, we shall bypass the latter issues and concentrate on the former questions.

In recent decades new mathematical manuscripts from early imperial China were discovered: works produced in the third and second centuries BCE. The full text of two of them was recently published. They allow us to start addressing the question of the historical process through which the body of knowledge regarding operations as recorded in *The Nine Chapters* was obtained. The sources are more or less of the same nature as *The Nine Chapters.* Once again, in this case, only an examination of clues found in the sources, and the practices they reveal, will enable us to draw conclusions.[25] Rather unexpectedly, what these new documents show is a mathematical landscape very different from what was uncovered above. I shall now demonstrate this.

III.   THE NATURE OF THEORETICAL WORK ON DIVISION IN ANCIENT CHINA

The previous section argued that *The Nine Chapters* testifies to a specific state of knowledge with respect to a set of operations. This set includes division, and square and cube root extraction, as well as an operation-equation (i.e., a form of quadratic equation.) The purpose of the present section is to show that the indirect evidence the earliest extant texts provide about these operations before the period of *The Nine Chapters* indicates a different state of knowledge with respect to the same operations. The conclusion we shall derive is that we can get some idea—albeit very sketchy—about the time period of development of this body of knowledge. More importantly, we can suggest hypotheses about the process that shaped the complex of operations described.

Let us begin by presenting the available documentary evidence. The first manuscript found, the *Book of mathematical procedures* (筭數書 *Suanshushu*), was excavated in 1984 from a tomb at Zhangjiashan 張家山 (Jiangling county, Hubei province).[26] Peng Hao gives *ca.* 186 BCE as a *terminus ante quem* for its composition. Since 2007, two new mathematical manuscripts have been excavated. The earlier of the two books dates from the Qin period (3rd century BCE), and actors titled it *Shu* 數 (*Mathematics*). Its entire text was recently published (Xiao Can, 2011; Zhu Hanmin 朱漢民 & Chen Songchang 陳松長主編, 2011). This manuscript, the product of illegal excavation, was bought in December 2007 on the Hong Kong antiques market. However, the second mathematical book found, titled 算術 *Suanshu* (*Mathematical procedures*), was excavated at Shuihudi (Yunmeng county, Hubei province) during regular excavations. Archeologists working on this source material estimate that the book was copied at the beginning of the Han dynasty in China, before 157 BCE. They have yet to publish the text (Chemla & Ma, 2011).

The evidence from the manuscripts yielded by archeology differs in nature from what we know through texts like *The Nine Chapters,* which were handed down by the written tradition. On the one hand, this relates to the different modes of transmission and the corresponding types of changes the books underwent in these different transmission channels. On the other hand, the mathematical books from ancient China handed down by the written tradition were usually passed on with ancient commentaries attached. When these commentaries have survived they help us understand the evidence provided by the book.

In what follows, it will be useful to bear in mind the testimony of another book that was handed down and was apparently composed before *The Nine Chapters—The Gnomon of the Zhou* (*Zhou bi* 周髀).[27] Opinions differ regarding its completion date. Some scholars date *The Gnomon of the Zhou* from the 1st century BCE (Qian Baocong dates it from *ca.* 100 BCE); whereas others argue it was completed in the early 1st century CE (see Cullen, 1996). The book provides information on mathematical knowledge and practices used in the context of astronomical activity. In 656, together with *The Nine Chapters* and other writings, *The Gnomon of the Zhou* was selected as one of the books that formed the collection *Ten Mathematical Classics.* In relation to the classic status of *The Gnomon of the Zhou*, commentaries on it had been composed in the preceding centuries, and those selected for the compilation in 656 were handed down with the text from then on. We shall refer below to the commentary on this classic composed by Zhao Shuang in the 3rd century.[28]

What do these other writings show in comparison to *The Nine Chapters?*

First, and importantly, all the documents older than *The Nine Chapters,* including *The Gnomon of the Zhou,* testify to the difficulty that doing division presented in that period.

According to my interpretation of the texts, an important part of the *Book of Mathematical Procedures* as well as *Mathematics* is devoted to problems arising from division.[29] The difficulties included, first and foremost, dividing quantities expressed with respect to different systems of measurement (capacity, volume, areas, weight, and so on), including fractions of measurement units.[30] The problems apparently also covered the actual process of dividing, when actors had to divide quantities expressed with respect to systems of measurement and achieve a result of the same type.

In fact, the only mathematical topics that *The Gnomon of the Zhou* treats at length are precisely these two aspects of division.[31] Accordingly, the difficulties associated with dividing are reflected in the wealth of details *The Gnomon of the Zhou* gives about the overall execution of a process of division. The procedures for the division of quantities expressed with respect to various units of measure begin by showing how to transform dividend and divisor into abstract integers, so that the quotient yielded corresponds to a given measurement unit. They then prescribe how to break down the execution of the division into a sequence of elementary divisions, each of which yields the component of the quotient corresponding to a given order of magnitude. To do this, the successive remainders are transformed, before being divided by the same divisor, so as to obtain the related part of the quotient. Finally, collating these results yields the overall result of the division in the form of a quantity expressed with respect to a series of measurement units and orders of magnitude. In this context, *The Gnomon of the Zhou* provides key information that will enable us to derive the clues we need and to reach conclusions. Let us now consider a piece of evidence in the book in greater detail.[32]

A given length, expressed as an integral number of the measurement unit *li*, is to be divided by a number of days, which consists of an integer increased by a fraction. In the first step, both values are correlatively transformed into abstract integers yielding the same result (952,000 and 1,461), before the first division is prescribed. The first prescription is carried out as follows: ". . . makes the dividend (*shi*), . . . makes the divisor. Eliminating this (*chu zhi* 除之), (each time it is) like the divisor (*ru fa* 如法), it yields one *li* (*de yi li* 得一里)". Here, I interpret the same character *chu* that we interpreted earlier as meaning "to divide", as referring to a subtraction. Accordingly, in this context I translate *chu* as "to eliminate". In other words, I claim that between the two contexts, the meaning of this key term has changed. I shall return below to this suggestion and the overall expression prescribing division. The prescription clearly specifies the measurement unit to be associated with each unit obtained for the quotient each time a quantity equal to the divisor is subtracted. As we already saw in the previous section, the formulation evokes one of the terms *The Nine Chapters* uses to refer to a division. We shall return to this point later as well.

The key issue for us now is the following part of the execution of the division. At this point we have obtained 651 *li* and 889 remains in the dividend, which is smaller than the divisor 1,461. The procedure goes on in order to obtain the remaining part of the quotient, expressed with respect to the measurement unit for length that is smaller than the *li*, i.e., the *bu*.

A *li* is equivalent to 300 *bu*. If the practice of computation were the same as that recorded later, the remainder 889 would be multiplied by 300, and the product would then be divided by the divisor 1,461 to yield the additional number of *bu* in the quotient, 182 *bu* and 798/1,461 *bu*. This is not merely my own perception. It also conforms to the estimate of the 3rd century commentator Zhao Shuang, who repeatedly begins his commentary on the related segment of the computation with the words: "One should multiply by 300 . . .". However, in *The Gnomon of the Zhou* the subsequent part of the operation was not executed in this way, which provides us with important clues.

In contrast to Zhao Shuang's expectation, the remainder is first multiplied by 3, computing what Zhao Shuang interprets as "the *dividend of the hundreds* (*bai shi* 百實)" (my emphasis). *The Gnomon of the Zhou* then prescribes a second division of this dividend by the same divisor. The prescription runs as follows: "(each time it is) like the divisor (*ru fa* 如法), it *yields a hundred bu* (得百步 *de bai bu*)" (my emphasis). This division yields the value of the part of the sought quantity corresponding to the hundreds of *bu*. The remainder from that division is then multiplied by 10, determining what Zhao Shuang calls "the dividend of the tens". After this transformation of the dividend, *The Gnomon of the Zhou* prescribes a third division by means of the following expression: "(each time it is) like the divisor (*ru fa* 如法), it *yields ten bu* (*de shi bu* 得十步)"—my emphasis). Lastly, the same procedure is repeated to determine the units in the part of the quotient which is expressed in *bu,* and it is concluded by the following statement: "(each time it is) like the divisor (*ru fa* 如法), it yields one *bu* (*de shi bu* 得一步). The (part of the dividend) which does not fill up the divisor is named by the divisor".[33]

What clues can we derive from this method of performing the division?

First, the procedure reflects features of the number system on which it is based and about which we have no explicit information. It also casts light on features of the computing device used. As mentioned above, *The Gnomon of Zhou,* like the newly discovered manuscripts, clearly refers to computations carried out on a surface, with numbers represented by counting rods. Some of the quantities mentioned in the text of the procedure sketched above appear to have their values modified throughout the calculation. This confirms that they were written in a way that allowed these changes. Rods certainly fit this assumption. However, this preliminary conclusion is not enough for us to determine the kind of number system on which the procedures operated at the time. We should be wary of assuming anything further about the practice of calculation.

The procedure we have examined in *The Gnomon of Zhou* fully justifies this caution.[34] Although it clearly makes use of a decimal conception of the quantity in determining the amount of *bu* in the required result, the calculation method makes it unlikely that it is based on a place-value notation, as we have seen in the case of *The Nine Chapters.* Is this conclusion correct?

The question prompts us to mine our corpus of manuscripts looking for possible clues about the number system. The first important conclusion is that, as far as I can tell, in these early writings there is no positive evidence, such as we find in *The Nine Chapters* and later texts, that a place-value system was used. No procedure, for instance, seems to introduce the elementary operations of moving quantities forward or backward, in order to multiply or divide them by a power of 10.[35] By contrast, as mentioned in the previous section, these operations occur frequently in later texts, exploiting the fact that they operate on the basis of such a number system. Observing the practice of algorithms, and the operations they involve, thus gives us insight into the number system used.

If this is correct, it suggests that a major change occurred between the period of the production of the manuscripts and composition of the procedures examined in *The Gnomon of the Zhou,* and the time when *The Nine Chapters* was compiled. This conclusion is supported by another indirect piece of evidence about the practice. The manuscripts contain tables of multiplication between powers of 10, which would seem useless if the number system had the property of being place-valued (Chemla & Ma, 2011). In fact, tables of this kind do not appear in mathematical writings such as *The Nine Chapters*. As a preliminary conclusion, our first set of clues thus suggests that the number system on which such divisions were carried out was decimal, but not place-valued. Accordingly, rods may have been used on the surface in a different way from that outlined in the previous section.

The second set of clues we can derive from the procedure of *The Gnomon of the Zhou* that we have described above comes from the terms used to prescribe divisions. To begin with, let us note that the context of this procedure makes clear the origin and meaning of expressions we find in all the writings under examination—that is, "when the dividend is like the divisor, then one" or, "then it yields one" or alternatively, "then one (name of a measuring unit)" (*shi ru fa er yi* 實如法而一, *shi ru fa de yi* 實如法得一, *shi ru fa er yi* (name of a measuring unit) 實如法而一 (name of a measuring unit). The clarification comes from the fact that now these expressions occur in the same context as other expressions such as, "(each time it is) like the divisor (*ru fa* 如法), it *yields a hundred bu* (得百步 *de bai bu*)"—my emphasis—or, "(each time it is) like the divisor (*ru fa* 如法), *it yields ten bu* (*de shi bu* 得十步)"—my emphasis. All these expressions seem to reveal a focus on the meanings of—that is, the orders of magnitude of and units to be attached to—the successive parts of the result, whether they are hundreds of *bu*, tens of *bu*, or units of *bu*. This emphasis on determining the meanings of the units produced by the operation on the dividend and the divisor may indicate the difficulty caused by the fact that the operands were modified in several ways during the process of the division. Such expressions possibly also cast light on the actual procedure of division used, which is based on successive subtractions. In fact, the expressions used in the manuscripts and in *The Gnomon of the Zhou* are all similar in inspiration.[36]

Consequently, it is striking that the verb *chu* occurs in the texts, but in these early writings it never refers to division.[37] There, *chu* only means "subtraction". Incidentally, it is this observation that led me to interpret the first prescription of division in *The Gnomon of the Zhou* as follows: "Eliminating this (*chu zhi* 除之), (each time it is) like the divisor (*ru fa* 如法), it yields one *li* (*de yi li* 得一里)". In fact, this use of *chu* as "subtraction" in the statement of division occurs in the manuscripts, but never in *The Nine Chapters*. This shows the strong relationship between *The Gnomon of the Zhou* (i.e., the earliest mathematical text handed down) and the earliest known mathematical documents discovered by archeology. The conclusion echoes, and supports, the assumption that we derived above, according to which these early writings share a similar number system, different from the one to which *The Nine Chapters* refers.

We have reached a crucial point: As far as the extant evidence permits us to draw conclusions, we can see that the verb *chu*, which refers to division in *The Nine Chapters*, and constitutes the structural pivot of the set of operations that we observed in the previous section, seems to have acquired this meaning of "division" between the last century BCE and the 1st century CE.[38] More precisely, its earlier meaning of "subtraction", evidenced in the most ancient texts, gave way to that of "division". In fact, not only did meaning of the *term chu* change, but the practice of computation yields clues showing that the *process* of division also underwent a key change. Let me broadly indicate the main clue. The algorithm of "division *chu*" between integers which is evidenced in *The Nine Chapters*, and which we can definitely reconstruct on the basis of observation of mathematical practices, proceeds through a progressive diminution of the dividend and a decimal shift of the divisor. The latter feature exploits the place-value feature of the number system. By contrast, the procedure for division recorded in *The Gnomon of the Zhou*, which we have described in some detail above, does not refer to any change of the divisor, a point the commentator Zhao Shuang regularly emphasizes in his commentary on the lengthy procedures describing divisions. These procedures only modify the dividend, alternately reducing and multiplying it. This method of division may have created the necessity to clarify at each step, as the terminology prescribing the successive divisions does, the nature of the result segment obtained.

In conclusion, we can see that the two sets of documents—first, the manuscripts and *The Gnomon of the Zhou,* and second, *The Nine Chapters* and later texts—reveal a change that apparently occurred between the two respective time periods. This change seems to have involved several correlated features: the number system, the way of prescribing division by means of a verb or not, and the method of performing a division process on integers. Essential for us in relation to the topic of this chapter is that all these facts can be obtained by indirect reflection of what the practices of computation evidenced in the books show.

We shall conclude the chapter with a final set of clues. It deals with square root extraction. The manuscripts contain procedures for determining the side of a square. However, the terms referring to this operation vary.[39] In addition, none of these terms refers to the division *chu* or to an algorithm for division. Lastly, the procedures differ among themselves and are different from division. In particular, in contrast to the procedure included in *The Nine Chapters*, they do not offer any clues about reliance on a place-value number system.

A fourth correlated feature can thus be added to the set of changes listed in the previous paragraph and attested by *The Nine Chapters*. The manuscripts seem to indicate that algorithms for square root extraction were not standardized in that period. Further, they do not show evidence of any inquiry into the relationship of these procedures and division that corresponds to what is recorded in later texts. By contrast, in the state of knowledge evidenced by *The Nine Chapters*, the name of the operation, the method for doing square root extraction, and the way the text of the algorithm was written down all indicate a reshaping that brought square root extraction into close relationship with the algorithm for the division *chu* and the place-value number system.

IV. CONCLUSION: MATHEMATICAL PRACTICES AND CONCEPTUAL HISTORY

Which conclusions can be drawn from the previous analyses? I shall answer this question from two perspectives.

To begin with, let us recapitulate the conceptual changes we have revealed with respect to operations in early imperial China. In Section II, I have argued how an observation of various practices allowed us to perceive knowledge about operations in *The Nine Chapters*. For us to perceive a body of knowledge regarding operations, rather than interpreting sustained theoretical discourse, we analyzed clues drawn from our sources and interpreted them in the light of a description of related practices. In particular, we have seen how, at the time of *The Nine Chapters*, the division *chu* appears to have been a key element of, and to have played a central part for, the set of known arithmetical operations. This operation was the reference point for writing down the texts for algorithms for square and cube root extraction. The execution of these two operations on the computing instrument, and the way of prescribing them, also referred to division *chu*. In addition, a new operation, i.e., the quadratic equation, was introduced on the basis of the method for square root extraction evidenced in *The Nine Chapters*. According to the evidence derived from *The Nine Chapters*, knowledge about these four operations included knowledge about the structure of the set they formed.

These observations raised the question we addressed in Section III, of understanding which historical processes led to the shaping of this knowledge. By relying on sources produced centuries earlier, we were able to suggest that the process was not a progressive derivation, using the division *chu,* of new operations that were naturally linked to it. On the contrary, we have shown that the operations of division and square root extraction were done by other means in the period before the composition of *The Nine Chapters.* Both operations underwent a key transformation.

This transformation appears connected to a change in the number system the algorithms were applied to. Perhaps the place-value decimal number system, written down in China with counting rods, was introduced in this context. Moreover, the transformation not only reshaped the algorithms that carried out the operations but also established relationships between them, giving rise to the structure that we illuminated. In this process, the operations were made to converge towards one another. These conclusions thus strongly support the claim that operations were a topic of inquiry in ancient China—a fact that has remained unremarked so far. It means that practitioners in ancient China did not limit themselves to creating methods to perform operations. They also devoted some attention to the operations as such, and developed practices to work on them, yielding new knowledge about their set. As a result, *The Nine Chapters* testifies to a radical shift in the understanding of, and practice with, the four operations examined.[40] The shift was reflected most clearly in the new terminology that was introduced and that we have analyzed above.

I believe we have shown clearly how the observation of mathematical practices can provide new means to do conceptual history. In the present case, it simply allows us to consider a history of the theoretical dimensions of the inquiry into arithmetical operations and of the means of working with these operations. In my view, this would otherwise remain out of reach.

This remark leads me to my second set of conclusions, focusing on the part the description of practices can play in the history of science. Historiography of mathematics mainly focused in the past on results stated and theories expounded by the actors. The need to widen our interests and include the description of practices goes beyond merely describing them. It is an indispensible tool for enriching interpretation of our sources and deepening our examination of conceptual history. What I have endeavored to show by example in this chapter is *how* the observation of practices—in the broad sense illustrated above—can help us achieve such aims.

It is not by chance, I have suggested, that history of mathematics in the ancient world sheds light on how our knowledge of practices can be a tool for mining our sources more profitably. These sources are scarce, and require elaborate treatment. Reconstructing practices in relation to which sources were produced enables us to uncover and interpret clues that give indirect insight into the results known to actors (e.g., how to divide) and more generally, into the nature of actors' knowledge (e.g., a way of structuring a set of operations), when these bodies of knowledge only leave traces in the documents. Our knowledge of practices also allows us to detect relevant clues.

What does the exercise tell us about the opposition between practices and results that Léna Soler employs to determine one of the meanings of the word "practice" in present-day science studies?[41] One outcome of the argument developed in this chapter is to highlight the various ways the opposition between results or knowledge, on the one hand, and practice in this sense, on the other, may have to be rethought.[42]

To begin with, the case study analyzed in this chapter illustrates that, even in a field like mathematics, practices have their own history. We were actually able to see that in the main time periods examined above, the methods of working with operations and executing them presented differences. For instance, in *The Nine Chapters,* the layout of computations and elementary operations such as moving rods representing numbers forward and backward were obviously important features of the practice of computing. In *The Gnomon of the Zhou* and the *Book of Mathematical Procedures,* apparently no importance was given to these features in the way division was conducted. This observation about the historical dimension of practices has a direct bearing on our theme: The types of clues discussed in relation to our two sets of sources were of a completely different nature. I would argue that practices are shaped in close relation to the questions addressed, and they are shaped in the process of knowledge making. As a result, the two aspects of scientific activity are intertwined.

We can detect one reflection of that intimate connection in the fact that practices made use of knowledge about the topics examined. For instance, computation practices rely on knowledge about the number system they operate on. This is one reason why such practices can reflect these pieces of knowledge and testify to their existence.

In addition, the fact that practices and bodies of knowledge are shaped conjointly explains why results and concepts can show correlation with practices. This idea is illustrated in the present chapter by the example of the quadratic equation. We have seen how the concept of quadratic equation recorded by *The Nine Chapters* can be correlated to the practice of computation on the surface on which rods represented numbers. This differs from other concepts of quadratic equation evidenced in other sources, and we can observe the sequence of syntheses of these distinct concepts carried out at different moments of history.[43] This remark about the joint production of practices and bodies of knowledge also explains why reconstructing practices enables historians to make sense of clues.

For the sake of the analysis, one can thus distinguish between practices and the knowledge produced. Ultimately, however, one can only observe their intimate relationship. More generally, practices developed to inquire into knowledge and carry out operations, far from being disconnected from the results produced, can be considered as belonging to the knowledge produced in the framework of a knowledge activity.[44] These practices are not formed by spontaneous generation. Like the results, concepts, and theories, they are conscious products of normed activity. Knowledge about their ability to guide action and inquiry is transmitted along with the other outcomes of scientific activity. This accounts for their relative stability and their collective dimensions. These features also explain why the description of practices can assist in the interpretation of sources from the past.

NOTES

1. The research that culminated in this chapter received funding from the European Research Council under the European Union's Seventh Framework Programme (FP7/2007–2013)/ERC Grant agreement no. 269804 SAW, "Mathematical sciences in the ancient world". I would like to thank Léna Soler and the group around her in the project PractiSciens for sharing their reflections with me. I am particularly grateful to Mélissa Arneton and Amirouche Moktefi for their comments on an initial version of this chapter. I would also like to thank my colleagues in the group around the SAW project, and the participants in our seminars, for their commentaries on the research, of which this chapter is merely an initial result. Last, but not least, many thanks to Karen Margolis for sharing her thoughts with me about the formulation of this chapter.

2. Florence Bretelle-Establet (2010) constitutes an attempt to address this issue.

3. Carlo Ginzburg (2012, pp. 3–4), summarizes the gist of the method, referring the reader to the theoretical insights of Marc Bloch, *Apologie pour l'histoire, ou Métier d'historien* (1949). This posthumous book by Bloch offers reflections on the key role of traces in the approach of the historian of his time. In particular, he writes:

   > Jusque dans les témoignages les plus résolument volontaires, ce que le texte nous dit expressément a cessé aujourd'hui d'être l'objet préféré de notre attention. Nous nous attachons ordinairement avec bien plus d'ardeur à ce qu'il nous laisse entendre, sans avoir souhaité le dire. . . . Dans notre inévitable subordination envers le passé nous nous sommes donc affranchis du moins en ceci que, condamnés toujours à le connaître exclusivement par ses traces, nous parvenons toutefois à en savoir sur lui beaucoup plus long qu'il n'avait lui -même cru bon de nous en faire connaître. C'est, à bien le prendre, une grande revanche de l'intelligence sur le donné. (Bloch, 1949, p. 25)

4. In my view, the traces a text leaves and what historians can deduce from them are still questions worthy of theoretical investigation. However, I shall leave this for discussion elsewhere.

5. How we can rely on our sources to reconstruct practices is another topic I have addressed in several other publications that I shall mention below when I need to make use of their results for my argument.

6. When I speak of division in this chapter, I mean only division between integers or between integral amounts of measurement units. I have dealt with the same pieces of evidence from *The Nine Chapters* in Karine Chemla, "Changing mathematical cultures . . ." *(forthcoming)*. There, the focus was to highlight a correlation between the concepts and the mathematical practice in relation to which these concepts were developed. Moreover, this conclusion served as a basis to establish, using an argument that considered *The Nine Chapters* in the light of later texts, that concepts are not determined by the scholarly culture

to which they belong. The purpose in the present chapter is different, since I intend to illuminate actors' knowledge with respect to operations and to focus on the method with which this can be done. In the following section, I shall rely on the conclusions obtained in the present section to advance an argument regarding earlier documents.

7. I have described the evidence we have on this system of computation in Karine Chemla (1996).

8. Rods have also been found in archeological excavations of tombs sealed in the Qin and Han dynasty. However, this evidence is difficult to interpret because we cannot establish with certainty for what activities these rods were used. Reference in mathematical writings is more reliable evidence in this case.

9. I shall return to this issue in greater detail in another publication. Authors who have claimed that the instruments were the same include Christopher Cullen (2004, p. 24) and Joseph W. Dauben (2008, p. 96). Before new mathematical manuscripts were excavated from tombs, archeologists had discovered rods. Moreover, some early written documents mention ways of computing. The prevalent assumption was that the rod system stayed unchanged. See, for example, Lam (1988).

10. Incidentally, this is the expression prescribing the division used in the text of the procedure mentioned earlier for computing the volume of the half-parallelepiped.

11. For remarks on terminology, unless otherwise stated I refer the reader to the glossary I published in Karine Chemla and Guo Shuchun, *Les neuf chapitres. Le Classique mathématique de la Chine ancienne et ses commentaires* (2004). In the glossary, I discuss the terms used to prescribe operations and give information on the syntax of the sentences employing these terms. Moreover, I refer to evidence supporting my conclusions. For the last set of expressions I shall return below to their original meaning on the basis of evidence provided in the next section. Li 李繼閔 (1998, pp. 144–148) discusses the terminology of division in *The Nine Chapters*.

12. We shall see below that division has cognate operations and that they are *only* prescribed by means of the verb *chu*, to which qualifications are added. In what follows, when I speak of "division" in the context of *The Nine Chapters*, I usually mean division and its cognate operations.

13. This description is not entirely correct. Sometimes, for instance, only the dividend is explicitly designated, and the expression prescribing division has a complement referring to the value or the magnitude taken as divisor. A broad description is enough for our purpose here.

14. For a critical edition, see (Qian 錢寶琮, 1963, vol. 2, pp. 282–283). Qian Baocong argues that the book was composed around 400 CE, but stresses that the received version displays hints of later changes in the Tang period. Lam Lay Yong and Ang Tian Se (2004, pp. 194–195) provide a rough translation. The algorithm is also explained in (Chemla & Guo Shuchun, 2004, pp. 16–19). I discuss the key features and terms of these texts in Chemla (1996).

15. As I suggested in the glossary published in Chemla and Guo Shuchun (2004, p. 945), the term *kai* is probably a synonym of the term *qi* 啟 "to open, to detach", which it replaced at some juncture, probably in the mid-1st century BCE. However, one can note a shift, which is interesting in terms of the overall transformation described in this chapter: *qi* is a verb whose complement is the object to be detached (as in "detach the side of the square"), whereas *kai* takes as its complement the area for which the side is sought ("open the area of the square"). As discussed below, it is precisely the latter area that, in *The Nine Chapters*, is the operand of the operation designated as "dividend", *shi* 實.

16. See Chapter D in Chemla and Guo Shuchun (2004, especially pp. 99–116), which deals with this question.

17. In Chemla and Guo Shuchun (2004, pp. 19–20, 322–335, 362–368, 370–377 and related footnotes), I give all the necessary details on these algorithms and the interpretation of the texts. I refer the reader to this publication for a bibliography on the topic, limiting myself here to the issues specific to this chapter.

18. Chemla (1989) describes this practice. It also contains a translation of the texts for root extraction included in *The Nine Chapters*. An English translation can be found in the appendices in Chemla (1994b). I refer the reader to these translations to facilitate following the argument.

19. In fact, the *Mathematical Classic by Master Sun* contains texts for procedures for multiplications, divisions, and square root extractions. It describes explicitly the positions and the basic operations required to execute these arithmetical operations. Clearly, during the process of doing a root extraction as described in the *Mathematical Classic by Master Sun*, the positions named "dividend", "quotient", and "divisor" undergo transformations that can be correlated with those that positions with the same names undergo in the process of doing a division. The same conclusion holds true for *The Nine Chapters*, even though the actual correlation exhibited changes. In other words, the practice is the same, even though there is a difference in what is expressed. In Chemla (1993), I interpreted an assertion by Li Chunfeng in his commentary on *The Nine Chapters* as referring precisely to both this commonality of practice and the difference in meaning expressed. In this context I would like to emphasize that Li Chunfeng's commentary deals more specifically with the practice of naming positions (位 *wei*). This point will prove meaningful below. What is also important here is that in the *Mathematical Classic by Master Sun*, we encounter the same phenomenon for the processes of multiplication and division, except that in these cases the relationship between the processes executing the operations on the computing instrument is that of an opposition, not a correlation. However, in this case, the *Mathematical Classic by Master Sun* inserts one of those extremely rare second-order comments that reveal actors' awareness of the practice. The beginning of the text for division reads: "The method for any division is *exactly opposed* to that of multiplication", (Qian, 1963, vol. 2, p. 282, my emphasis). This statement can be read as referring to the result of the practice of writing down computation processes on the computing instrument. To recapitulate what we have observed so far, *The Nine Chapters* and the *Mathematical Classic by Master Sun* display several common features. First, they share the same algorithm for the operation of division, designated in both cases by the term *chu*. Second, they both attest to the same practices of expressing the relationship between operations using texts of algorithms that execute them as well as a material inscription of the processes on the computing instrument. Let me repeat that, despite these continuities of practice, the relationships expressed differ.

20. (Li Jimin 李繼閔, 1990, pp. 112–114). The text and an interpretation can be found in (Chemla & Guo Shuchun, 2004, pp. 671–672, 698–699, 734–735, and related footnotes).

21. I have discussed the identity of various types of algebraic equations in the ancient world in (Chemla, 1994a). Moreover, I have devoted another article to the discussion of this concept of equation and its subsequent history in China: "Changing Mathematical Cultures, Conceptual History and the Circulation of Knowledge" (Chemla, *Forthcoming*). These two publications offer a more detailed discussion of the interpretation of the text, especially its

mathematical dimensions. Here I am focusing on the knowledge of the structure of a set of operations as evidenced in *The Nine Chapters*.

22. At that time in China, an equation of this kind was thought to have a single root. Such details, however, are not relevant to my present aim.

23. To avoid distraction, I shall not comment on the terminological differences between "fixed divisor", "divisor", and "joined divisor".

24. Chemla (Forthcoming) discusses how concretely the operation was detached. In that article I emphasize *how*, in the succeeding centuries, knowledge about algebraic equations in China appears to have been sought within the same conceptual framework.

25. In what follows, for simplicity and to concentrate on the topic of the chapter, I state results I have obtained without presenting the underlying arguments in full. I shall provide these in other publications.

26. A first critical edition with annotations was published in Peng Hao 彭浩 (2001). It was first translated into Japanese in Jochi Shigeru 城地茂 (2001). We have already referred to the two translations into English that have appeared: Cullen (2004) and Dauben (2008). A critical edition and new translations into Japanese and Chinese have also appeared: (Chôka zan kankan Sansûsho kenkyûkai. 張家山漢簡『算數書』研究会編 Research group on the Han bamboo strips from Zhangjiashan *Book of Mathematical Procedures*, 2006).

27. Qian Baocong 錢寶琮 (1963) contains a critical edition of the book. It was the basis for the translation into English published in Cullen (1996). In what follows, I shall rely on the same critical edition.

28. The anthology provided the editions of books that were used as textbooks to teach mathematics in the context of the College of Mathematics (*Suan xue* 算學), and prepare candidates for the newly established state examinations on this subject. Compare Man-Keung Siu and Alexei Volkov (1999).

29. Another reflection of this could be the importance given to the expression and treatment of proportions which, according to Peng Hao 彭浩 (2001, pp. 17–19), represent one half of the *Book of Mathematical Procedures*.

30. Some of the procedures created to address this issue are of the type discussed in Chemla (2006). Historians' publications on the manuscripts since they were released have so far underestimated the importance of these procedures. I shall return to them more systematically elsewhere. The difficulty posed by division is also reflected in the number of procedures listed for dividing between fractions or integers increased by fractions.

31. See, for instance, Qian Baocong 錢寶琮 (1963, p. 52, 61).

32. I am relying on Qian Baocong 錢寶琮 (1963, vol. 1, p. 52). The same comments apply to the passage on p. 61 referred to above.

33. The last sentence is the usual prescription when the remainder of the dividend is taken as the numerator of a fraction whose corresponding denominator is the divisor.

34. Note that a procedure containing steps similar to those we focus on in *The Gnomon of the Zhou* in the context of a decimal system of measuring units is evidenced in the *Book of Mathematical Procedures*, slip 42, (Peng Hao 彭浩, 2001, p. 56).

35. For instance, the procedure of *The Gnomon of the Zhou* prescribes multiplying by 10 twice, to yield first what Zhao Shuang interprets as "the dividend of the tens" and then "the dividend" that corresponds to the units. In neither case is the multiplication prescribed as a displacement of the rods. Exactly the same feature characterizes the procedure in the *Book of Mathematical Procedures* mentioned in the preceding footnote.

36. Xiao Can 肖燦 (2011, pp. 121–124) gives an overview of the expressions referring to division in the manuscripts. Guo Shuchun 郭書春 (2002, pp. 525–527) surveys the expression of division in the *Book of Mathematical Procedures*. In what follows, the statement on the expression of division in *The Gnomon of the Zhou* and the other claims made on the book are my own conclusions. They require a philological argument that I shall develop elsewhere at the time when I present a full analysis of the expressions for division in early Chinese mathematical texts.

37. A passage from *The Gnomon of the Zhou* seems to contradict this assertion (Qian, 1963, vol. 1, pp. 77–79). It is thus interesting to read Zhao Shuang's view on this passage "非周髀本文。蓋人問師之辭。其欲知度之所分，法術之所生。 This is not from the original text of *The Gnomon of the Zhou*. Probably these are the words of someone asking a teacher. He wants to know the division of the *du* and that from which procedures originate". We shall return to the use of the term *chu* in the various early mathematical texts in another publication.

38. The use of *chu* "subtraction" occurs in *The Nine Chapters*, although rarely. For instance, we encounter it in the procedures attached to problems 3.17 and 6.16. In the former case, the use of the term "there remains" immediately afterwards makes the sense of *chu* clear. However, in the latter case, the gloss attributed to the 3rd-century commentator Liu Hui makes the meaning of *chu* as "subtraction" explicit, as if he saw it as an archaism that could cause problems for the readers. By contrast, the commentator does not comment on any occurrence of *chu* as division. This indicates that in the other occurrences the commentator interprets *chu* as referring to division. Moreover, it shows that the commentator did not see division as a subtraction to be repeated. The different uses of the same term are one of many indications that *The Nine Chapters* was produced by compilation, a topic I shall return to in a future publication. In addition, I would argue that among the extant mathematical documents to date, the meaning of *chu* as "division" only occurs from *The Nine Chapters* onwards. Finally, the fact that commentators such as Liu Hui seem to regard *chu* "subtraction" as an archaism is an important sign supporting the idea that, compared to other earlier sources, *The Nine Chapters* testifies to a change in the understanding of the set of operations. More generally, mathematical sources from ancient China do not record wholly independent traditions. This assertion does not mean, however, that the change shown in *The Nine Chapters* that we discuss in the present chapter obliterates earlier bodies of knowledge.

39. See, for instance, the procedure presented in slips 185–186, entitled "Squaring a field (or finding the side of the square for a field) *fangtian* 方田", of the *Book of mathematical procedures* (Peng Hao 彭浩, 2001, pp. 124–125). Peng Hao notes (fn 1) that the algorithm differs from that of *The Nine Chapters*. Cullen (2004, p. 88) suggests that perhaps the latter procedure was not yet discovered at that time.

40. It can also be shown that during approximately the same time span the reflection about division yielded other theoretical insights (Chemla, 2006). The use of terms like "divide in turn *baochu* 報除" or the theoretical role devoted to the division *chu* more generally, for instance in relation to the algorithm "Measures in square" (*fangcheng* 方程), testify to yet other developments.

41. I refer to the meaning 1 identified in Léna Soler (2012c).

42. The aim of the conference "From practice to results in logic and mathematics", organized by the research group PratiScienS, led by Léna Soler in Nancy, 21–23 June, 2010, was to contribute to this program. The same issue

is addressed in Chapter IX of Léna Soler (2009). This chapter, especially pp. 300–302, includes a discussion of theses close to those I present here. My analysis here focuses on the issue of clues.

43. I outline this thesis in Chemla (1992).
44. In fact, like the concepts and results on which we concentrated in this chapter, ancient practices can be restored mainly through an examination of clues. I have not addressed the issue in this chapter to avoid overloading the argument. I refer the reader to my other publications on these topics.

The Interplay Between Mathematical Practices and Results

Commentary on "Observing Mathematical Practices as a Key to Mining Our Sources and Conducting Conceptual History: Division in Ancient China as a Case Study", by Karine Chemla

Mélissa Arneton, Amirouche Moktefi, and Catherine Allamel-Raffin

This comment addresses issues regarding the interplay between practices and results in Karine Chemla's contribution to this volume. Chemla explains how a historian of ancient mathematics can get further information by restoring practices and deriving results that are indirectly evidenced in the rare surviving sources that the historian has. It is noteworthy that Chemla discusses directly the relation between practices and results, where results might be understood as the body of knowledge that one can gather in a scientific outcome (e.g., a publication), while practices are the activities and conceptions that led to those results. This issue is still insufficiently addressed in the recent rising of studies on mathematical practices, as far as we can tell from many writings that we consulted and meetings that we attended (Giardino, Moktefi, Mols, & Van Bendegem, 2012).

Chemla argues that "reconstructing mathematical practices involved in producing our sources yields key resources for inquiring into issues of conceptual history that our sources do not tackle in detail" (p. 239). In order to defend this thesis, Chemla appeals to a case study from an ancient mathematical culture—namely, ancient China. We are told that the practitioners of mathematics in early imperial China worked assiduously on procedures (i.e., algorithms, in modern terminology) such as how to compute the volume of a half-parallelepiped. Such procedures make use of blocks that stand for operations (e.g., division, square root extraction, etc.). Executing these operations requires a procedure in itself. However, not all the documents that we have do include such procedures on how to execute an operation. Moreover, even when the documents contain such procedures, this does not make them "treatises about operations" (p. 247). Most conceptual work on operations can be gathered only indirectly. In her paper, Chemla shows how studying practices helps to reveal *hidden* knowledge about these operations.

To show this, Chemla relies mostly on a classic Chinese book: *The Nine Chapters on Mathematical Procedures*. The corresponding work was probably composed in the 1st century CE, but circulated and survived together with two commentaries composed subsequently. The book refers to a computing instrument used to perform procedures, probably by placing and moving counting rods on a simple surface. Among the various operations involved in *The Nine Chapters,* division plays a central role. Unlike other operations (addition, subtraction, and multiplication), division is clearly understood as an operation because technical terms are introduced for its operands. Examining how operations are prescribed provides additional clues about operations. For instance, root extractions are frequently prescribed as follows: "Divide this by extraction of the square/cube root". This practice shows a link between the operation of root extraction and division, and attests that division is more fundamental. Chemla also identifies a new operation which corresponds—in modern terminology—to solving a quadratic equation. Here again, the way this operation is prescribed shows that it derives from root extraction. Altogether, these examples show how the understanding of naming practices is used effectively by historians to acquire knowledge that was only indirectly evidenced in the way *The Nine Chapters* presented the operations.

Similarly, Chemla worked with other clues to restore more "practices of naming, practices of writing down texts of procedures, and practices of computing" (p. 240). For instance, thanks to the study of computing and executing procedures, combined with clues she extracted from a later text—*The Mathematical Classic by Master Sun,* completed in 400 CE—Chemla argues that the number system used in *The Nine Chapters* was a place-valued decimal system, a fact that is not directly recorded in the surviving text. Hence, understanding practices enables one to improved interpretation of existing clues, to detect additional clues and, consequently, to derive knowledge that was "evidently possessed" by the authors of *The Nine Chapters* (p. 253).

I.   RESTORING PRACTICES AND DERIVING RESULTS

In her case study, Chemla is in the position of a historian who handles a precious text from an ancient mathematical culture. Since such textual sources are rare, it is crucial to derive from them the greatest possible amount of information. For this purpose, Chemla implements a method that can be carried out in two steps:

Step 1: The historian has sources which directly report mathematical results. She collects and interprets clues in the sources to restore the practices of the mathematicians who produced those results.

Step 2: The historian has identified mathematical practices. She uses them to collect and interpret further clues in order to derive new results that were not directly evidenced in the sources.

Chemla uses this method as described in the first section of this commentary. The reader is invited to examine her paper carefully to observe the numerous clues she collected, and to grasp how she uses them efficiently to derive knowledge that was not explicitly stated in her sources. For the sake of brevity, let us consider a hypothetical simple case to illustrate the whole process. Suppose that you are examining an ancient manuscript from some mysterious mathematical culture.

Step 1: You face some formulae where an operation noted " * " is used, but you do not know what it operates yet. All you know is that when this operation is applied to 2 and 3, it gives 5, and when applied to 4 and 5, it gives 9. These clues highly suggest that operation " * " stands for addition. As such, you have restored the notational practice of representing the operation of addition with symbol " * ".

Step 2: Now, you use this information to better understand the text and interpret other passages. For instance, if you observe that the manuscript substitutes "2*3" for "3*2" and "4*5" for "5*4", you might reasonably assume that the author of the manuscript knew that addition was commutative. Also, if you find that "2§3" (§ being an unknown operation) is replaced by "2*2*2", then it is likely that symbol "§" stands for multiplication and that the relation of this operation to addition was known to the author of the manuscript.

This hypothetical example shows how a historian works on restoring practices used by the mathematicians of the past (notations, substitutions, etc.) and how these practices can be used with benefit to provide more clues and improve the interpretation of the very same text from which those practices were extracted. It might be noted that this process involves some circularity, as shown in the diagram above. Indeed, one first restores practices from the clues in the source. Then, one re-reads the very same source in order to derive more results. Actually, one must keep in mind that the process does not stop after one round. Indeed, when the historian has restored some practices and derived new results, she continues to look for other practices and to derive further results. The circularity, however, involves no fallacy, since the restored practices are not just used to interpret the clues that first suggested them. In a way, the processes of restoring practices and deriving results complement each other the same way cryptology and cryptanalysis do. The more one understands encryption practices, the more one is likely to decrypt. And the more keys one gets in decoding a message, the more likely it is that one will get further keys. Every crossword or Sudoku player is familiar with the interplay just described, and also knows how its circularity makes any misjudgment at some stage of the interplay affect subsequent stages. Consequently, one has to pay attention to the constraints that are inherent to each step of the historiographical method described above.

The possibility of restoring practices is not disputable in itself. Historians do that all the time. An important lesson of the practice turn in philosophy and sociology of science, however, is precisely the recognition that the final output of research alone does not suffice to understand the activity that led to that output; this is why the study of science-in-the-making is so interesting. To get around the difficulty of reconstructing activities, historians usually exploit several intermediary and related sources (notebooks, drafts, private correspondence, diaries, etc.). When one works on an ancient mathematical culture with only few surviving sources, however, the restoration of activities is both more difficult and more crucial to achieve. Such are the "working conditions of the historian of ancient mathematics" (p. 238). Once the historian has restored a few practices, she attempts to derive from them results that were not directly stated in the sources at hand. Logical analysis helps at this stage, but is insufficient. Mathematical practitioners might not have looked for further results or might have reached erroneous ones. All we can say is that "Since they operated according to *this* practice, they might have reached *that* result". This situation strengthens Chemla's plea for the study of practices in order to get clues that enable the historian to go further, in her reconstruction of past knowledge, than what is permitted by logical analysis alone.

The constraints that we discussed above make us wonder what chances a historian of ancient mathematics has to *accurately* restore implicit practices and derive hidden results from the sources. In this respect, the historian might be regarded as a player who assembles a jigsaw puzzle, without having all the pieces and with no certainty about the final picture. In such a situation, accuracy refers merely to the sense the historian makes of her pieces. With the clues she has, she tries to make the best possible idea of what that final picture could be. There might be disputes about it and about the place of different pieces. But since there is no other way to come to know how the pieces *should* fit together, all the historian is doing is finding the most meaningful reconstruction. In case a new source is discovered and suggests that the present state of the puzzle is inaccurate, the historian reworks it in order to get again the most plausible picture. This analogy should not make the reader think that no historical work is trustworthy and reliable. In many cases, we have enough pieces to restore a satisfactory picture.

II.   LOCAL PRACTICES *VERSUS* UNIVERSAL RESULTS

In the third section of her paper, Chemla identifies two sets of manuscripts. The first set contains several early works prior to *The Nine Chapters*, notably a mathematical classic, *The Gnomon of the Zhou*, and several other works that have been recently discovered and that were mostly produced in the third and second centuries BCE. The second set is formed by the *Nine Chapters* and subsequent works such as *The Mathematical Classic by Master Sun* to which we alluded in the first section of this commentary. Her comparison of the two sets "reveal[s] a change that apparently occurred between the two respective time periods. This change seems to have involved several correlated features: the number system, the way of prescribing division by means of a verb or not, and the method of performing a division process on integers" (p. 259). Though not directly stated by Chemla, this argument shows that she considers the two sets of work to belong to the same mathematical tradition. As such, substantial differences between the two sets suggest a conceptual change in the meantime.

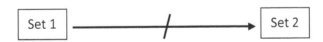

It might be noted that an alternative reading is possible: The two sets belong to distinct mathematical traditions. Given that we have no late manuscript of the first-set-tradition and no early manuscript of the second-set-tradition, we could not tell whether any change ever occurred.

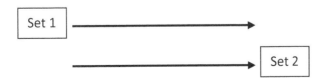

Since Chemla did not opt for this second reading, we conclude that she considers the two sets to belong to the same (and single) mathematical culture. Chemla does not make explicit what this mathematical culture would be; she just refers to "Chinese mathematics". But this leaves us with questions about how a conceptual tradition can be identified within a cultural area. In this respect, Chemla's paper offers further inspiration on an issue that did not receive the attention it deserves, namely the role of cultural elements in mathematical practices. The case studied here concerns mathematics of a cultural area other than the western (Greek-based) tradition, which is usually explored in practice-turn scholarship (e.g., Schatzki, 2001). Paying attention to different cultures provides an opportunity to discuss the universality of scientific results in spite of the local and social roots of the practices that led to those results.

It is common within practice-based studies to insist on the social context of scientific practices, including the mathematical ones (see Van Bendegem's contribution to this volume). The educational setting, in particular, is an obvious instance of a context that should not be ignored. For instance, the scientific practices developed in a given educational context at a given historical time are linked to the choices of didactic formats (see, for instance, Andrea Woody's contribution to this volume on the emergence of the chemistry periodic table), and also to the social conditions in which pupils learn (Norenzayan, Smith, Kim, & Nisbett, 2002). Hence, the prescriptions and practices of *The Nine Chapters* would not be adequately understood by a historian who would miss the fact that this work was used for educational purposes. Although in the present text, Chemla does not give extensive information on the cultural and educational contexts, she has provided plenty of elements on this same topic in previous writings (Chemla, 2011).

Introducing the cultural dimension might prove valuable for the study of scientific practices, if we consider that practices are a behavior socially shared by a group whereas culture stands for the representations which orient the collective cognition of the group (e.g., DiMaggio, 1997; Kozma, 2003). Some authors deny, however, the usefulness of culture as a supra level of explanation of values and discourse systems, because certain practices "are anchoring practices which play a key role in reproducing larger systems of discourse and practice" (Swiddler, 2001, p. 99). Accordingly, the study of practices alone would be sufficient. In her contribution to the present book, Chemla does not explore such kinds of issues, but elsewhere, she argued that "scientific activity feeds on the cultures where it occurs" (Chemla, 1998, p. 74). In her present paper, Chemla insists on the historicity of mathematical practices, which, thus, should be considered within their historical, social, and cultural contexts.

This local shaping of practices contrasts with the (assumed) universality of mathematical knowledge, as considered, for instance, by cognitive psychologists (Dehaene, Molko, Cohen, & Wilson, 2004), even if individual differences are observed (Reuchlin, 1978). For instance, one might consider the opposition between a holistic world conception in East Asia and an analytic western tradition (Dasen, 2004; Nisbett, Peng, Choi, & Norenzoyan, 2001), and ask whether this opposition influences the cognitive understanding of mathematics. More generally, the question is: How would singular, culturally located practices lead to universal, culture-free results? In the present paper and many others, Chemla stresses that regarding this question, the challenge is "to satisfy two requirements: showing how a scientific practice is singular and still produces universal results, some of which do circulate all over the world and contribute to the shaping of modern knowledge" (Chemla, 1998, p. 75).

Although Chemla does not provide a direct and definitive solution to this challenge, she proposes subtle methodological remarks in her paper that can be taken to offer some preliminary answers. A first step in this direction would be not to consider the distinction of practices and results to be strict. Indeed, this distinction depends to a large extent on one's perspectives and interests. Making use of procedures is in itself a practice; accordingly, one might consider each procedure to be a result of such practice. If one is interested in one specific problem that can be solved using different procedures, however, then each procedure might be considered as a different practice. Chemla considers practices to be part of the body of knowledge possessed by scientists, and, as such, practices can be seen as results that have been shaped and produced according to a given historical process. Even if the distinction between practices and results is valuable for the sake of analysis, one should keep in mind their "intimate relationship" and the importance of considering them jointly (p. 262).

A second step in getting to grips with the tension between the locality of practices and the universality of the results of those practices is to question what is meant by locality and universality. On the one hand, if locality refers to the shaping of practices within singular conditions, then one has to admit that results are local, too, because they are produced within the same local conditions. On the other hand, the claim that results are universal might be understood as the claim that results are culture-free and thus belong to an international mathematical culture (Chemla, 1998, p. 88). Nothing prevents us, however, from considering practices, too, to be universal in the same sense—that is, to also consider them as part of the international mathematical culture. For instance, even if we consider Chinese mathematics to be a Chinese way to do mathematics, there is no need to be Chinese in order to do Chinese mathematics. Local (cultural) factors determine only what part of that international mathematical culture will be locally used or favored (e.g., the choice of a specific number system). Considered in this perspective, local (cultural) factors do not harm the universality of that international culture and its components, be they practices, results, or practice-result connections. Studying mathematical practices from different cultural areas certainly contributes to a better understanding of the fascinating and still open problem of the tension between the locality of mathematical practices and the universality of their results.

ACKNOWLEDGMENTS

This commentary benefited from discussions with the members of the Prati-ScienS group. We would like to acknowledge our debt to them here, and also thank Karine Chemla for her contribution to this volume and her insightful answers to our inquiries.

9   Scientific Practice and the Scientific Image

Joseph Rouse

> In some remote corner of the universe, poured out and glittering in innumerable solar systems, there once was a star on which clever animals invented knowledge. That was the haughtiest and most mendacious minute of "world history"—yet only a minute. After nature had drawn a few breaths the star grew cold, and the clever animals had to die. One might invent such a fable and still not have illustrated sufficiently how wretched, how shadowy and flighty, how aimless and arbitrary, the human intellect appears in nature. There have been eternities when it did not exist; and when it is done for again, nothing will have happened.
>
> —Friedrich Nietzsche, "On Truth and Lie in an Extra-Moral Sense", 1873

This passage from Nietzsche's early manuscripts both draws upon and ironically denigrates our capacities for scientific understanding. The tension underlying Nietzsche's fable and commentary has been taken up in a different way in an influential essay by Wilfrid Sellars (2007). Sellars framed one of the central issues in 20th-century philosophy as the relation between two influential "images" or conceptions of ourselves and our place in the universe. The "manifest image" understands us as *persons,* rational, sentient agents accountable to norms: "to think is to be able to measure one's thoughts by standards of correctness, of relevance, of evidence" (Sellars 2007, p. 374). The "scientific image" is the composite formed from the explanatory theoretical postulations of the sciences: a systematic, scientific representation of nature that explains its manifold appearances. For philosophical naturalists, the scientific image takes priority: as Sellars put it, "in the dimension of describing and explaining the world, science is the measure of all things, of what is that it is, and of what is not that it is not" (1997, p. 83). Critics of naturalism respond that scientific claims are only authoritative for us because they answer to rational norms of understanding and justification. Sellars sought to do justice to the comprehensiveness and apparent autonomy of both images, combining them in what he called a "stereoscopically unified" vision of ourselves in the world.

I am sympathetic to Sellars' broadly naturalistic approach to such a ste-reoscopic vision of ourselves in the world, but further work is needed to make it defensible. In a forthcoming book, *Articulating the World,* I argue that a revised conception of both images is needed to reconcile our scien-tific understanding of ourselves as natural beings with our philosophical sense of ourselves as answerable to conceptual norms. Here I develop an overview of how this larger project requires revision to our understanding of "the scientific image". Although Sellars intended the phrase to apply pri-marily to how the sciences understand the natural world, it also embodied a conception of scientific understanding. Implicit in his conception of "the scientific image" was a repudiation of the then-predominant empiricist-foundationalist conception of scientific knowledge. In its place, Sellars offered a conception of "[science's sophisticated extension of] empirical knowledge as rational, not because it has a foundation, but because it is a self-correcting enterprise" (1997, p. 79). Yet Sellars' conception of the sci-entific image is still a conception of a body of scientific knowledge. We need to replace that conception in turn with one that emphasizes understanding scientific practice.

How would an emphasis upon science as practice revise the scientific image, in its dual sense of a scientific understanding of the world and a con-ception of scientific understanding? Many advocates of a "practice turn" in philosophy deny that the sciences produce, or even aim to produce, a single, unified conception of the world. Scientific understanding is supposedly a disunified patchwork that need not even aspire to the ideal of the "Perfect Model Model" (Teller, 2001) implicit in Sellars's vision. Scientific disunity has itself been conceived in various ways: as theoretical understanding embod-ied in diverse models whose cross-classifications are useful and informative for some purposes and not others (Teller, 2001; Giere, 1988, 2006a); in laws of limited scope whose gerrymandered domains are circumscribed by where they admit of even approximately accurate models (Cartwright, 1999); in the mutual adjustment of highly specialized theory and instrumentation such that "the several systematic and topical theories that we retain . . . are true to different phenomena and different data domains" (Hacking, 1992a, p. 57); in the collective inferential stability of different sets of laws across different ranges of counterfactual perturbation, corresponding to dif-ferent scientific domains governed by different interests (Lange, 2000); or as a recognition of the metaphysical "disorder" of things (Dupre, 1993), a dis-order perhaps most radically expressed in the speculative vision of scientific understanding as like "a Borgesian library, each book of which is as brief as possible, yet each book of which is inconsistent with every other [such that] for every book, there is some humanly accessible bit of Nature such that that book, and no other, makes possible the comprehension, prediction and influencing of what is going on" (Hacking, 1983, p. 219).

Advocates of the disunity of science are onto something important, but it is not sufficient to see the import of the practice turn as primarily a rejection of the unity of science. For one thing, Sellars himself was attentive to the plurality of scientific theories, disciplines, and domains, and thought it could be accommodated within his account of the scientific image, understood as an idealization drawn from a more complex and messy practice:

> Diversity of this kind is compatible with intrinsic "identity" of the theoretical entities themselves, that is, with saying [for example] that biochemical compounds are "identical" with patterns of sub-atomic particles. For to make this "identification" is simply to say that *two* theoretical structures, each with its own connection to the perceptible world, could be replaced by *one* theoretical framework connected at *two levels of complexity* via different instruments and procedures to the world as perceived. (2007, p. 389)

Yet I think there is a more fundamental issue underlying Sellars's disagreements with the advocates of scientific disunity. Sellars shares with the disunifiers a conception of scientific understanding as representing the world, whether or not these various representations can be unified into a single, idealized, systematic "image". Scientific understanding is supposedly embodied in scientific knowledge. Whether that knowledge primarily takes propositional form, or is realized to a substantial extent through mathematical, material, visual, or computational models, scientific understanding is mediated in whole or part by a representational simulacrum of the world it seeks to understand.[1] This representationalist vision of scientific understanding has been especially influential among naturalists in the philosophy of mind and language, who have often taken mental or linguistic representation as the key to accommodating scientific understanding itself within a scientific conception of the world (Dennett, 1987; Dretske, 1981; Fodor, 1979; Millikan, 1984). If scientific understanding is representational, then a naturalistic account of mental or linguistic representation might suffice at a single stroke to incorporate scientific understanding of the world within the world so understood. I think this aspiration is misplaced, and if so, we need to look elsewhere to grasp what a scientific understanding of the world could amount to. A central part of the difficulty is the aspiration to identify the shape or form of scientific knowledge as a whole, whether conceived as a systematic, unified "image" or as a less systematically integrated set of partial representations for different purposes. Indeed, I shall argue, a more adequate account of scientific understanding must do justice *both* to its disunified practices and achievements, and to the ways in which those divergences nevertheless remain mutually accountable within an interconnected discursive practice.

Sellars himself offers a key formulation that points toward the possibility of a naturalistic alternative to representationalist conceptions of scientific understanding. In a justly famous passage from "Empiricism and the Philosophy of Mind", he argued that "in characterizing an episode or a state as that of knowing, we are not giving an empirical description of that episode or state; we are placing it in the logical space of reasons, of justifying and being able to justify what one says" (Sellars, 1997, p. 76). Representationalist conceptions have identified scientific understanding with some specific position or set of positions within the space of reasons, that is, as a body of knowledge that represents the world or aspects of it in particular ways. I respond that scientific understanding should instead be conceived as the constitution and reconfiguration of the entire space. The sciences change the terms and inferential relations through which we understand the world, which aspects of the world are salient and significant within that understanding, and how those aspects of the world matter to that overall understanding. Scientific research also brings aspects of the world into the space of reasons by articulating them conceptually, so as to allow them to be discussed, understood, recognized, and responded to in ways that are open to reasoned assessment.

The sciences thus mark a continuation and expansion of a characteristic feature of human life more generally. We human beings are organisms whose way of life configures our surroundings as an environment with which we interact perceptually and practically. Yet our environment is the outcome of both intensive and extensive niche construction (Odling-Smee, Laland, & Feldman, 2003), with linguistic performances as salient features of our inherited environment.[2] Language functions within a broader pattern of material and social interaction, in which our way of life and the environment it discloses are themselves at issue for us. Conceptually contentful engagement with the world only emerges in this context. Specifically linguistic performances thus play a central role in conceptual articulation, but only as integral components of a larger field of practical involvement.

This approach emphasizes that the sciences only emerge historically within an already well-established pattern of discursive and material niche construction, and need to be understood as extensions of that pattern. Scientific inquiry takes place within the pervasive setting of human life as a conceptually articulated understanding and engagement with the world. It is a commonplace to think of the sciences' contribution as the attainment and growth of knowledge, which also discredits false belief and undercuts the acceptance of mysterious or incomprehensible powers. That commonplace lies behind a Sellarsian conception of the scientific image, and suitably qualified, that commonplace thought is surely correct. Yet a more fundamental achievement underlies the acquisition and refinement of knowledge through scientific inquiry. Scientific knowledge is possible only because of ongoing practical work within the sciences and a broadly scientific culture,[3] which enables relevant aspects of the world to show up at all within the space of conceptually articulated understanding. Through scientific inquiry, human beings are able to talk about, act upon, recognize, and reason about aspects of the world that had previously been inaccessible. Similarly, the sciences show how to understand interrelations among aspects of the world that were previously disconnected. Of course, these efforts also close off or render dubious other conceptual relations and even whole domains that once seemed accessible and intelligible. The critical dimension of science is not merely a matter of falsifying claims, but also of reconfiguring the conceptual space within which they are intelligible.

Being able to say what others cannot, and to talk about things not within their ken, is not just a matter of learning new words; it requires being able to *tell* what you are talking about with those words.[4] As John Haugeland once noted,

> Telling [in the sense of telling what something is, telling things apart, or telling the differences between them] can often be expressed in words, but is not in itself essentially verbal. . . . People can tell things for which they have no words, including things that are hard to tell. (1998, p. 313)

The sciences allow us to talk about an extraordinary range of things, by enabling us to tell about them, and tell them apart. To pick a small sample of exemplary cases, people can now tell and talk about mitochondria, the pre-Cambrian Era, subatomic particles, tectonic plates, retroviruses, spiral galaxies, and chemical kinetics. One need not go back very far historically to find not error, but silence, on these and so many more scientific topics.

Despite widespread assumptions to the contrary, the conceptual articulation that enables such discursive achievements is not merely intralinguistic. The normative space of conceptually articulated understanding is not limited to a logical space of intralinguistic inferences that only engages the world at occasional observational and practical interfaces.[5] Practical skills, perceptual discriminations, material transformations of the world, and a socially articulated way of life (including scientific life) integrally contribute to opening and sustaining possibilities for conceptual understanding. In this respect, scientific understanding resembles the larger process of discursive niche construction to which it contributes. Yet in many ways, scientific concepts are especially intensively interdependent with practical skills, equipment, and the creation of new material arrangements in specially prepared work sites. Even though scientific concepts and theories aim to provide understanding of the world as we find it, their proximate application is typically to the world as we make it in the specialized setting of laboratories or field practices. The point goes far beyond the commonplace that the sciences conduct experiments to assess the accuracy of their claims. Experimental work in the sciences normally involves constructing, maintaining, and revising whole "experimental systems", in which simplified, purified, or constructed elements are brought together in regimented settings that enable their interactions to manifest themselves in clear patterns that enable a domain of objects and events to be articulated conceptually. These experimental systems constitute a kind of "microworld", within which the relevant concepts acquire exemplary uses and normative governance.[6] It is in this respect that the making of laboratories, experimental systems, and conceptually articulated domains of scientific understanding and research go hand in hand. Scientific research thereby introduces a deliberate expansion and re-configuration of the available possibilities for conceptual understanding. It seeks to make salient and comprehensible many aspects of the world that would otherwise be hidden and inaccessible to us perceptually, practically, or discursively. While the inferential relations among concepts and judgments are indispensable to that process of conceptual transformation, the sciences also require more extensive efforts to connect those concepts to newly accessible or salient features of the world.

Understanding conceptual articulation in the sciences thus requires a more inclusive conception of the sciences as social and material practices than has been common in most philosophical reflection upon scientific understanding and scientific concepts. Such an account begins with the recognition that the sciences are first and foremost research enterprises. "Scientists" in the primary sense are those people who are engaged in empirical inquiry, whether as primary investigators, or as active members of a larger research team. Yet the research enterprise, as a distinctive form of niche construction, extends far beyond the scientists whose work constitutes its primary focus. Scientific research depends upon an extensive institutional structure, ranging from disciplinary and other professional associations, journals and other publishers, to the universities, hospitals, institutes, government agencies, or corporations in which research activities are situated.[7] Scientific research has become extraordinarily expensive, and its changing sources of financial and other material support are also integral to the research enterprise, especially where they help shape the priorities and direction of research itself. A significant part of that expense is devoted to material resources. The equipment, materials, research sites, and other physical infrastructure of the sciences have become complex and sophisticated, but also sufficiently widely used to support their own supply networks.

Part of that wider use of scientific material and equipment reflects the extension of scientific concepts, instrumentation, materials, and practices beyond the research laboratory, as scientific understanding has become materially as well as conceptually embedded throughout modern industrial societies. Laboratories and their component apparatus and skills are not solely research facilities. Nor can the research enterprise be limited to its suppliers, consumers, and institutional context. Scientific research is of limited import unless its achievements are disseminated more broadly, and the skills, understanding, and research orientation of its participants are differentially reproduced in succeeding generations.[8] This pedagogical and disseminative role has grown to dwarf the primary research component, such that there are many more science teachers, writers, and administrators than there are scientists. Scientific education is not merely intended to produce more scientists, but also to embed scientific concepts and scientific understanding throughout a broad range of modern society and culture.

Even though the active contributors to research are comparatively few in number and fairly readily identifiable, the research enterprise that sustains them is thoroughly entangled with government, the economy, culture, and a wide range of professions and activities that are not themselves primarily scientific.

I emphasize the expansive scope and pervasive social entanglement of the scientific research enterprise as a reminder of the complexity of scientific understanding. Philosophers of science often seek to compartmentalize and narrow the phenomena we seek to understand, both because of our own parochial disciplinary interests in the sciences as intellectual achievements, and in order to make our subject matter tractable. The result has been a highly idealized, disembodied, and largely retrospective conception of scientific understanding and its conceptual content. Philosophers have typically identified science first and foremost with a systematic "body" of knowledge claims that is nowhere actually assembled and expressed in that form, and that is in any case extracted from the diverse and complex institutional and material settings within which those verbal and mathematical formulations are articulated, understood, and deployed. We have largely neglected the factors involved in determining which research questions are important, which ones are actually pursued, and what forms the pursuit takes. We have likewise neglected how the resulting achievements are actually understood, deployed, and implemented. Yet these developments of scientific understanding have substantially transformed the world we live in, and the ways we talk about and understand that world. Such an idealized and disembodied philosophical conception of scientific understanding is especially unsuited for naturalists. At the very least, naturalists who endorse a more traditional conception of the scientific image owe an account of how their idealized image is actually to be discerned among the more diverse, messy and complex practices in which scientific understanding is embedded. Even that might not be enough, however, if we take seriously an account of the evolution of conceptual understanding as a form of niche construction. If scientific understanding is itself a further development of our material and behavioral niche construction, then to abstract away from a more expansive conception of science as a research enterprise would be to give up on a naturalistic account of scientific understanding.

Understanding the sciences as research enterprises also requires a different conception of the temporality of scientific understanding. The predominant philosophical accounts of scientific understanding are retrospective, looking back at the structure and content of what has already been understood and codified as scientific knowledge. That retrospective orientation often persists even in thinking about possible future developments in the sciences. These possibilities are typically addressed in the future perfect tense, by looking forward speculatively to the further development of science, whether or not that speculation is carried all the way to the supposedly regulative ideal of a completed science. Philosophical conceptions of science still typically look forward to looking back from the vantage point of a scientific knowledge not yet achieved.

This retrospective philosophical orientation is in sharp contrast to the understanding that drives the practice of scientific research. Research workers have a more prospective understanding of their field, as oriented toward outstanding problems and opportunities. While they certainly rely upon what has already been achieved, their understanding of the content and significance of those achievements is transformed by their concern to move beyond it. What were once research topics in their own right are now often regarded not so much as achieved propositional knowledge, but instead as reliable effects and procedures that can be used as tools to explore and articulate new possibilities (Rheinberger, 1997). The concepts employed are understood as open-textured in ways that both permit and encourage further articulation and possible correction of previous patterns of use. Not just which scientific claims to accept, but what those claims say about the world, is always open to further transformation. Empirical inquiry involves the articulation of concepts, including the refined skills, practices and material reconfigurations of the world through which those concepts become both meaningful and informative about the world. Above all, the significance of various topics, claims, tools, and issues is organized around their place in the configuration of available and promising research opportunities, rather than their role in the systematic reconstruction of current knowledge.

Yet it is not just an orientation toward outstanding issues and opportunities for research that makes for a divergence between scientific understanding of a research field and philosophical conceptions of the scientific image as an idealized retrospective reconstruction. Even the retrospective assessment of what has already been accomplished in a scientific research field is re-structured by an orientation toward research. Researchers do, after all, engage in retrospective assessments, and those compilations are actually produced in specific forms, ranging from review articles in journals to textbooks and handbooks. Yet none of these summations, either individually or collectively, exemplifies anything like philosophers' conception of the scientific image. Each compilation is intended for specific audiences, and their content is selected and organized for their prospective significance for the research enterprise. Review articles are oriented toward assessing the significance of recent work in the field for subsequent research. Textbooks focus upon the skills and knowledge likely to be needed by the next generation of scientists. Handbooks and atlases are similarly selective and prospective.[9] Yet one also cannot take the primary journal literature as a distributed repository of the scientific image, since it includes conflicting reports as well as preliminary findings understood as open to, and even oriented toward, subsequent further development and correction. Moreover, in many fields, research develops rapidly enough that researchers' grasp of the field typically forges well ahead of the published literature. Actual compilations of scientific knowledge are thus always partial and oriented toward what is significant for specific projects, and the notion of an all-purpose or no-specific-purpose compilation may be an oxymoron. Indeed, the suspicion arises that the conception of a unified "scientific image", a systematic, idealized compilation of scientific knowledge as a whole apart from any specific purposive use of it, is intended to serve the specifically first-philosophical purposes of epistemology and metaphysics. Naturalists ought to be worried about this aspiration.

Perhaps the clearest indication of the divergence between researchers' understanding and anything like a Sellarsian "scientific image" is provided by the unusual occasions for sustained efforts to identify a scientific consensus on the current state of knowledge. Such efforts have recently been undertaken with extraordinary care and thoroughness across the multi-disciplinary domain of climate science, in the reports of the Intergovernmental Panel on Climate Change (IPCC, 1990, 1995, 2001, 2007). The research literature and the opinions of research scientists have been carefully and comprehensively vetted, with a thorough review process aiming to correct errors and accommodate critical assessment of preliminary drafts. Disagreement has been recognized and accommodated by incorporating estimates of degrees of confidence within the reporting of results and predictions. Yet the inherent conservatism of the process of consensus-formation, alongside the cautions provoked by awareness of vigilant critics and skeptics even among the governments responsible for the review process, strongly suggest that most researchers' own understanding of climate science often diverges from the IPCC conclusions, even when they endorse the process and its outcome as an expression of scientific consensus. The point is not to suggest flaws in the IPCC process or reports, but instead to highlight the possible divergence between scientific understanding and even the most diligent and thorough determination of a "scientific consensus". Both the idea of a scientific consensus, and that of the "scientific image" as a characterization of scientific understanding, are idealized composites. Yet they may not be the same idealized composite. The issue is whether scientific understanding is adequately expressed as a collective consensus or composite representational "image", even when its constituent judgments are qualified by estimates of confidence or reliability.

These conceptions of the scientific image as a comprehensive representation of the world, and of the space of reasons as an intra-linguistic domain, nevertheless retain the virtues of familiarity and sophisticated philosophical articulation. Philosophers have developed a rich vocabulary for talking about knowledge and inference in these terms, and a good understanding of how to use that vocabulary to understand the methods and achievements of the sciences. What alternative can I offer to these familiar accounts of the scientific image as an empirically justified, systematic representation of the world?[10] Without a serious alternative conception of scientific understanding, philosophical naturalists will continue to fall back upon the familiar presumption that science aspires to a systematic representation of the world, justified within an intralinguistic space of reasoning.[11]

In what follows, I sketch an alternative way to think about a scientific conception of the world and the naturalistic philosophical stance that it helps constitute. Familiar accounts of the scientific image, and the space of reasons within which it is expressed and justified, assume that these are relatively self-contained linguistic expressions or activities. I argue instead that conceptual understanding in the sciences involves material, social, and discursive transformations of the human environment, taken together. These transformations amount to extensive forms of what evolutionary biologists call niche construction (Odling-Smee, Laland, & Feldman, 2003). An environmental niche is not something specifiable apart from the way of life of an organism, which in turn cannot be understood except in its specific patterns of interdependence with its environment. A niche is a configuration of the world itself as relevant to an ongoing pattern of activity. Yet organismic activities in turn affect their environment, in ways that bear upon the subsequent development of the organism and its way of life. Such processes become all the more consequential to the extent that an environmental niche is transformed by the organism's own forms of niche construction. The emergence of discursive practice, conceptual normativity, and ultimately scientific understanding within our evolutionary lineage places our own way of life at issue for itself in its continuing reproduction and development.

Thomas Kuhn was widely criticized for claiming that "after discovering oxygen, Lavoisier worked in a different world" (1970, p. 118), but in a quite straightforward sense, even ordinary "normal" scientific research is world-transforming. Scientific practices re-arrange things so that novel aspects of the world show themselves, and familiar features are manifest in new ways and new guises. They develop and pass on new behaviors and skills (including new patterns of talk), which also require changes in prior patterns of talk, perception, and action to accommodate these novel possibilities. These developments thereby introduce new ways of understanding ourselves and living our lives, while reconfiguring or even closing off some previously familiar possibilities. Overall, they reconfigure the world we live in as a normative space, a field of meaningful and significant possibilities for living a life and understanding ourselves and the world.

The sciences thereby conceptually articulate the world itself (and not just our thought or talk about it). We are often inclined to say that the world itself does not change, but only our patterns of talk, thought, and social relations. The world is thereby conceived as already articulated into entities and properties, which may or may not be discernible to us, with nothing we can say or do to change that, except by adding new kinds of human-made artifact. Yet such dismissive claims presume that changes in how we talk, think, and relate to one another and things around us are *not* themselves changes in the world. They certainly are changes in our practical, perceptual, and socially interactive environment. More important, however, these changes wrought by scientific work also allow the world to show itself in new patterns, with newly discriminable elements, and new significance. These patterns are not themselves intelligible as patterns, except in relation to the correlative forms and norms of pattern-recognition (Haugeland, 1998, ch. 12; Dennett, 1991).

The idea that scientific understanding and other conceptual transformations are also world-transforming has sometimes been dismissed as a kind of fuzzy-minded, unscientific, idealist metaphysics (Scheffler, 1967). How could changes in our thoughts or utterances change the world they describe and not just our descriptions of it? I instead take the world-transforming character of scientific inquiry to be a straightforward commonplace for any naturalistic understanding of ourselves and the world. Naturalists should instead reject that notion that conceptual understanding is merely a matter of thoughts or utterances in isolation. Conceptual content and authority incorporate patterns of material interaction within an environment. Claiming that changes in conceptual understanding do not change the world implicitly presupposes that changes in conceptual content can take place and be recognized intra-linguistically. Changes in language then need not involve changes in the world, because changes in language would be contained within language. If instead we understand language and conceptual understanding as integral to larger patterns of interaction with the world that constitute our environment and our biological way of life, then the notion that the development and ongoing revisions of language are integral to changes in the world should be unsurprising.

Part of a reconception of scientific understanding as niche construction emphasizes its mediation by experimental practice and theoretical modeling. The sciences transform the world around us and the capacities through which we encounter and live in it, and only thereby allow it to be intelligible to us in revealing ways. The result is a conception of scientific practice as an ongoing reconfiguration of our discursively articulated environmental niche. Although I do not explicitly address this point here, the first part of *Articulating the World* argues that a naturalistic account of conceptual understanding within the context of biological evolution requires understanding language more generally as integral to socially, discursively, and materially articulated niche construction. In bringing these two lines of argument together, I thereby seek to satisfy a central coherence condition for any philosophical naturalism: A naturalism that could not account for scientific understanding as part of nature as scientifically understood is fundamentally incoherent. Meeting this condition is not merely an obligation that must be met in order to sustain a viable philosophical naturalism, however. We gain a richer and more detailed grasp of scientific understanding and scientific practice by recognizing it to be an ongoing process of niche construction. Scientific niche construction involves coordinated changes that create new material phenomena, new patterns of talk and skillful performance, the opening of new domains of inquiry and understanding, and transformations in what is at issue and at stake in how we live our lives and understand ourselves. The sciences thereby transform the world we live in and our place and possibilities within it. In doing so, they articulate the world to allow its conceptual intelligibility. Neither merely "made up" by us, nor found to have been already there, conceptual articulation is the outcome of new ways of interacting with the world that mutually reconstitute us as organisms and the world around us as our biological environment.[12]

In working out this reconception of the scientific image, several aspects of scientific practice receive heightened attention and reinterpretation as contributions to the conceptual articulation of our biological niche. Scientific research often requires creating novel phenomena (Hacking, 1983, 2009), prototypically in laboratories, but in my expanded sense, the field and observational sciences also bring new phenomena into play. These phenomena introduce new patterns into the world, which make different aspects of the world salient within our overall way of life. My discussion thus emphasizes a shift in how we think of the broadly empirical side of the sciences, from asking about what we can observe in the world to focusing upon what various phenomena can show us. Scientifically significant phenomena are structured events that allow patterns or relations to stand out as salient and significant. Observation may seem to be a private, experiential event, but phenomena are public, mutually accessible features of the world.[13]

In addition to arranging or uncovering new phenomena, scientists also build models of various sorts: analytical-mathematical, computational, physical, pictorial, diagrammatic, verbal, and more. Models are themselves internally structured systems, often ones that produce reliable responses to operations performed upon them. Juxtaposing these model systems to laboratory phenomena and more complex events also changes what is salient within the modeled events. Phenomena and models highlight the ways in which scientific conceptualization is a public, material process, in which meaning arises from patterned interactions within the world rather than from the internal, inferential relations among mental or linguistic representations. The point is not to deny the role of explicit judgments and inferences in scientific understanding, but instead to assimilate discursive articulation to the kinds of worldly patterns manifest in natural and experimental phenomena, and model systems of various kinds. Inferentially interconnected judgments and their constituent concepts are especially powerful model systems in just this way.

The modal character of conceptual understanding and scientific practice also requires heightened attention. Recent work on scientific practice has often de-emphasized the role of laws and nomological necessity (Beatty, 1995; Bechtel, 2006; Cartwright, 1999; Giere, 2006a; Teller, 2001). This shift away from conceiving scientific understanding in terms of laws has had multiple sources: a recognition that theoretical understanding is embedded in multiple, partial models rather than unifying principles; renewed attention to biology and other sciences that study historical contingencies rather than physical necessities; the waning of Humean skepticism about causal relations unmediated by lawlike regularities; the decline of the deductive-nomological theory of explanation and with it, a partial eclipse of the aspiration for a general philosophical theory of explanation at all. Yet these concerns turn out to be objections not to laws, but to specific conceptions of natural laws and their necessity. I argue in the book (following and building upon Lange, 2000, 2007, and Haugeland, 1998, 2013) that a more adequate understanding of laws highlights their indispensable role in scientific practice, and their important connections to measurement, inductive reasoning, and conceptual articulation, even in sciences such as biology, geology, or psychology that have often been mistakenly thought to lack laws of their own, on a narrower conception of what laws are.

Along with the alethic modalities, this account also gives greater emphasis to the normativity of scientific understanding. It is now a commonplace that the European history of modern science replaced an irreducibly normative conception of the world inherited from Greek philosophy with conceptions of nature in terms of causes, mechanisms, laws, or symmetries that leave no obvious place for normativity. In philosophical domains from logic to ethics and politics to aesthetics, normative considerations have consequently been construed as originating with us, as rational agents, social beings, affective perceivers, or makers of meaning. For naturalists who incorporate human life within scientifically understood nature, however, such relocations of normativity as instituted within our way of life only postpone the problem. If human beings are natural entities who can be understood physically, biologically, or psychologically, then our role as sources of normative authority and force must also be situated within a scientific understanding of nature.

Once conceptual understanding is situated within a biological context, any supposed abyss between nature and normativity seems to close. The way of life of an organism as a normative configuration of an environment is significant for the maintenance and reproduction of that way of life. In our case, what that way of life is, and thus how things show up as significant within and for that way of life, is itself at issue for us, rather than fixed as a relatively stable pattern. Our conceptually articulated way of life as human thereby allows aspects of the world to show up as significant in novel ways, and for other seemingly intelligible possibilities to be closed off or reconfigured. The sciences have been powerful examples of such conceptually articulated niche construction. Scientific research discloses new aspects of the world, new interrelations among familiar aspects, and new possibilities for our own self-understanding and way of life.

Normative considerations are central to scientific practice, since scientific understanding is highly selective. Which aspects of the world matter scientifically, which phenomena are worth exploring and understanding, and thus which inquiries are scientifically significant, are all integral to scientific understanding. Many responses to this recognition of scientific normativity have worked within the canonically modern conception that situates norm-instituting human activity within an anormative nature. Such traditional approaches either take scientific significance to be determined objectively by anormative nature (e.g., determined by the generality of explanatory laws or the specificity of causal relations), or humanistically by our practices and interests. I argue that when we look more closely at scientific inquiry and the world it discloses, we find them to be normatively constituted in ways that cannot be reduced either to objective features of the world or to human imposition or institution.

As a naturalist, I am committed to understanding the normativity of scientific practice from within the horizons of the natural world disclosed within scientific research. One manifestation of this commitment has been the recognition of conceptual understanding in science as niche construction, a material transformation of the world that allows the world to show itself and affect us in new ways. Our conceptual understanding of nature does not and cannot take place from an imaginary standpoint outside nature that would allow us to represent it as a whole in an intra-linguistically articulated image. Scientific understanding is intra-worldly, and cannot transcend its own involvement in the world. Yet that involvement extends outward from scientific practices in the narrowest sense to encompass the place of scientific understanding within human life more generally. Conceptually articulated niche construction extends throughout the entirety of human life, and the sciences are important to us because of their integration within that larger set of issues, rather than because they are separate from it and relatively self-contained. In this respect, scientific understanding has to be understood within the contingencies of human history and culture. I thus take naturalism to be opposed to essentialist conceptions of science or scientific understanding. Scientific understanding is not a perennial possibility always available throughout human history, or even available to rational or intelligent beings of different biological species or different planetary ecologies. Sciences are historically specific practices emerging within human history. Moreover, that historical specificity reflects the biological specificity of language and conceptual understanding more generally.

This recognition of the historical specificity of scientific understanding brings us, at long last, back to the passage from Nietzsche with which I began. A naturalistic commitment to situating our self-understanding within the scientific image has sometimes seemed to challenge the authority and significance of scientific understanding, self-destructively. Philosophical understanding of science has often focused upon the sciences' supposed transcendence of the local and the human, precisely in order to understand the normative authority of a scientific understanding of the world. I think this aspiration to transcendence of our historical and ecological embeddedness has been erroneous, and has in any case been at odds with a naturalistic standpoint. *Articulating the World* concludes with a discussion of the contingency and locality of conceptual articulation generally and scientific understanding specifically. My aspiration is to show how science matters, and makes authoritative claims upon us, *because* of rather than *despite* its historical and cultural specificity. Science, as a powerful but historically specific extension of the conceptually articulated way of life that is our biological heritage, is not an essential possibility perennially available to any entities with sufficient intellect and social support. It is likewise not an aspiration to transcend our historical contingency through taking up a "god's-eye view" of ourselves and the world. Sciences are instead precarious and risky possibilities that only emerged in specific circumstances, and could disappear.[14] The contingency of conceptual understanding generally and scientific understanding specifically does not thereby undercut the authority or significance of the sciences, but instead calls attention to what is at stake in whether and how those practices continue and develop. The contingent historical emergence and open-ended future possibilities for the subsequent development of scientific understanding should not be regarded as just one historical possibility among many, whose fate might be a matter of arbitrary indifference from the standpoint of the universe. We do not and cannot occupy such a standpoint. From the standpoint of those living a life in the midst of that history, these possibilities are the horizons for our lives, and for how our possibilities matter. Who we are and shall be, what our world is like and how it might further reveal itself, and what possibilities it might thereby open to us and our descendants or close off, are at stake in the subsequent development of our social-discursive way of life and the conceptually articulated manifestation of the world that it makes intelligible. Nothing could matter to us more, or be less arbitrary from a naturalistic standpoint, from *within* the world we aspire to understand.

NOTES

1. In talking about the Sellarsian Scientific Image as representational, I do not have in mind Sellars's more specific notion of representational "picturing", but the more general notion of scientific understanding embedded in its metaphorical characterization as an "image". Sellars to some extent disavows some of those connotations, but still holds onto a conception of scientific understanding as embedded in the epistemic product of inquiry as a body of knowledge. In a similar vein, he explicitly characterized the ideal limit of philosophical understanding:

   > To press the metaphor to its limit, the completion of the philosophical enterprise would be a single model . . . which would reproduce the full complexity of the [conceptual] framework in which we were once unreflectively at home. (1985, p. 296)

2. *Articulating the World* will include a much more extensive discussion of niche construction and its role in human evolution, in Chapters 8 and 9. Here is a brief summary of the conception for those unfamiliar with the basic idea:
   Niche construction theory in evolutionary biology emphasizes how the developmental and selective environment of an organism is transformed by the ongoing and cumulative interactions of other organisms with that

environment; an organism's biological environment is not something given, but is instead dynamically shaped by its ongoing intra-action with the organism. Such transformations are not limited to enduring physical effects on the abiotic environment, for there can also be persistent forms of behavioral niche construction. Behavioral niche construction requires only that the presence of behavioral patterns and their selective significance for individual organisms' evolutionary fitness is reliably reproduced in subsequent generations. The emergence of communicative-cooperative practices that evolve into language can then be understood as a preeminent example of niche construction. Language is a persisting public phenomenon, which develops and co-evolves with human beings. Human beings normally develop in an environment in which spoken language is both pervasive and salient, and languages only exist in the gradually changing forms in which they can be thus learned and reproduced. Our ability to acquire and take up the skills and discriminations that enable the ongoing reproduction of that phenomenon is then integral to our overall practical/perceptual responsiveness to our environment, which has become a discursively articulated environment. The evolutionary emergence of this capacity, and its ontogenetic reconstruction in each generation, relies on the close coupling of organisms' capacities for perceptual and practical responsiveness to their environment. There is nothing mysterious or even discontinuous about the gradual development of the linguistic capacities and performances that enable conceptual understanding. Yet the partial autonomy of discursive practice due to its systematic interconnectedness with other aspects of our perceptual-practical immersion in an environment allows for the emergence of symbolic displacement and conceptual understanding.

Conceptual understanding thus emerges biologically as a highly flexible, self-reproducing and self-differentiating responsiveness to cumulatively constructed aspects of our selective environment. Discursive niche construction is not limited to our abilities to perceive and produce linguistic expressions. Other expressive capacities (pictorial, musical, corporeal, and more) have emerged in the wake of language. More important, however, is that the resulting capacities for symbolic displacement also come to incorporate our broader practical-perceptual immersion in an environment. Our perceptual and practical capacities are not themselves different in kind from those of other organisms, but they are transformed by their uptake within discursive practice. The possible discursive significance of everything we do has been prominently characterized by McDowell (1994) as "the unboundedness of the conceptual". Our discursively articulated practical/perceptual involvements are pervasive throughout and integral to the world in which we develop as and into adult human beings. Their cumulative effects have dramatically transformed us as a species, and indirectly affected many others, including many driven to extinction. Yet the proximal marks of the discursive articulation of the world, however salient and significant for us, are environmental features to which our various "companion species" (Haraway, 2008) are almost entirely insensitive. Our inherited responsiveness and massive ongoing contribution to this peculiar cumulative history of niche construction, and not any general cognitive capacities, are what differentiate us as concept users from any other known organism.

3. In the book, I discuss the ways in which scientific practice draws upon the conceptual and practical resources of its broader cultural setting. In the context of the present discussion, I would only note that the establishment of controlled experimental systems as the locus for the articulation of conceptual domains draws upon the much broader effort to establish and maintain standard units of measurement, and to extend beyond the laboratory setting the isolations,

purifications, shielding and so forth that enables experimental phenomena to manifest clear and intelligible patterns in the world.

4. Strictly speaking, as Putnam (1975) prominently called to philosophical attention, the division of linguistic labor allows people to talk intelligibly about all sorts of things that they are not themselves capable of telling about in this sense, or telling apart. Yet someone must be able to tell what is being talked about in some domain of discourse, if such talk is not to become a "frictionless spinning in a void" (McDowell, 1994). Enabling and sustaining such conceptual engagement with the world is a central part of what the sciences accomplish.

5. For a useful discussion of what is at issue in understanding a domain as a system in which distinct components interact at well-defined interfaces, see Haugeland, 1998, ch. 9, and Simon, 1969.

6. For more extended discussion of experimental systems, and their role in conceptual development, see Rheinberger, 1997; Rouse, 1987 (esp. ch. 4), 2009, 2011.

7. Recognition of the ways in which the development of scientific institutions has been integral to our conception of scientific understanding affects how we think about the history of science. Recent work in the history of science (Biagioli, 1993, and Shapin & Schaffer, 1985, are especially influential examples) highlights the ways in which very important scientific accomplishments arose in contexts with a quite different sense of the aims and significance of scientific research, which were also contested at the time. The point of such work is not to reject our contemporary sense of Galileo's or Boyle's accomplishments as anachronistic, but instead to highlight the ways in which what it is to do science, and to understand the natural world, has been shifting in the course of ongoing inquiry.

8. Emphasis upon the differential reproduction of scientific skills and understanding is important. Although scientific practices and understanding are sufficiently continuous over time to talk about their reproduction, we should recognize that concepts, skills, practices, and materials are continually being reconfigured and re-conceived in the course of their transmission.

9. Daston and Galison's (2007) important study of changing conceptions of objectivity expressed in scientific atlases and handbooks usefully highlights the fact that epistemologically potent concepts such as "objectivity" are not merely terms for philosophical reflection upon science, but also terms that function within scientific practice to shape the direction, form, and content of scientific work. Yet they do not explicitly call attention to the prospective orientation of such works, as preparation for encountering and making sense of novel cases, even though the importance of this role for such works shows up throughout their discussion.

10. I use "justification" here in a broad sense that incorporates reliabilist accounts of the authority of scientific knowledge. My point is not to distinguish internalist accounts of reasoning and justification from externalist accounts of reliable methods and strategies of inquiry, but to highlight the focus of both upon a mostly intralinguistic domain of statements or propositions. Reliabilists are usually as much concerned with the reliability of scientific knowledge claims (even if that reliability is grounded in scientific methods and the perceptual or instrumental detection of entities and their properties), as internalists are with the (more narrowly conceived) justification of knowledge claims by other statements or propositions.

11. There is a crucial role for reasoning and justification even in reliabilist accounts of knowledge, since the reliability of various methods or procedures is always indexed to a reference class within which they are or are not

reliable, and the determination of the relevant reference class cannot itself be understood in reliabilist terms (Brandom, 1994, ch. 4; 2000, ch. 3).

12. Strictly speaking, a naturalistic position understood in this way cannot talk about the world as a whole, which is only then differently configured as a meaningful environment by the way of life of various organisms. There are only these interlocking environments that allow meaningfully configured manifestations of a world. Yet there is another way to vindicate talk of the world as it is apart from its significance for us. The "environment" of an organism is not an enclosed space with an "outside" to it. Entities that do not matter to an organism's way of life are not "outside" its environing world, but are only relatively opaque to its life activities. These forms of opacity and transparency are also mediated by other organisms, since what does not affect an organism directly may nevertheless figure prominently in the developmental and selective environments of other organisms that do matter to it. Moreover, these significance relations are open to change, since environments are dynamic. An organism like us, whose way of life is discursively and thus conceptually articulated, thematizes this possibility of disclosing aspects of the world previously hidden from it. Thus, from within our way of life, we understand and comport ourselves toward possibilities of the world being more or other than it appears selectively within our own way of life. The notion of an "objective" world, a world as it is apart from its meaningful configuration within any specific organism's way of life, thus should not refer to an already determinate configuration of entities, but to an issue we confront within our ongoing conceptually articulated way of life.

13. In fact, observation is never merely a private, experiential event, but always situates perceptual and practical responsiveness within a larger pattern of material-discursive practice, including the ways in which we call others to share in our observational discoveries (Kukla & Lance, 2009).

14. There is no necessary tension between essentialist conceptions of science as a perennial possibility within human life, and a recognition of the vulnerability of a scientific ethos and the way of life it sustains. One might think of the conceptual and epistemic norms of the sciences as always making claims upon us, even though recognition and uptake of those claims is at risk. I will nevertheless be making a stronger claim: the normative authority of scientific practices, concepts and claims only emerges within an historically and biologically specific context, such that maintaining that authority requires also sustaining the way of life within which those practices, concepts and claims could be authoritative for us. Recognizing the contingency of scientific practices and norms does not undercut their authority, I argue, but instead intensifies the significance of what is at stake in sustaining a scientific way of life. There is nevertheless an important insight in essentialist conceptions of the normative authority of the sciences. They are best understood not at face value as descriptions of the "nature" of science, but are instead efforts to focus what is at issue in specific conflicts or tensions over the maintenance of the intelligibility of a scientific way of life and a scientific culture as we know it. They should thereby be themselves understood as an important aspect of a scientific way of life. They are situated, reflective efforts to articulate who we are, how we live, and why it matters to sustain that way of life, from within its horizons. In doing so, they help sustain, and to some extent transform, that way of life in bringing its normative claims and their authority to reflective attention.

Is Rouse's Scientific Image Really Scientific?

Commentary on "Scientific Practice and the Scientific Image", by Joseph Rouse

Emiliano Trizio

In his complex paper, Joseph Rouse takes up Wilfrid Sellars' claim that one of the central issues in 20th-century philosophy lies in the challenge of recombining the image of the world developed by science, the "scientific image", with the so-called "manifest image", in which we appear as personal subjects answerable to norms. Rouse believes, however, that both images must be revised in order to succeed in developing the wished-for stereoscopic vision in a satisfactory naturalistic framework. In particular, and this is the topic of his paper, it is our understanding of the scientific image that must be deeply modified in order to serve this purpose. In this vein, Rouse's paper addresses the theme of this volume by asking: "How would an emphasis upon science as practice revise the scientific image, in its dual sense of a scientific understanding of the world and a conception of scientific understanding"? (Chapter 9, this volume, p. 278). Let us stress at once that: 1) According to Rouse (and coherently with his naturalistic stance[1]), the scientific image must include a conception of scientific understanding itself; 2) the practice turn allows revising the scientific image in such a way that it includes a correct "scientific" understanding of scientific understanding itself.

Rouse begins by addressing possible qualms coming from the side of the so-called "disunifiers"—that is, those among the advocates of the practice turn who argue that science should not be seen as producing an image of the world at all. Yet the dis-unifiers, just like Sellars, believe that scientific understanding is based on representation. The dis-unifiers thus conclude, contra Sellars, that there is no "scientific image" of the world, in the sense of a single unified scientific representation of it. Rouse points out, instead, that what the practice turn really teaches us is that scientific understanding is not primarily representational, while naturalists themselves have so far tried to naturalize scientific understanding by naturalizing mental or linguistic representations. The outline of Rouse's anti-representationalist account of science can be captured in terms of a series of oppositions between the old philosophical understanding of science and the new understanding issued from the practice turn. According to the old view: 1) Scientific understanding consists in an empirically justified fabric of beliefs, or in a position in an intralinguistic space of reason; 2) scientific understanding is disembodied from social, historical, technological, biological contexts; 3) correlatively, the world, as the objective correlate of scientific knowledge, is endowed with a fixed unchangeable structure; 4) scientists strive to reach consensus; 5) scientific understanding expresses itself through retrospective compilations enjoying the consensus of scientists. In contrast, according to the view of scientific understanding based on the study of scientific practices: 1) Scientific understanding is essentially "research activity/enquiry" and, as such, it is incorporated in social, technical, practical, institutional contexts; 2) scientific understanding, along with its practical, interactional, and linguistic performances, is part of a perceptual and practical involvement with the world; 3) the world itself is involved in this interaction and changes in a way that does not amount only to the invention and construction of material artifacts in and out of the laboratory, because the very structure of what is phenomenally salient, of what the world is for us (à la Kuhn), is continuously modified; 4) scientists strive less for the construction of consensus than for making progressive development possible; 5) nowhere do we find the recapitulation of an alleged scientific worldview shared by scientists, for scientific literature is, at all levels, future-oriented in the sense that its principal aim is to foster further research. In particular, results matter primarily for their being a source of promising future development (Chapter 9, this volume, p. 284).

These brief indications should suffice to appreciate what lesson, according to Rouse, naturalists should draw from the practice turn. In other words, Rouse has explained why naturalism needs the practice turn. Has he also explained if, and why, the practice turn needs naturalism, and, in particular, a biologistic variant of naturalism? The second part of the paper indeed addresses this issue, where Rouse tries to argue in what way the notion of scientific image (or scientific understanding) can be improved by situating it in a biological context. This is a crucial point here, for it concerns precisely the way in which scientific practices are to be analyzed and understood.

As we know, Rouse believes that the way in which scientific practices disclose the world, must be understood in scientific terms, and, more specifically, in biological terms. In other words, scientific practices along with science studies must be naturalized. Let us try to understand the implications of this project by taking a quick look at the methodological status of the current researches on scientific practices. What seems fairly uncontroversial is that the practice turn marks a movement away from grand tentatively all-embracing philosophical accounts of science, and, a fortiori, from first-philosophical reconstructions of science; but this shift can be best captured, as Andrea Woody has done in her contribution to this volume (Woody, Chapter 4, this volume), as a shift from the "a priori" to the "empirical". It does seem to be a methodological feature shared by most advocates of this trend that what is said about science is supported by empirical evidence; but this empirical evidence is mainly derived from historical, sociological, and ethnological case studies. Perhaps one could even try to argue that the practice turn is analyzable in several components such as a historical turn, a sociological turn, an ethnological turn, and so forth, perhaps in as many components as there are social sciences in the broad sense of the word. The methodological import of biological or cognitive evidence has been so far rather meager, and it is unlikely to acquire a more prominent role in the foreseeable future. Even most of the items from 1) to 5) characterizing, according to Rouse, the new understanding of science, have been brought about by researches ranging from the analyses of new experimentalists to historical and social studies of science, which one can hardly see as situating science in a biological context. It is therefore doubtful that the actual research activity within science studies needs a biologistic framework in order to gather more evidence about its object.[2] At any rate, one might wonder whether biological naturalism has any real heuristic value for science studies and case-study-based philosophy of science, in the sense that it can provide otherwise unattainable evidence about science in general and about scientific practices in particular.

However, the value of Rouse's attempt to develop a stereoscopic vision of the world as disclosed by science and of the scientific understanding disclosing it (which in turn includes an account of the manifest image) does not rest, I would contend, on the heuristic virtues of naturalism, but on the way it provides a general framework in which the results issued from the practice turn can be combined in a coherent picture, that is, in a naturalistic stereoscopic vision. I take this to be Rouse's own idea when he insists on the benefits of the notion of biological niche for a correct understanding of scientific practices: "We gain a richer and more detailed grasp of scientific understanding and scientific practice by recognizing it to be an ongoing process of niche construction" (Chapter 9, this volume, p. 287).

A biologistic naturalism would thus provide an ontological and conceptual scaffolding for the otherwise largely social science-based understanding of scientific practices. Scientific practices, which are identified by the historian or by the social analyst, would thus need to be naturalized, that is, reconceptualized in such a way that their being a continuation of our species' niche construction is thematized and fully developed. In particular, Rouse lists five consequences of the introduction of the notion of niche construction for the understanding of scientific practice: 1) Seeing new phenomena as the introduction of new experimental patterns that make aspects of the world salient within our way of life; 2) seeing models as implying a shift from representation to interaction for "Models are themselves internally structured systems, often ones that produce reliable responses to operations performed upon them" (Chapter 9, this volume, p. 288); 3) renewed attention to science that studies historical contingency rather than nomological necessities; 4) elimination of the abyss between nature and normativity;[3] 5) understanding scientific understanding within the contingency of human history and culture. Hereafter, I will focus on the way in which the concept of niche bears on the notion of world and on the relation between the world and the scientific understanding of it.

Rouse's proposal appears as an impressive naturalistic synthesis of the results of the last three decades of researches on scientific practices. One may say that Rouse's aim is to do for the contemporary practice-based account of science what Quine did for the account of science previously developed by the logical empiricists and focused on linguistic representations.[4] Quine had replaced more traditional forms of empiricism with a physicalistic-naturalistic account of knowledge centered on the notion of a web of belief, which is, in turn, interpreted in the light of his behavioristic semantics. For Quine each element of a web of belief is potentially involved in a holistic and never ending process of revision whose constraints, or checkpoints, are imposed by experience—that is, by physical events occurring at the speakers' nerve endings. Quine's physicalism consists in accepting the ontology of physics as the guarantee of the objectivity of the constraints imposed upon us by the external world.[5] I suggest that to appreciate the shift from Quine's to Rouse's naturalism, one should stress that the latter replaces the notion of a web of belief with the notion of an ecological niche, while at the same time dropping physicalism. According to Rouse, the problem with the concept of a web of belief is that it is a naturalistic version of a representational account of scientific understanding (what Rouse defined as an intralinguistic space), whose holistic plasticity is allowed by the reconfiguration of linguistic items only. Coherently with the practice turn, the dynamic of scientific understanding is instead viewed by Rouse as a process involving not only the space of reason but material, practical, and social factors as well. The holistic entanglement becomes much wider than previously foreseen by traditional epistemology. Scientific research involves readjustments that cannot be reduced to changes to the web of belief, for they imply transformations in which what counts as a salient phenomenon is modified, just as are beliefs, material performances, social institutions, and cognitive values. In this sense, I believe, Rouse's naturalism ends up assimilating constructivist motives that have been common in recent epistemology, while "urbanizing" them in a naturalistic framework. Examples are: incommensurability, world-change, world-multiplicity, and the emphasis on the historically and socially contingent character of scientific research.[6] Indeed, the concept of niche construction includes, besides the series of elements entering in the previously mentioned dynamics of scientific understanding, all the elements characterizing a certain cognitive form of life, and also the patters of phenomena that become salient within a certain form of life; moreover, a niche is transformed by the species inhabiting it, and co-evolves with them, as Kuhn already pointed out while proposing what he saw, less boldly, as a "parallel" between biological evolution and the history of science (Kuhn, 2000, pp. 101–104). In sum, the notion of ecological niche should capture that the world itself is at stake in scientific understanding, and not just our representation of it. In this respect, we should stress how Rouse defines the concept of niche: "A niche is a configuration of the world itself as relevant to an ongoing pattern of activity" (Chapter 9, this volume, p. 286). The notion of world as existing as an absolute invariant beyond and independently of the actual forms of life of the various organisms, such as the one of the physicalist, is thus rejected in the framework of Rouse's naturalism.[7]

Rouse's appeal to the notion of niche, which is crucial for his naturalistic reconceptualization of scientific practices, deserves careful analysis. The general problem that needs further discussion is, of course, whether notions such as ecological niche and niche construction enable us to describe what is encompassed by the term of "scientific practices".[8] Here I limit myself to pointing what I consider to be a problematic aspect of Rouse's approach: the definition of niche, central to Rouse's reconceptualization of scientific practice, is, as it stands, at odds with the standard biological use. There are many different definitions of ecological niche (some stress the function of a species within an ecological system, others refer to the part of an environment that a species occupies or to an abstract space characterized by environmental parameters[9]); however, they do not (I would contend, cannot) refer to the notion of world, nor to its ontological status, as is implied by an expression such as "configuration of the world itself". Rouse seems to use a kind of Uexküllian view of environment as a world literally existing but "for a species", in contrast to the notion that there is a single encompassing reality which different species inhabit and understand differently. But biology, by itself, is committed to the view that a niche is a configuration of a very limited "part of the world", not of the world itself, even if some biologists venture into this kind of highly speculative territory. A niche has a beginning and an end in time; its emergence is made possible by pre-existing causally relevant conditions that were in no way part of it, and its existence can, at any moment, come to end as a result of external events (such as a natural catastrophe), whether the species occupying the niche has any understanding of them or not. This situation here looks different from the familiar case in which a physical theory receives an ontological interpretation that can be seen as simply added to it, for Rouse literally modifies a central concept of ecological theory, namely that of niche.

In conclusion, Rouse aims to develop a biologistic-naturalistic stereoscopic vision of the world and of the scientific practices disclosing it, a vision that, as I said earlier, assimilates and reconfigures many constructivist motives brought about by the advocates of the practice turn. However, I wonder whether Rouse achieves this aim only at the price of deforming the ontology of evolutionary biology and ecology, in such a way that the rather mundane notion of ecological niche is turned into some kind of world-constituting structure, however dynamical and contingently characterized. Is this deformation compatible with Rouse's own formulation of naturalism, which implies the commitment "to avoid arbitrary impositions upon the development of science"? The question, in other words, is the following one: Is Rouse's scientific image of scientific understanding really scientific?

NOTES

1. Rouse (2002) defines his naturalistic stance as follows:

   Think of these two concerns, expressed in the Quinean metaphilosophical commitment to avoid arbitrary impositions upon the development of science and the Nietzschean philosophical commitment not to accept or rely upon what is mysterious or supernatural, as articulating what is at stake in a commitment to philosophical naturalism. Naturalists can disagree about how these concerns should be articulated and applied, but they should agree that it matters to get them both right, and to hold one another accountable to their appropriate application. I suggest that the tradition of naturalism is held together by an underlying commitment to be responsible to these two concerns, and it is in this sense that I take this book to have been conceived in the spirit of naturalism. (Rouse, 2002, p. 4)

   It must also be noted, that, more recently, Rouse's naturalism becomes more overtly biologistic, for biology and ecology provide the framework for Rouse's attempt to give a naturalistic account of scientific practices.

2. Let us add, in passing, that a form of methodological naturalism about social sciences stressing their lack of autonomy with respect to the natural ones will not help in this case. Indeed the thesis that there is no specifically historical or sociological understanding to be contrasted with the form of knowledge produced by natural science hardly implies that social scientists need notions and methodological frameworks issued from one or another natural science in order to make advances in their research activity. The job of historians, for instance, consists mainly in reading documents, and in order to do that historians don't need to be reminded at every step that they are members of a given animal species studying the way in which other members of the same species produce some sort of broadly conceived adaptive responses to their environment.

3. Under this perspective, normativity is part of what sustains a certain way of life and is itself constantly at issue for that way of life rather than fixed once and for all.

4. In spite of the fact that first philosophy was not an important issue for the advocates of the practice turn.

5. Elsewhere, Rouse has convincingly argued that Quine's physicalism still presupposed a dualism between theory and empirical content, the former being subject to the indeterminacy of translation and the latter being fixed by the ontology of physics (Rouse, 2002, pp. 106–134).

6. This does not mean that Rouse is committed to any variant of constructivism, as standardly defined.

7. Unless it is interpreted as an issue for potential new disclosures in the evolutionary course of the scientific understanding of our species, (Rouse, Chapter 9, this volume, note 12).

8. I am confident that Rouse's forthcoming book, *Articulating the World*, will not fail to further clarify this point.

9. This is the approach initiated in Hutchinson (1957).

Contributors

Catherine Allamel-Raffin is associate professor in philosophy and history of science at the University of Strasbourg, France. She is a member of the Institut de Recherches Interdisciplinaires sur les Sciences et la Technologie (IRIST, Strabourg, France), and a member of the Laboratoire d'Histoire des Sciences et de Philosophie—Archives Henri Poincaré (Nancy, France). She recently worked on the production and functions of images in scientific investigation processes, especially in astrophysics, material physics, and pharmacology. She is also interested in issues related to the interactions between science and fiction. Her recent publications include "The Meaning of a Scientific Image: Case Study in Nanoscience. A Semiotic Approach", in *Nanoethics* (2011), and "Robustness and Scientific Images", in L. Soler et al., (Eds.), *Characterizing the Robustness of the Sciences After the Practical Turn in Philosophy of Science* (Springer, 2012).

Hanne Andersen is professor of science studies and head of the Centre for Science Studies at Aarhus University, Denmark. She is PI of the research group "Philosophy of Contemporary Science in Practice" and works on philosophy of interdisciplinarity, expertise, social epistemology, and research integrity.

Mélissa Arneton has a PhD in psychology and is associate researcher at IRIST in Strasbourg, France. Her research focuses on how different practitioners develop knowledge, attitudes, and identities through sociocultural processes. She is currently a research engineer in the French national institute for disabilities and adapted teachings (INSHEA).

Louis Bucciarelli is a professor of Engineering and Technology Studies, Emeritus, at MIT. His early publications were in engineering—the analysis of static and dynamic stability of structures—and in the history of the development of theories describing the behavior of structures (e.g., *Sophie Germain: An Essay in the History of the Theory of Elasticity*, with N. Dworsky; Reidel, 1980). He is the author of *Designing Engineers* (MIT Press, 1994), an ethnography of engineers at work, and *Engineering Philosophy* (IOS Press, 2004).

Regis Catinaud is completing a PhD in philosophy of science and epistemology at the University of Geneva University (History and Philosophy of Science Unit, Switzerland) and at the University of Lorraine (Laboratoire d'Histoire des Sciences et de Philosophie—Archives Henri Poincaré, Nancy, France). He is currently working in the PratiScienS group on the analysis of the notion of *theorizing practice* and its application, through an *activity-based* perspective, on Maxwell's electrodynamics. In Geneva, he is working on the historical emergence of theoretical physics in western Switzerland.

Hasok Chang is Hans Rausing professor of History and Philosophy of Science at the University of Cambridge. Previously, he taught for 15 years at University College London, after receiving his PhD in Philosophy at Stanford University and his undergraduate degree at the California Institute of Technology. He is the author of *Is Water H₂O? Evidence, Realism and Pluralism* (Springer, 2012), and *Inventing Temperature: Measurement and Scientific Progress* (Oxford University Press, 2004), which was a joint winner of the 2006 Lakatos Award. He is also co-editor (with Catherine Jackson) of *An Element of Controversy: The Life of Chlorine in Science, Medicine, Technology and War* (British Society for the History of Science, 2007), a collection of original work by undergraduate students at University College London. He is a co-founder of the Society for Philosophy of Science in Practice (SPSP) and the International Committee for Integrated History and Philosophy of Science. Currently he is the President of the British Society for the History of Science.

Karine Chemla is currently senior researcher at the French National Center for Scientific Research (CNRS), France, in the research group SPHERE (UMR 7219, CNRS & University Paris Diderot). Her interest is in the history of mathematics in ancient China within the context of a world history. She also studies modern European mathematics. In both cases, she focuses, from a historical anthropology viewpoint, on the relationship between mathematics and the various cultures in the context of which it is practiced and developed. Karine Chemla and Guo Shuchun published *Les Neuf Chapitres* (2004), which was granted the Hirayama Prize for 2006 (Académie des Inscriptions et Belles-Lettres). In 2006, Karine Chemla also received the Prize Binoux (Académie des sciences) and, in 2008, the silver medal from CNRS. Since 2011, together with Agathe Keller and Christine Proust, she is the head of the project "Mathematical Sciences in the Ancient World" (SAW), for which she obtained an Advanced Research Grant from the European Research Council. In 2012, she edited *The History of Mathematical Proof in Ancient Traditions* (Cambridge University Press). She serves on editorial boards of several scholarly journals.

Jean-Luc Gangloff is teaching philosophy in high school and is a member of the Institut de Recherches Interdisciplinaires sur les Sciences et la

Technologie (Strasbourg, France) and of the Laboratoire d'Histoire des Sciences et de Philosophie—Archives Henri Poincaré (Nancy, France). He is presently interested in issues related to the interactions between science and fiction. He has recently published a paper in collaboration with Catherine Allamel-Raffin, "Robustness and Scientific Images", in L. Soler et al., (Eds.), *Characterizing the Robustness of the Sciences After the Practical Turn in Philosophy of Science* (Springer, 2012).

Cyrille Imbert is a CNRS research fellow at the Archives Poincaré (Nancy, France). His main areas of research are philosophy of science and social epistemology. He has co-edited special issues of *Synthese* and a book published by Routledge about models and simulations, as well as another special issue of *Synthese* about the collective dimension of science. He has published on issues related to computational science (like the nature and epistemology of computer simulations or how computers change scientific practices and computational physics), thought-experiments, scientific explanation (in particular, the problem of explanatory relevance), as well as complexity theory and its implications for the philosophical analysis of science.

Vincent Israel-Jost has studied applied mathematics at Université de Strasbourg and philosophy of science at Université de Paris 1, Panthéon-Sorbonne, both at doctoral level. He is particularly interested in the status of contemporary imaging instruments and has written an essay on scientific observation that will be published early in 2014 (Garnier Classiques, Paris). He currently focuses on the role that science plays in society, both at the individual and at the collective levels, and works at the Université Catholique de Louvain (Belgium) on a project that aims to redefine the notion of "authority" of science.

Caroline Jullien is an associate member of the Laboratoire d'Histoire des Sciences et de Philosophie—Archives Henri Poincaré (Nancy, France). After having been educated in mathematics, she turned toward philosophy of mathematics. She works on the meaning and the function of rhetoric and aesthetics in mathematics. In the book *Esthétique et Mathématiques— Une Exploration Goodmanienne*, Jullien described historically, then systematically, the problems that are linked to the question of aesthetics in mathematics. She has developed the cognitive aspects of aesthetics in mathematics, in a Goodman sense, in several articles. Some of her publications are: *Esthétique et mathématiques—Une exploration goodmanienne* (PUR, 2008), "Rôle cognitif de l'esthétique en mathématiques", in J. Przychodzen et al. (Eds.), *L'esthétique du beau ordinaire dans une perspective transdisciplinaire* (L'Harmattan, 2010), "Densité syntaxique et densité sémantique en mathématiques", in P. Lombard and E. Bardin (Eds.), *La figure et la Lettre* (PUN, 2011), and "From the Languages of Art to mathematical languages, and back again", *Enrahonar 49*, 2012.

Katherina Kinzel studied philosophy and political science in Vienna and Berlin and is now a PhD student and course instructor at the University of Vienna. She currently works in a project on "Contingency and Inevitability in the History and Philosophy of Science" funded by the Austrian Science Fund (FWF). Her research interests include the general philosophy of science, the sociology of scientific knowledge, the relations between the history and the philosophy of science, historical epistemology, and the philosophy and methodology of historiography.

Peter Kroes is professor in the Philosophy of Technology at the Delft University of Technology, The Netherlands. He has an engineering degree in physics (1974) and wrote a PhD thesis on the notion of time in physical theories (University of Nijmegen, 1982). He has been teaching courses in the philosophy of science and technology and the ethics of technology, mainly for engineering students. His research in the philosophy of technology focuses on the nature of technical artifacts and engineering design, the modeling of socio-technical systems, and the nature of technological knowledge. His book publications include *Technical Artefacts: Creations of Mind and Matter* (Springer, 2012), *A Philosophy of Technology; From Technical Artefacts to Socio-Technical Systems* (together with Pieter Vermaas, Ibo van de Poel, Maarten Franssen, and Wybo Houkes; Morgan and Claypool, 2011), *Functions in Biological and Artificial Worlds; Comparative Philosophical Perspectives* (editor with Ulrich Krohs; MIT Press, 2009), and *The Empirical Turn in the Philosophy of Technology* (editor with Anthonie Meijers; JAI/Elsevier Science, 2000). He also edited the section on Philosophy of Engineering Design of the *Handbook of Philosophy of Technology and Engineering Sciences* (editor, Anthonie Meijers; Elsevier, 2009).

Michael Lynch is professor in the Department of Science & Technology Studies at Cornell University. He studies discourse, visual representation, and practical action in research laboratories, clinical settings, and legal tribunals. He received the 1995 Robert K. Merton Professional Award from the Science, Knowledge and Technology Section of the American Sociological Association for his book *Scientific Practice and Ordinary Action*. His most recent book, *Truth Machine: The Contentious History of DNA Fingerprinting* (with Simon Cole, Ruth McNally, & Kathleen Jordan), examines the interplay between law and science in criminal cases involving DNA evidence. The book received the 2011 Distinguished Publication Award from the Ethnomethodology/Conversation Analysis section of the American Sociological Association. He was Editor of *Social Studies of Science* from 2002 until 2012 and President of the Society for Social Studies of Science from 2007 to 2009.

Amirouche Moktefi is research fellow at Ragnar Nurkse School of Innovation and Governance, Tallinn University of Technology, Estonia. He is also an associate member of the Institut de Recherches Interdisciplinaires sur les Sciences et la Technologie, University of Strasbourg, France. His research interests include visual reasoning, theory of opposition, Euclidean geometry, and the history of symbolic logic. He recently co-authored with Prof. Sun-Joo Shin "A History of Logic Diagrams", in *Logic: A History of its Central Concepts* (North-Holland, 2012). He also published "Geometry: The Euclid Debate", in *Mathematics in Victorian Britain* (Oxford University Press, 2011).

Joseph Rouse is the Hedding professor of Moral Science in the Philosophy Department, and Chair of the Science in Society Program, at Wesleyan University, in Middletown, Connecticut, USA. He is the author of *Articulating the World: Conceptual Understanding and the Scientific Image (forthcoming)*, *How Scientific Practices Matter: Reclaiming Philosophical Naturalism* (2002), *Engaging Science: How to Understand its Practices Philosophically* (1996), *Knowledge and Power: Toward a Political Philosophy of Science* (1987), and is the editor of John Haugeland's posthumously published *Dasein Disclosed* (2013). His research program encompasses the philosophy of scientific practice, interdisciplinary science studies, naturalism in philosophy, and the history of 20th-century philosophy.

Jean-Michel Salanskis is professor of Philosophy of Science, Logic and Epistemology at University Paris Ouest Nanterre La Défense. He has worked in philosophy of mathematics, phenomenology, contemporary philosophy in general, and on Jewish tradition. He is the author of about 130 papers and 20 books. Among them: *Le Temps du Sens* (Orléans, Editions Hyx, 1997), *Modèles et Pensées de L'action* (Paris, L'Harmattan, 2000), *Talmud, Science et Philosophie* (Paris, Les Belles Lettres, 2004), *Philosophie des Mathématiques* (Paris, Vrin, 2008), *L'herméneutique Formelle* (Paris, Klincksieck, 2013).

Léna Soler has been, from 1997 onwards, associate professor at the University of Lorraine and a member of the Laboratoire d'Histoire des Sciences et de Philosophie—Archives Henri Poincaré, in Nancy, France. Her areas of specialization are philosophy of science and philosophy of physics. She first studied physics and has an Engineering Degree in materials science from the Formation d'Ingénieurs de l'Université Paris Sud Orsay, France. She went on to study general philosophy and philosophy of science at the University of Paris I, Sorbonne, France, and in 1998 completed her PhD on Einstein's early work on light quanta and theorizing processes in physics. In 2008, she received a four-year research grant from the French Agence Nationale de la Recherche for the project "Rethinking Sciences from the Standpoint of Scientific Practices". She wrote an *Introduction à l'épistémologie* (Ellipses, first edition in 2000; second edition, revised and enlarged, in 2009), and has been the main editor of *Rethinking Scientific*

Change and Theory Comparison: Stabilities, Ruptures, Incommensurabilities? (2008, Springer; co-edited with H. Sankey & P. Hoyningen); *Characterizing the Robustness of Science: After the Practice Turn in Philosophy of Science* (2012, Springer; co-edited with E. Trizio, T. Nickles, & W. Wimsatt); a special issue on contingency (*Studies in History and Philosophy of Science, 39,* 2008; co-edited with H. Sankey); and a special issue on *Tacit and Explicit Knowledge: Harry Collins's Framework* (*Philosophia Scientiæ, 17*(3), 2013; co-edited with S. D. Zwart & R. Catinaud).

Emiliano Trizio studied physics at the University of Padua and philosophy of science at the London School of Economics. After completing his PhD on Husserl's phenomenology at the University of Paris X and at the University of Venice Ca' Foscari, he has held a post-doc at the University of Nancy 2, working within the research group PratiSciens. He has taught introduction to physics at the Universities of Panthéon-Sorbonne and Venice Ca' Foscari, as well as philosophy at the Universities of Lille 3 and Seattle. At the moment, he is full time instructor at the philosophy department of the University of Seattle. Most of his research activity has focused on phenomenology and philosophy of science.

Jean Paul Van Bendegem is full time professor at the Vrije Universiteit Brussel (Free University of Brussels) where he teaches courses in logic and philosophy of science. He has been a Dean of the Faculty of Arts and Letters from 2005 to 2009. He is director of the Center for Logic and Philosophy of Science. He has been president of the National Center for Research in Logic (founded in 1955 by, among others, Chaïm Perelman and Leo Apostel). He is one of the co-founders of the Association for the Philosophy of Mathematical Practice (APMP). He is the editor of the journal *Logique et Analyse*. His research focuses on two themes: the philosophy of strict finitism and the development of a comprehensive theory of mathematical practice. Among his recent academic publications are *Perspectives on Mathematical Practices: Bringing Together Philosophy of Mathematics, Sociology of Mathematics, and Mathematics Education* (jointly edited with Bart Van Kerkhove; Springer, 2006), *Philosophical Dimensions in Mathematics Education* (jointly edited with Karen François; Springer, 2007) and *Philosophical Perspectives on Mathematical Practice* (jointly edited with Jonas De Vuyst and Bart Van Kerkhove; College Publications, 2010).

Frédéric Wieber received his PhD in History and Philosophy of Science from Université Paris Diderot in 2005 and is now associate professor of History and Philosophy of Science at the Université de Lorraine (Nancy, France). He is a member of the Laboratoire d'Histoire des Sciences et de Philosophie—Archives Henri Poincaré, Université de Lorraine. His primary interest is in the history and philosophy of biology and chemistry. His works include papers on computational protein chemistry in the 1970s and 80s. He is also interested in practice-based analysis of science and has participated in the PratiScienS research project.

Andrea Woody is associate professor of Philosophy at the University of Washington. Her research concerns methodological issues in the sciences, especially the physical sciences and chemistry, and frequently focuses on issues concerning modeling strategies, representational choices, and explanatory discourse. These interests intersect with issues in social epistemology and the social nature of science—in particular, the roles of community formation and disciplinarity in modern scientific practice. She also has research interests in science education, history of chemistry, feminist perspectives in philosophy, and philosophy of the performing arts.

Sjoerd Zwart, is assistant professor in the Philosophy of Technology and Engineering Sciences at the Delft (1997) and Eindhoven (2002) Universities of Technology, The Netherlands. He has been teaching courses in logic, argumentation, the philosophy of science and technology, and engineering ethics. He has studied mathematics and has a master's degree in the formal philosophy of science. He wrote a PhD thesis on verisimilitude distance measures on Lindenbaum algebras that bear a considerable similarity to classical Belief Revision (*Refined Verisimilitude*, Kluwer, 2001). His research efforts have shifted towards subjects within methods and techniques in engineering design and norms and values within engineering. Most recently he has taken up studies into practices and tacit knowledge in the engineering sciences and technology. Besides the present volume, some relevant publications and editorships include the section on modeling in the *Handbook of Philosophy of Technology and Engineering Sciences* (editor, Anthonie Meijers; Elsevier, 2009), a *SHPS* special issue on values and norms in modeling, co-edited with Martin Peterson (*forthcoming*), and a special issue of Philosophia Scientiae (2013) on Collins' account of tacit knowledge, co-edited with L. Soler and R. Catinaud.

References

Achinstein, P., & Hannaway, O. (Eds.). (1985). *Observation, experiment and hypothesis in modern physical science*. Cambridge, MA: MIT Press.

Ackermann, R. J. (1985). *Data, instruments and theory: A dialectical approach to understanding science*. Princeton, NJ: Princeton University Press.

Ackermann, R. (1989). The new experimentalism. *British Journal for the Philosophy of Science, 30*(2), 185–90.

Akeroyd, F. M. (2003). Prediction and the periodic table: A response to Scerri and Worrall. *Journal for General Philosophy of Science, 34,* 337–355.

Andersen, H. (2007, August). *Demarcating misconduct from misinterpretations and mistakes*. Talk delivered at the First Biennial SPSP Conference, University of Twente, the Netherlands, Retrieved from http://philsci-archive.pitt.edu/4153/1/Andersen_Scientific_Misconduct.pdf

Anderson, W. (1992). The reasoning of the strongest: The polemics of skill and science in medical diagnosis. *Social Studies of Science, 22*(4), 653–684.

Arendt, H. (1958). *Human condition*. Chicago: University of Chicago Press.

Argyris, C. & Schön, D. (1974). *Theory in practice: Increasing professional effectiveness*. San Francisco: Jossey-Bass.

Ascher, M. (2002). *Mathematics elsewhere*. Princeton, NJ: Princeton University Press.

Ascher M., & Ascher R. (1986). Ethnomathematics. *History of Science, 24,* 125–144.

Ashmore, M. (1989). *The reflexive thesis. Writing sociology of scientific knowledge*. Chicago & London: University of Chicago Press.

Ashplant, T. G., & Wilson, A. (1988). Present-centered history and the problem of historical knowledge. *The Historical Journal, 31*(2), 253–274.

Aspray, W., & Kitcher, P. (Eds.). (1988). *History and philosophy of modern mathematics*. Minneapolis: University of Minnesota Press.

Austin, J. L. (1961). A plea for excuses. In J. O. Urmson & G. J. Warnock (Eds.), *Philosophical papers* (pp. 123–170). Oxford: Oxford University Press.

Austin, J. L. (1962). *How to do things with words: The William James lectures delivered at Harvard university in 1955*. Oxford: Clarendon Press.

Baigrie, B. S. (1995). Scientific practice, the view from the tabletop. In J. Z. Buchwald (Ed.), *Scientific practice: Theories and stories of doing physics* (pp. 87–122). Chicago: University of Chicago Press.

Bailer-Jones, D. M. (2003). When scientific models represent. *International Studies in the Philosophy of Science, 17,* 59–74.

Baker, G. P., & Hacker, P. M. S. (1985). *Wittgenstein: Rules, grammar and necessity*. Oxford: Blackwell.

Barad, K. (2011). Erasers and erasures: Pinch's unfortunate "uncertainty principle". *Social Studies of Science, 41*(3), 443–454.

Barber, B., & Fox, R. (1958). The case of the floppy-eared rabbits: An instance of serendipity gained and serendipity lost. *American Journal of Sociology, 64,* 128–136.

Barnes, B. (1974). *Scientific knowledge and sociological theory.* London: Routledge and Kegan Paul.

Barnes, B. (1977). *Interests and the growth of knowledge.* London: Routledge and Kegan Paul.

Barnes, B. (1992). Realism, relativism and finitism. In D. Raven & L. Van Vucht Tijssen (Eds.), *Cognitive relativism and social science* (pp. 131–147). New Brunswick: Transaction Publishers.

Barnes, B., Bloor, D., & Henry, J. (1996). *Scientific knowledge: A sociological analysis.* Chicago: University of Chicago Press.

Barnes, B., & Shapin, S. (Eds.). (1979). *Natural order: Historical studies of scientific culture.* London: Sage.

Barron, J. (1991). "Ask Dr. Science" passes a landmark: Puncturing "experts" 2,000 times. *New York Times* (2 July). Retrieved from http://www.nytimes.com/1991/07/02/arts/ask-dr-science-passes-a-landmark-puncturing-experts-2000-times.html

Barth, E. M., & Krabbe, E. C. W. (1982). *From axiom to dialogue. A philosophical study of logics and argumentation.* Berlin: Walter de Gruyter.

Batens, D., & Van Bendegem, J. P. (Eds.). (1988). *Theory and experiment, recent insights and new perspectives on their relation.* Dordrecht, Boston, Lancaster, & Tokyo: D. Reidel Publishing Company.

Baudouin, J.-M., & Friedrich, J. (Eds.). (2001). *Théories de l'action et education.* Bruxelles: De Boeck université.

Baumhauer, H. (1870). *Die Beziehungen zwischen dem Atomgewichte und der Natur der chemischen Elemente.* Braunschweig: Vieweg.

Beatty, J. (1995). The evolutionary contingency thesis. In G. Wolters & J. Lennox (Eds.), *Concepts, theories, and rationality in the biological sciences* (pp. 45–81). Pittsburgh: University of Pittsburgh Press.

Beaver, D. (2001). Reflections on scientific collaboration (and its study): Past, present, and future. *Scientometrics, 52,* 365–377.

Beaver, D. & Rosen, R. (1978). Studies in scientific collaboration. Part I. The professional origins of scientific co-authorship. *Scientometrics, 1,* 65–84.

Beaver, D. & Rosen, R. (1979a). Studies in scientific collaboration. Part II. Scientific co-authorship, research productivity and visibility in the French scientific elite, 1799–1830. *Scientometrics, 1,* 133–149.

Beaver, D. & Rosen, R. (1979b). Studies in scientific collaboration Part III. Professionalization and the natural history of modern scientific co-authorship. *Scientometrics, 1,* 231–245.

Bechtel, W. (2005). Explanation: A mechanistic alternative. *Studies in History and Philosophy of the Biological and Biomedical Sciences, 36,* 421–441.

Bechtel, W. (2006). *Discovering cell mechanisms.* Cambridge: Cambridge University Press.

Beer, F. P., Johnston, E. R., Jr., & DeWolf, J. T. (2006). *Mechanics of materials* (4th ed.). New York: McGraw-Hill.

Beguyer de Chancourtois, A. E. (1862). Sur un classement naturel des corps simples ou radicaux appelé vis tellurique. *Comptes-Rendus de l'Academie des sciences, 54,* 757–761.

Bensaude-Vincent, B. (1986). Mendeleev's periodic system of chemical elements. *British Journal for the History of Science, 19,* 3–17.

Biagioli, M. (1993). *Galileo, courtier.* Chicago: University of Chicago Press.

Bijker, W. E., Hughes, T. P., & Pinch, T. J. (Eds.). (1987). *The social construction of technological systems: New directions in the sociology and history of technology.* Cambridge, Massachusetts: MIT Press.

Bird, A. (2008a). The historical turn in the philosophy of science. In S. Psillos & M. Curd (Eds.), *Routledge companion to the philosophy of science* (pp. 67–77). Abingdon: Routledge.

Bird, A. (2008b). Incommensurability naturalized. In L. Soler, H. Sankey, & P. Hoyningen-Huene (Eds.). *Rethinking scientific change and theory comparison* (pp. 21–39). Dordrecht, NL: Springer.

Bishop, A. (1988). *Mathematical enculturation*. Dordrecht: Kluwer.

Blais, M. (1987). Epistemic "tit for tat". *The Journal of Philosophy, 84*(7), 363–375.

Bloch, M. (1949). *Apologie pour l'histoire, ou Métier d'historien*. Paris: Armand Colin.

Bloor, D. (1973). Wittgenstein and Mannheim on the sociology of mathematics. *Studies in the History and Philosophy of Science, 4*, 173–191.

Bloor, D. (1976/1991). *Knowledge and social imagery* (second ed.). London: Routledge and Kegan Paul. (First edition, 1976)

Bloor, D. (1981). The strengths of the strong programme. *Philosophy of the Social Sciences, 11*, 199–213.

Bloor, D. (1999). Anti-Latour. *Studies in History and Philosophy of Science, 30*(1), 81–112.

Bloor, D. (2011). *The enigma of the aerofoil*. Chicago & London: University of Chicago Press.

Boghossian, P. (2006). *Fear of knowledge. Against relativism and constructivism*. Oxford: Oxford University Press.

Bourdieu, P. (1977) *Outline of a theory of practice* (R. Nice, Trans.). Cambridge, UK, New York: Cambridge University Press.

Bratman, M. E. (1992). Shared cooperative activity. *Philosophical Review, 101*, 327–341.

Bratman, M. E. (1999). Shared intention. In M. E. Bratman (Ed.), *Faces of intention. selected essays on intention and agency* (pp. 109–129). Cambridge: Cambridge University Press.

Bratman, M. E. (2009). Modest sociality and the distinctiveness of intention. *Philosophical Studies, 144*, 149–165.

Brandom, R. (1994). *Making it explicit*. Cambridge: Harvard University Press.

Brandom, R. (2000). *Articulating reasons*. Cambridge: Harvard University Press.

Braverman, H. (1976). *Labor and monopoly capital*. New York: Monthly Review Press.

Bretelle-Establet, F. (Ed.). (2010). *Looking at it from Asia: The processes that shaped the sources of history of science* (Boston Studies in the Philosophy of Science, Vol. 265). Dordrecht & New York: Springer.

Bridgman, P. W. (1954). The present state of operationalism. In P. G. Frank (Ed.), *The validation of scientific theories* (pp. 75–80). New York: Collier Books.

Bridgman, P. W. (1959). *The way things are*. Cambridge, MA: Harvard University Press.

Broad, W. J. (1981). Fraud and the structure of science. *Science, 212*, 137–141.

Broad, W. J. & Wade, N. (1982). *Betrayers of the truth: Fraud and deceit in the halls of science*. New York: Simon & Schuster.

Brunschvicg, L. (1912). *Les étapes de la philosophie mathématique*. Paris: Alcan.

Brush, S. G. (1996). The reception of Mendeleev's periodic law in America and Britain. *Isis, 87*, 595–628.

Brush, S. G. (2007). Predictivism and the periodic table. *Studies in History and Philosophy of Science, 38*, 256–259.

Bucciarelli, L. (1988). An ethnographic perspective on engineering design. *Design Studies, 9*(3), 159–168.

Bucciarelli, L. (1994). *Designing engineers*. Cambridge, MA: MIT Press.

Bucciarelli, L. (2002, May). Between thought and object in engineering design. *Design Studies 23*(3), 219–232.

Bucciarelli, L. (2004). *Engineering philosophy*. Delft, the Netherlands: DUP Satellite.

Bucciarelli, L. (2012) *Bachelor of arts in engineering—The full proposal*. Retrieved from http://hdl.handle.net/1721.1/71008

Buchwald, J. Z. (Ed.). (1995). *Scientific practice: Theories and stories of doing physics*. Chicago: University of Chicago Press.

Bush, V. (1945/1995). *Science: The endless frontier* (Reprint). North Stratford, NH: Ayer.

Butterfield, H. (1931). *The Whig interpretation of history*. London: Bell & Sons.

Button, G. (Ed.). (1991). *Ethnomethodology and the human sciences*. Cambridge, UK: Cambridge University Press.

Button, G., & Sharrock, W. (1993). A disagreement over agreement and consensus in constructionist sociology. *Journal for the Theory of Social Behaviour, 23*(1), 1–25.

Callebaut, W. (1993). *Taking the naturalistic turn. How real philosophy of science is done*. Chicago: University of Chicago Press.

Callendar, C., & Cohen, J. (2006). There is no special problem about scientific representation. *Theoria, 55*, 7–25.

Canguilhem, G. (1968). L'objet de l'histoire des sciences. In G. Canguilhem (Ed.), *Études d'histoire et de philosophie des sciences* (pp. 9–23). Paris: Vrin.

Canguilhem, G. (1970). *Introduction à l'histoire des sciences* (Vol. 1). *Eléments et instruments*. Paris: Hachette.

Canguilhem, G. (1971). *Introduction à l'histoire des sciences* (Vol. 2). *Objet, méthode, exemples*. Paris: Hachette.

Carter, J. (2010). Diagrams and proofs in analysis. *International Studies in the Philosophy of Science, 24*(1), 1–14.

Cartwright, N. (1983). *How the laws of physics lie*. Oxford: Clarendon Press.

Cartwright, N. (1999). *The dappled world*. Cambridge: Cambridge University Press.

Cartwright N., Shomar T., & Suarez, M. (1996). The tool box of science: Tools for building of models with a superconductivity example. In W. E. Herfel et. al (Eds.), *Theories and models in scientific processes* (pp. 137–149). Amsterdam: Editions Rodopi.

Caudill, D., & LaRue, L. (2006). *No magic wand: The idealization of science in law*. Lanham, MD: Rowman & Littlefield.

Cavaillès, J. (1962). *Philosophie mathématique. Collection histoire de la pensée VI*. Paris: Hermann.

Chang, H. (2004). *Inventing temperature: Measurement and scientific progress*. Oxford: Oxford University Press.

Chang, H. (2011a). The philosophical grammar of scientific practice. *International Studies in the Philosophy of Science, 25*, 205–221.

Chang, H. (2011b). Compositionism as a dominant way of knowing in modern chemistry. *History of Science, 49*, 247–268.

Chang, H. (2012). *Is water H_2O?: Evidence, realism and pluralism*. Dordrecht, Heidelberg, New York, London: Springer.

Chang, H., & Fisher, G. (2011). What the ravens really teach us: The intrinsic contextuality of evidence. In P. Dawid, W. Twining, & M. Vasilaki (Eds.), *Evidence, inference and enquiry, Proceedings of the British Academy 171* (pp. 341–366). London: British Academy.

Chemla, K. (1989). Qu'apporte la prise en compte du parallélisme dans l'étude de textes mathématiques chinois ? Du travail de l'historien à l'histoire du travail. In F. Jullien (Ed.), *Parallélisme et appariement des choses* (pp. 53–80), Extrême-Orient, Extrême-Occident 11. Retrieved from http://persee.cines.fr/web/revues/home/prescript/article/oroc_0754–5010_1989_num_11_11_948

Chemla, K. (1992). De la synthèse comme moment dans l'histoire des mathématiques, *Diogène, 160*, 97–114.

Chemla, K. (1993). Cas d'adéquation entre noms et réalités mathématiques. Deux exemples tirés de textes chinois anciens. In K. Chemla & F. Martin (Eds.), *Le juste nom* (pp. 102–138). Extrême-Orient, Extrême-Occident 15. Retrieved from http://per see.cines.fr/web/revues/home/prescript/issue/oroc_0754–5010_1993_num_15_15

Chemla, K. (1994a). Nombres, opérations et équations en divers fonctionnements: Quelques méthodes de comparaison entre des procédures élaborées dans trois mondes différents. In I. Ang & P.-E. Will (Eds.), *Nombres, astres, plantes et viscères: Sept essais sur l'histoire des sciences et des techniques en Asie orientale* (pp. 1–36). Paris: Institut des Hautes Etudes Chinoises.

Chemla, K. (1994b). Similarities between Chinese and Arabic mathematical writings. I. Root extraction, *Arabic Sciences and Philosophy. A Historical Journal*, 4(2), 207–266.

Chemla, K. (1996). Positions et changements en mathématiques à partir de textes chinois des dynasties Han à Song- Yuan. Quelques remarques. In K. Chemla & M. Lackner (Eds.), *Disposer pour dire, placer pour penser, situer pour agir. Pratiques de la position en Chine* (pp. 115–147, 190, 192, Extrême-Orient, Extrême-Occident 18).

Chemla, K. (1998). Aperçu sur l'histoire des mathématiques en Chine ancienne dans le contexte d'une histoire internationale. In D. Tournès (Ed.), *L'océan indien au carrefour des mathématiques arabes, chinoises, européennes et indiennes*, Actes du colloque à Saint-Denis de la Réunion (pp. 71–90). Saint-Denis : IUFM de La Réunion.

Chemla, K. (2006). Documenting a process of abstraction in the mathematics of ancient China. In C. Anderl and H. Eifring (Eds.), *Studies in Chinese language and culture—Festschrift in honor of Christoph Harbsmeier on the occasion of his 60th birthday* (pp. 169–194), Oslo: Hermes Academic Publishing and Bookshop A/S. Retrieved from http://halshs.archives-ouvertes.fr/halshs-00133034, http://www.instphi.org/Festschrift.html

Chemla, K. (2008). *Classic and commentary: An outlook based on mathematical sources* (Preprint/Max-Planck-Institut für Wissenschaftsgeschichte, Vol. 344). Berlin: Max-Planck-Institut für Wissenschaftsgeschichte.

Chemla, K. (2009). On mathematical problems as historically determined artifacts. Reflections inspired by sources from ancient China. *Historia Mathematica, 36*(3), 213–246.

Chemla, K. (2010). Mathematics, nature and cosmological inquiry in traditional China. In G. Dux & H.-U. Vogel (Eds.), *Concepts of nature in traditional china: Comparative approaches* (pp. 255–284). Leiden: Brill.

Chemla, K. (2011). Une figure peut en cacher une autre: Reconstituer une pratique des figures géométriques dans la Chine du XIIIe siècle. *Images des mathématiques* (CNRS). Retrieved from http://images.math.cnrs.fr/Une-figure-peut-en-cacher-une.html

Chemla, K. (2012). Reading proofs in Chinese commentaries: Algebraic proofs in an algorithmic context. In K. Chemla (Ed.), *The history of mathematical proof in ancient traditions* (pp. 423–486). Cambridge: Cambridge University Press.

Chemla, K. (*forthcoming*). Changing mathematical cultures, conceptual history and the circulation of knowledge. A case study based on mathematical sources from ancient China. In K. Chemla & E. Fox-Keller (Eds.), *Cultures without culturalism*.

Chemla, K., & Guo, S. (2004). *Les neuf chapitres. Le Classique mathématique de la Chine ancienne et ses commentaires*, Paris: Dunod.

Chemla, K., & Ma, B. (2011). Interpreting a newly discovered mathematical document written at the beginning of Han dynasty in China (before 157 B.C.E.) and excavated from tomb M77 at Shuihudi 睡虎地, *Sciamvs, 12*, 159–191.

Chôka zan kankan Sansûsho kenkyûkai. 張家山漢簡『算數書』研究会編 (2006). Research group on the Han bamboo strips from Zhangjiashan *Book of Mathematical Procedures. 漢簡『算數書』Kankan Sansûsho. The Han bamboo strips from Zhangjiashan Book of Mathematical Procedures*, 京都 Kyoto: 朋友書店 Hôyû shoten.

Collins, H. M. (1974), The TEA set: Tacit knowledge and scientific networks, *Science Studies, 4*, 165–86.

Collins H. M. (1975). The seven sexes: A study in the sociology of a phenomenon, or the replication of experiments in physics. *Sociology, 9*(2), 205–24.

Collins, H. M. (1981a). Stages in the empirical programme of relativism. *Social Studies of Science, 11*, 3–10.

Collins, H. M. (1981b). Son of seven sexes: The social destruction of a physical phenomenon. *Social Studies of Science, 11*, 33–62.

Collins, H. M. (1981c). The place of the core-set in modern science: Social contingency with methodological property in science, *History of Science, 19*(1), 6–19.

Collins, H. M. (1984). When do scientists prefer to vary their experiments? *Studies in History and Philosophy of Science, 15* (2), 169–174.

Collins, H. M. (1985/1992). *Changing order: Replication and induction in scientific practice* (2nd ed.). Chicago: University of Chicago Press.

Collins, H. M. (1994). A strong confirmation of the experimenters' regress. *Studies in History and Philosophy of Modern Physics, 25*(3), 493–503.

Collins, H. M. (2001a). Tacit knowledge, trust and the Q of sapphire. *Social Studies of Science, 31*(1), 71–85.

Collins, H. M. (2001b). What is tacit knowledge?. In T. R. Schatzchi, K. Knorr-Cetina, & E. Von Savigny (Eds.), *The practice turn in contemporary theory* (pp. 107–119). London, New York: Routledge.

Collins, H. M. (2004). *Gravity's shadow*. Chicago: University Press.

Collins, H. M. (2010). *Tacit and explicit knowledge*. Chicago: University of Chicago Press.

Collins, H. M., & Evans, R. (2002). The third wave of science studies: Studies of expertise and experience. *Social Studies of Science, 32*(2), 235–296.

Collins, H. M., & Evans, R. (2007). *Rethinking expertise*. Chicago: University of Chicago Press.

Collins, H. M., & Evans, R. (2008, July 15). A response from Harry Collins and Robert Evans to Michael Lynch's review of *Rethinking Expertise*. *American Scientist*. Retrieved from http://www.americanscientist.org/bookshelf/pub/a-response-from-harry-collins-and-robert-evans-to-michael-lynchs-review-of-rethinking-expertise

Collins, H. M., & Kusch, K. (1998). *The shape of actions: What humans and machines can do*. Cambridge, MA: MIT Press.

Collins, H. M., & Yearley, S. (1992). Epistemological chicken. In A. Pickering (Ed.), *Science as practice and culture* (pp. 301–326). Chicago & London: University of Chicago Press.

Constant II, E. W. (1987). The social locus of technological practice: Community, system, or organization? In W. E. Bijker, T. Hughes, & T. Pinch (Eds.), *The social construction of technological systems* (p.223–242). Cambridge, MA: MIT Press.

Conway, E., & Oreskes, N. (2010), *Merchants of doubt: How a handful of scientists obscured the truth on issues from tobacco smoke to global warming*, London: Bloomsbury Press.

Coulter, J. (2001). Human practices and the observability of the "macro social". In T. Schatzki, K. Knorr-Cetina, & E. Von Savigny (Eds.), *The practice turn in contemporary theory* (pp. 29–41). New York: Routledge.

Couzin, J. (2006). Breakdown of the year: Scientific fraud. *Science, 314*, 1853.

Couzin, J., & Unger, K. (2006). Cleaning up the paper trail. *Science, 312*, 38–43.

Craver, C. (2007). *Explaining the brain: Mechanisms and the mosaic unity of neuroscience*. Oxford: Oxford University Press.

Crombie, A. (1994). *Styles of scientific thinking in the European tradition*. London: Duckworth.

Cullen, C. (1996). *Astronomy and mathematics in ancient China: The Zhou bi suan jing* (Needham Research Institute Studies). Cambridge, UK, & New York: Cambridge University Press.

Cullen, C. (2004). *The Suan shu shu* 筭數書 *"Writings on reckoning"* (Needham Research Institute Working Papers, Vol. 1). Cambridge: Needham Research Institute.

Dahlberg, J. E., & Mahler, C. C. (2006). The Poehlman case: Running away from the truth. *Science and Engineering Ethics, 12*, 157–173.

D'Ambrosio, U. (1985). Ethnomathematics and its place in the history and pedagogy of mathematics. *For the Learning of Mathematics, 5*(1), 44–48.

D'Ambrosio, U. (1990). The history of mathematics and ethnomathematics. How a native culture intervenes in the process of learning science. *Impact of Science on Society, 40*(4), 369–377.

D'Ambrosio, U. (2007). Peace, social justice and ethnomathematics. *The Montana Mathematics Enthusiast. Monograph 1, 25–34.*

Danto, A. (1973). *Analytical philosophy of action*. Cambridge, MA: Cambridge University Press.

Dasen, P. R. (2004). Education informelle et processus d'apprentissage. In A. Akkari & P. R. Dasen (Eds.), *Pédagogues et pédagogies du sud* (pp. 23–52). Paris: L'Harmattan.

Daston, L. (2009). Science studies and the history of science. *Critical Inquiry, 35*, 798–813.

Daston, L. & Galison, P. (2007). *Objectivity*. New York: Zone.

Dauben, J. W. (2008) 算數書 Suan Shu Shu (A book on numbers and computations; English Translation with Commentary). *Archive for History of Exact Sciences 62*, 91–178.

Daubert v. Merrell Dow Pharmaceuticals, Inc., 509 U.S. 579 (1993).

Dauenhauer, B. P. (Ed.). (2011). *At the nexus of philosophy and history*. University of Georgia Press.

Davidson, D. (1963). Actions, reasons and causes. *The Journal of Philosophy, 60*(23), 685–700.

Davidson, D. (2001). *Essays on actions and events* (2nd ed.). Oxford: Oxford University Press.

De Chadarevian, S., & Hopwood, N. (Eds.). (2004). *Models: The third dimension of science*. Stanford: Stanford University Press.

Dear, P. (2001). Science studies as epistemography. In J. Labinger & H. Collins (Eds.), *The one culture? A conversation about* science (pp. 128–141). Chicago: University of Chicago Press,.

Dear, P. (2006). *The intelligibility of nature: How science makes sense of the world*. Chicago: University of Chicago Press.

Dear, P., & Jasanoff, S. (2010). Dismantling boundaries in science and technology studies. *Isis, 101, 759–774.*

Dehaene, S., & Brannon, E. (Eds.). (2011). *Space, time and number in the brain. Searching for the foundations of mathematical thought*. New York: Academic Press.

Dehaene, S., Molko, N., Cohen, L., & Wilson, A. H. (2004). Arithmetic and the brain. *Current opinion in neurobiology, 14*, 218–224.

Dennett, D. (1987). *The intentional stance*. Cambridge: MIT Press.

Dennett, D. (1991). Real patterns. *Journal of Philosophy, 89*, 27–51.

Desanctis, G., & Poole, M. S. (1994). Capturing the complexity in advanced technology use: adaptive structuration theory. *Organization Science, 5*(2): 121–147.

De Vries, R., Anderson, M., & Martinson, B. C. (2006). Normal misbehavior: Scientists talk about the ethics of research. *Journal of Empirical Research on Human Research Ethics, 1,* 43–50.

Dewey, J. (1917). The need for a recovery of philosophy. In J. Dewey (Ed.), *Creative intelligence: Essays in the pragmatic attitude* (pp. 3–69). New York: Holt.

Dieudonné, J. (1982). Mathématiques vides et mathématiques significatives. In F. Guénard & G. Lelièvre (Eds.), *Penser les mathématiques* (pp. 15–38). Paris: Editions du Seuil.

DiMaggio, P. (1997). Culture and cognition. *Annual review of sociology, 23,* 263–287.

Dorst, K. (1997). *Describing design: A comparison of paradigms* (unpublished doctoral dissertation). University of Technology, Delft.

Dorst, K., & van Overveld, K. (2009). Typologies of design practice. In D.M. Gabbay, A. Meijers, P. Thagard, & J. Woods (Eds.), *Philosophy of technology and engineering sciences* (pp., 455–488). Burlington, MA: North Holland.

Downey, G. L. (1988). Structure and practice in the cultural identities of scientists: Negotiating nuclear wastes in New Mexico. *Anthropological Quarterly, 61*(1), 26–38.

Downey, G. L. (1998). *The machine in me. An anthropologist sits among computer engineers.* New York, London: Routledge.

Downey, G. L. & Beddoes, K. (Eds.). (2011). *What is global engineering education for?: The making of international educators.* San Rafael, CA: Morgan and Claypool Publishers.

Dretske, F. (1981). *Knowledge and the flow of information.* Cambridge: MIT Press.

Dreyfus, H. L. (1991). *Being-in-the-world: A commentary on Heidegger's* Being and Time (Division I). Cambridge, MA: MIT Press.

Dreyfus, H. L. & Dreyfus, S. E. (1986). *Mind over machine: The power of human intuition and expertise in the era of the computer.* New York: Free Press.

Dung, P. M. (1995). On the acceptability of arguments and its fundamental role in non-monotonic reasoning, logic programming and n-person games. *Artificial Intelligence, 77*(2), 321–358.

Dupre, J. (1993). *The disorder of things.* Cambridge: Harvard University Press.

Edmond, G., & Mercer, D. (2002). Conjectures and exhumations: Citations of history, philosophy and sociology of science in US federal courts. *Law & Literature, 14,* 309–366.

Edmond, G., & Mercer, D. (2004). Daubert and the exclusionary ethos: The convergence of corporate and judicial attitudes towards the admissibility of expert evidence in tort litigation. *Law & Policy, 26*(2), 231–257.

Eggert, L. D. (2011). Best practices for allocating appropriate credit and responsibility to authors of multi-authored articles. *Frontiers in Psychology, 2,* article 196.

Fanelli, D. (2009). How many scientists fabricate and falsify research? A systematic review and meta-analysis of survey data. *PLoS ONE, 4,* 1–11.

Feinberg, J. (1968). Collective responsibility. *Journal of Philosophy, 65,* 674–688.

Feldman, R. (2012, Summer). Naturalizing epistemology. In E. N. Zalta (Ed.), *The Stanford Encyclopedia of Philosophy.* Retrieved from http://plato.stanford.edu/archives/sum2012/entries/epistemology-naturalized/

Ferreiros, J., & Gray, J. (Eds.). (2006). *Architecture of modern mathematics. Essays in history and philosophy.* Oxford: Oxford University Press.

Fichant, M., & Pêcheux M. (1969). *Sur l'histoire des sciences.* Paris: Maspero.

Fleck, L. (1935/1979). *Genesis and development of a scientific fact.* (T. J. Trenn, Trans., and R. K. Merton, Ed.). Chicago: University of Chicago Press. (Translation of *Entstehung und Entwicklung einer wissenschaftlichen Tatsache,* 1935, Frankfurt am Main: Suhrkamp TB)

Fodor, J. (1979). *The language of thought*. Cambridge: Harvard University Press.

Fortun, M., & Bernstein, H. J. (1998). *Muddling through: Pursuing science and truths in the 21st century*. Washington, DC: Counterpoint.

Foucault, M. (1972). *The archaeology of knowledge* (A. M. Sheridan Smith, Trans.). New York: Pantheon Books.

Foucault, M. (1997). The will to knowledge (R. Hurley, Trans.). In P. Rabinow, (Ed.), *Ethics: Subjectivity and truth. Essential works of Michel Foucault, 1954–1984* (Vol. 1; pp. 11–16). London: Allen Lane.

François, K., & Van Bendegem, J. P. (Eds.). (2007). *Philosophical dimensions in mathematics education*. New York: Springer.

Frankenberger, E., & Badke-Schaub, P. (1996). Modeling design processes in industry: Empirical investigations of design work in practice. In O. Akin & G. Saglamer (Eds.), *Proceedings of 1st International Symposium on the Descriptive Models of Design*, Istanbul: Technical University.

Franklin, A. (1979). The discovery & non-discovery of parity non-conservation. *Studies in History and Philosophy of Science, 10*, 201–252.

Franklin, A. (1981). What makes a good experiment? *British Journal for the Philosophy of Science, 32*, 367–74.

Franklin, A. (1986). *The neglect of experiment*. Cambridge: Cambridge University Press.

Franklin, A. (1994). How to avoid the experimenters' regress. Studies in the History and Philosophy of Science, 25, 97–121.

Franklin, A. (1997). Calibration. *Perspectives on Science, 5*(1), 31–80.

Franklin, A. (1998/2012, Winter) Experiment in physics. In E. N. Zalta (Ed.), *The Stanford Encyclopedia of Philosophy*. Retrieved from http://plato.stanford.edu/archives/win2012/entries/physics-experiment/

Franklin, A. & Howson, C. (1984). Why do scientists prefer to vary their experiments? *Studies in History and Philosophy of Science, 15*(1), 51–62.

Friedman, M. (1974). Explanation and scientific understanding. *Journal of Philosophy, 71*, 5–19.

Fuller, S. (1988/2002). *Social epistemology* (2nd ed.). Indiana University Press.

Fuller, S. (2009). Response to Lynch. *Spontaneous generations: A journal for the history and philosophy of science, 3*(1), 220–222.

Fuller, S., De Mey, M., & Shinn, T. (Eds.). (1989). *The cognitive turn: sociological and psychological perspectives on science* (Selected papers of the Sociology of Sciences Yearbook conference, held at the University of Colorado at Boulder, Nov. 23–25, 1987). Dordrecht: Kluwer Academic Publishers.

Galison, P. (1983). How the first neutral-current experiments ended. *Review of Modern Physics, 55*(2), 477–509.

Galison, P. (1987). *How experiments end*. Chicago: University of Chicago Press.

Galison, P. (1995). Context and constraints. In J. Buchwald (Ed.), *Scientific practice, theories and stories of doing physics* (pp. 13–41). Chicago & London: University of Chicago Press.

Galison P., & Warwick A. (Eds.). (1998). Cultures of theory. *Studies in History and Philosophy of Modern Physics, Special Issue, 29B*(3).

Garfinkel, H. (1967). *Studies in ethnomethodology*. Englewood Cliffs, NJ: Prentice Hall.

Garfinkel, H., Lynch, M., & Livingston, E. (1981). The work of a discovering science construed with materials from the optically discovered pulsar. *Philosophy of the Social Sciences 11*, 131–158.

Giardino, V., Moktefi, A., Mols, S., & Van Bendegem, J-P. (Eds.). (2012). *From practice to results in mathematics and logic*. Special issue of *Philosophia Scientiae, 16*(1).

Giardino, V., Moktefi, A., Mols, S., & Van Bendegem, J-P. (2012). Introduction. In V. Giardino, A. Moktefi, S. Mols, & J-P. Van Bendegem (Eds.), *From practice to results in mathematics and logic.* Special issue of *Philosophia Scientiae, 16*(1), 5–11.

Giddens, A. (1984). *The constitution of society: Outline of the theory of structuration.* Cambridge: Polity Press.

Giere, R. (1988). *Explaining Science.* Chicago: University of Chicago Press.

Giere, R. (2006a). *Science without laws.* Chicago: University of Chicago Press.

Giere, R. (2006b). *Scientific perspectivism.* Chicago: University of Chicago Press.

Gieryn, T. F. (1983). Boundary-work and the demarcation of science from non-science. *American Sociological Review, 48*, 781–795.

Gilbert, G. N., & Mulkay, M. (1984/2003). *Opening Pandora's box: A sociological analysis of scientists' discourse.* Cambridge: Cambridge University Press.

Gillies, D. (Ed.). (1992). *Revolutions in mathematics.* Oxford: Clarendon Press.

Gingras Y. (1995). Following the scientists? Yes, but at arm's length! In J. Z. Buchwald (Ed.), *Scientific practice: Theories and stories of doing physics* (pp. 123–148). Chicago: Chicago University Press.

Ginzburg, C. (1989). Traces. Racines d'un paradigme indiciaire. In Carlo Ginzburg (Ed.), *Mythes, emblèmes, traces. Morphologie et histoire* (pp. 139–180). Paris: Flammarion.

Ginzburg, C. (2012). *Threads and traces. True false fictive.* Berkeley, Los Angeles, & London: University of California Press.

Godin, B. (2006). The linear model of innovation: The historical construction of an analytic framework. *Science, Technology and Human Values, 31*(6), 631–667.

Godin, B., & Gingras, Y. (2000). Impact of collaborative research on academic science. *Science and Public Policy, 27*, 65–73.

Goldman A. (1970). *A theory of human action,* Englewood-Cliffs, NJ: Prentice Hall.

Goldman, A. (1999). *Knowledge in a social world.* Oxford: Oxford University Press.

Goldman, A. (2001). Experts: Which ones should you trust? *Philosophy and Phenomenological Research, 63,* 85–110.

Goldman, A. (2010, Summer). Social epistemology. In E. N. Zalta (Ed.), *The Stanford Encyclopedia of Philosophy.* Retrieved from http://plato.stanford.edu/archives/sum2010/entries/epistemology-social/

Gooding, D. (1990). *Experiment and the making of meaning.* Dordrecht: Kluwer.

Gooding, D. (1992). Putting agency back into experiment. In Andrew Pickering (Ed.), *Science as practice and culture* (pp. 65–112). Chicago: University of Chicago Press.

Goodman, N. (1954). *Fact, fiction, and forecast.* Cambridge, MA: Harvard University Press.

Gordin, M. D. (2004). *A well-ordered thing: Dmitrii Mendeleev and the shadow of the periodic table.* New York: Basic Books.

Gordin, M. D. (2012). The textbook case of a priority dispute: D. I. Mendeleev, Lothar Meyer, and the periodic system. In J. Riskin & M. Biagoli (Eds.), *Nature engaged: Science in practice from the Renaissance to the present* (pp. 59–82). New York: Palgrave MacMillan.

Grasso, D., & Burkins, M. (Eds.) (2010). *Holistic engineering education: Beyond technology.* New York: Springer Verlag.

Gray, J. (2008). *Plato's ghost. The modernist transformation of mathematics.* Princeton: Princeton University Press.

Grene, M. (1974) *The Knower and the known.* Berkeley & Los Angeles: University of California Press.

Grosholz, E. (2007). *Representation and productive ambiguity in mathematics and the sciences.* Oxford: Oxford University Press.

Gross, P., & Levitt, N. (1994). *Higher superstition. The academic left and its quarrels with science.* Baltimore & London: John Hopkins University Press.

Gross, S. R., & Mnookin, J. L. (2004). Expert information and expert evidence: A preliminary taxonomy. *Seton Hall Law Review, 34*(1): 141–189.

Guo, S. (1992). *Gudai shijie shuxue taidou Liu Hui* 古代世界數學泰斗劉徽 (Liu Hui, a leading figure of ancient world mathematics, 1st ed.) Jinan: Shandong kexue jishu chubanshe.

Guo, S. (2002). Shilun *Suanshushu* de lilun gongxian yu bianzuan 試論算數書的理論貢獻與編纂 (On the theoretical achievements and the compilation of the *Book of Mathematical Procedures*), *Faguo hanxue* 法國漢學 *French Sinology, 6*, 505–537.

Haack, S. (2005). Trial and error: The Supreme Court's philosophy of science. *American Journal of Public Health, 95*(S1), S66–S73.

Hackett, E., Amsterdamska, O., Lynch, M., & Wajcman, J. (Eds). (2008). *Handbook of science and technology studies* (3rd ed.). Cambridge, MA: MIT Press.

Hacking, I. (1983). *Representing and intervening: Introductory topics in the philosophy of natural science.* Cambridge: Cambridge University Press.

Hacking, I. (1988). Philosophers of experiment. In A. Fine & J. Leplin (Eds.), *PSA Proceedings 1988* (Vol. 2, pp. 147–156) East Lansing, MI: Philosophy of Science Association.

Hacking, I. (1992a). The self-vindication of the laboratory sciences. In A. Pickering (Ed.), *Science as practice and culture* (pp. 29–64). Chicago: University of Chicago Press.

Hacking, I. (1992b). "Style" for Historians and Philosophers. *Studies in History and Philosophy of Science, 23*, 1–20.

Hacking, I. (1995). Introduction. In J. Z. Buchwald (Ed.), *Scientific practice: Theories and stories of doing physics* (pp. 1–10). Chicago: University of Chicago Press.

Hacking, I. (1999). *The social construction of what?* Cambridge, MA: Harvard University Press.

Hacking, I. (2009). *Scientific reason.* Taipei: National Taiwan University Press.

Hammersley, M., & Atkinson, P. (2007) *Ethnography—Principles in practice* (3rd ed.). New York, NY: Routledge.

Hampshire, S. (1959/1982). *Thought and action* (new ed.). London: Chatto and Windus. (First edition 1959)

Handling Misconduct, 42 C.F.R. § 50.102 (1989).

Hansson, S. O. (2012, Winter). Science and pseudo-science. In E. N. Zalta (Ed.), *The Stanford Encyclopedia of Philosophy.* Retrieved from http://plato.stanford.edu/archives/win2012/entries/pseudo-science/

Haraway, D. (2008). *When species meet.* New York: Routledge.

Hardwig, J. (1985). Epistemic dependence. *Journal of Philosophy, 82*, 335–349.

Hardwig, J. (1988). Evidence, testimony, and the problem of individualism—a response to Schmitt. *Social Epistemology, 2*, 309–321.

Hardwig, J. (1991). The role of trust in knowledge. *Journal of Philosophy, 88*, 693–708.

Harris, Z. S. (1952). Discourse analysis. *Language, 28*(1), 1–30.

Harrison, E. (1987). Whigs, prigs and historians of science. *Nature, 329*, 213–214.

Haugeland, J. (1998). *Having thought.* Cambridge: Harvard University Press.

Haugeland, J. (2013). *Dasein disclosed.* Cambridge: Harvard University Press.

Heidegger, M. (1927/1962). *Sein und Zeit.* Tübingen: Max Niemayer Verlag. (1962 Translation by J. Macquarrie & E. Robinson, New York: Harper)

Heinz, L. C., & Chubin, D. E. (1988). Congress investigates scientific fraud. *BioScience, 38*, 559–561.

Hempel, C. G. (1965a). Aspects of scientific explanation. In C. G. Hempel (Ed.), *Aspects of scientific explanation and other essays in the philosophy of science* (pp. 331–496). New York: The Free Press.

Hempel, C.G. (1965b). The logic of functional analysis. In C.G. Hempel (Ed.), *Aspects of scientific explanation and other essays in the philosophy of science* (pp. 297–330). New York: The Free Press.

Hempel, C.G., & Oppenheim, P. (1965). Studies in the logic of explanation. In C.G. Hempel (Ed.), *Aspects of scientific explanation and other essays in the philosophy of science* (pp. 245–290). New York: The Free Press.

Heritage, J. (1984). *Garfinkel and ethnomethodology.* Oxford, UK: Polity Press.

Hester, S., & Eglin, P. (Eds.). (1997). *Culture in action: Studies in membership categorization analysis.* Lanham, MD: University Press of America.

Hettema, H., & Kuipers, T. (1988). The periodic table—Its formalizations, status, and relation to atomic theory. *Erkenntnis, 28,* 87–408.

Hintikka, J. (1985). A spectrum of logics of questioning. *Philosophica, 35,* 135–150.

Hixon, J. (1976). *The patchwork mouse.* New York: Anchor Press.

Hornsby, J. (2004). Agency and actions. In J. Hyman & H. Steward (Eds.), *Agency and action* (pp. 1–23). Cambridge: Cambridge University Press.

Howson, C., & Franklin, A. (1991). Maher, Mendeleev, and Bayesianism. *Philosophy of Science, 58,* 574–585.

Hughes, R.I.G. (1997). Models and representation. *Philosophy of Science, 64,* S325–S336.

Hutchinson, G.E. (1957). Concluding remarks. *Cold Spring Harbor Symposia in Quantitative Biology, 22,* 415–427.

Intergovernmental Panel on Climate Change (IPCC). (1990). *First assessment report.* Retrieved from http://www.ipcc.ch/ipccreports/far/wg_I/ipcc_far_wg_I_full_report.pdf

Intergovernmental Panel on Climate Change (IPCC). (1995). *Second assessment report.* Retrieved from http://www.ipcc.ch/pdf/climate-changes-1995/ipcc-2nd-assessment/2nd-assessment-en.pdf

Intergovernmental Panel on Climate Change (IPCC). (2001). *Third assessment report.* Retrieved from http://www.grida.no/publications/other/ipcc_tar/

Intergovernmental Panel on Climate Change (IPCC). (2007). *Fourth assessment report.* Retrieved from http://www.ipcc.ch/pdf/assessment-report/ar4/syr/ar4_syr.pdf

Irwin, A. (2006). The politics of talk: Coming to terms with the "new" scientific governance. *Social Studies of Science, 36*(2), 299–320.

Jardine, N. (2003). Whigs and stories: Herbert Butterfield and the historiography of science. *History of Science, 41,* 125–140.

Jasanoff, S. (1992). What judges should know about the sociology of science. *Jurimetrics, 32,* 345–359.

Jasanoff, S. (1995). *Science at the bar.* Cambridge, MA: Harvard University Press.

Jasanoff, S. (2003). Breaking the waves in science studies. Comment on H.M. Collins & Robert Evans' the third wave of science studies. *Social Studies of Science, 3*(3), 389–400.

Jasanoff, S. (2005). *States of knowledge: The co-production of science and social order.* New York: Routledge.

Jochi, S. (2001). "Sansusho nihongo yaku 算數書日本語譯" (Japanese translation of the *Suan shu shu*). *Wasan kenkyuso kiyo 和算研究所紀要, 4,* 19–46.

Kaiser, D. (2005). *Drawing theories apart: The dispersion of Feynman diagrams in postwar physics.* Chicago: University of Chicago Press.

Kitcher, P. (1976). On the uses of rigorous proof. Book reviews: Proofs and Refutations—The logic of mathematical discovery, I. Lakatos. *Science, 196*(4291), 782–783.

Kitcher, P. (1983). *The nature of mathematical knowledge.* New York: Oxford University Press.

Kitcher, P. (1989). Explanatory unification and the causal structure of the world. In P. Kitcher & W. C. Salmon (Eds.), *Minnesota studies in the philosophy of science, Volume XIII: Scientific explanation* (pp. 410–505). Minneapolis: University of Minnesota Press.

Kitcher, P. (1992). Authority, deference, and the role of individual reason. In E. McMullin (Ed.), *The social dimension of science* (pp. 244–271). Notre Dame: University of Notre Dame Press.

Kitcher, P. (1993), *The advancement of science: Science without legend, objectivity without illusions.* New York & Oxford: Oxford University Press.

Klein, U. (2001a). Paper tools in experimental culture. *Studies in History and Philosophy of Science 32*(2), 265–302.

Klein, U. (Ed.). (2001b). *Tools and modes of representation in the laboratory sciences.* London: Kluwer Academic Publishers.

Klein, U. (2003). *Experiments, models, paper tools: Cultures of organic chemistry in the nineteenth century.* Stanford: Stanford University Press.

Knorr, K. (1977). Producing and reproducing knowledge: Descriptive or constructive? *Social Science Information, 16*(6), 669–696.

Knorr-Cetina, K. (1979). Tinkering toward success. Prelude to a theory of scientific practice. *Theory and Society, 8,* 347–76.

Knorr-Cetina, K. (1981). *The manufacture of knowledge. An essay on the constructivist and contextual nature of science.* Oxford, UK: Pergamon.

Knorr-Cetina, K., & Mulkay, M. (1983a). Introduction: Emerging principles in social studies of science. In K. Knorr-Cetina, & M. Mulkay (Eds.), *Science observed: Perspectives on the social study of science* (pp. 1–17). London: Sage.

Knorr-Cetina, K., & Mulkay M. (Eds.). (1983b). *Science observed: Perspectives on the social study of science.* London: Sage.

Koocher, G., & Keith-Spiegel, P. (2010). Peers nip misconduct in the bud. *Nature, 466,* 438–440.

Koyré, A. (1961/1973). Perspectives sur l'histoire des sciences. In *Etudes d'histoire de la pensée scientifique* (pp. 390–399). Paris: Gallimard.

Kozma, R. (2003). The material features of multiple representations and their cognitive and social affordances for science understanding. *Learning and instruction, 13,* 205–226.

Kroes, P., & Van de Poel, I. (2009). Problematizing the notion of social context of technology. In S. H. Christensen, B. Delabouse, & M. Meganck (Eds.), *Engineering in context* (pp 61–74). Aarhus: Academica.

Kroes, P. A., & Meijers, A. W. M. (Eds.). (2000). *The empirical turn in the philosophy of technology* (Vol. 20). New York: JAI/Elsevier Science.

Kuhn, T. S. (1970). *The structure of scientific revolutions* (2nd ed.). Chicago: University of Chicago Press.

Kuhn, T. (1977). Objectivity, value judgment, and theory choice. In *The essential tension: Selected studies in scientific tradition and change* (pp. 320–339). Chicago: University of Chicago Press.

Kuhn, T. (2000). *The road since "Structure".* Chicago: University of Chicago Press.

Kukla, R., & Lance, M. (2009). *Yo! And lo!* Cambridge: Harvard University Press.

Lakatos, I. (1970a). History of science and its rational reconstructions. *PSA: Proceedings of the Biennial Meeting of the Philosophy of Science Association* (Vol. 1970, pp. 91–136). Chicago: University of Chicago Press.

Lakatos, I. (1970b). Falsification and the methodology of scientific research programmes. In I. Lakatos & A. Musgrave (Eds.), *Criticism and the Growth of Knowledge* (pp. 91–196). Cambridge: Cambridge University Press.

Lakatos, I. (1976). *Proofs and refutations.* Cambridge: Cambridge University Press.

Lakoff, G., & Núñez, R. (2000). *Where mathematics comes from. How the embodied mind brings mathematics into being.* New York: Basic Books.

Lam, L. Y. (1988). A Chinese genesis: Rewriting the history of our numeral system, *Archive for History of Exact Sciences, 38*(2), 101–108.

Lam, L. Y., & Ang, T. S. (2004). *Fleeting footsteps: Tracing the conception of arithmetic and algebra in ancient China* (rev. ed.). River Edge, NJ: World Scientific.

Lambert, K. (2006). Fuller's folly: Kuhnian paradigms, and intelligent design. *Social Studies of Science, 36*(6), 835–842.

Lange, M. (2000). *Natural laws in scientific practice.* Oxford: Oxford University Press.

Lange, M. (2007). Laws and theories. In S. Sarkar & A. Plutynski (Eds.), *A companion to the philosophy of biology* (pp. 489–505). Oxford: Blackwell.

Larvor, B. (1998) *Lakatos. An introduction.* London: Routledge.

Latour, B. (1987). *Science in action: How to follow scientists and engineers through society.* Cambridge: Harvard University Press.

Latour, B. (1998). Sur la pratique des théoriciens. In J. M. Barbier (Ed.), *Savoirs théoriques et savoirs d'actions* (pp. 131–146). Paris: Presses Universitaires de France.

Latour, B. (2004). Why has critique run out of steam? From matters of fact to matters of concern. *Critical Inquiry, 30*(2), 225–248.

Latour, B., & Woolgar, S. (1979). *Laboratory life: The construction of scientific facts.* Princeton, NJ: Princeton University Press.

Laudan, L. (1981). The pseudo-science of science? *Philosophy of the Social Sciences, 11*(2), 173–198.

Laudan, L. (1992). Science at the bar: Causes for concern. *Science, Technology, & Human Values 7*(41), 16–19.

Lautman, A. (1977). *Essai sur l'unité des mathématiques et divers écrits.* Paris: Union générale des éditions.

Lave, R., Mirowski, P., & Randalls, S. (Eds.). (2010). Special Issue: STS and neoliberal science. *Social Studies of Science, 40*(5).

Lawrence, C. (1985). Incommunicable knowledge: Science, technology, and the clinical art in Britain, 1850–1914. *Journal of Contemporary History, 20,* 503–520.

Lee, S., & Bozeman, B. (2012). The impact of research collaboration on scientific productivity. *Social Studies of Science, 35,* 673–702.

Leontyev, A. (2006). "Units" and levels of activity. *Journal of Russian and East European Psychology, 44*(3), 30–46.

Li, J. (1990). *Dongfang shuxue dianji Jiuzhang suanshu ji qi Liu Hui zhu yanjiu* 東方數學典籍 ——《九章算術》及其劉徽注研究 (Research on the Oriental mathematical Classic *The Nine Chapters on Mathematical Procedures* and on its Commentary by Liu Hui). Xi'an: Shaanxi renmin jiaoyu chubanshe.

Li, J. (1998). *Jiuzhang suanshu daodu yu yizhu* 九章算術導讀與譯註 (Reading guide and annotated translation of *The Nine Chapters on Mathematical Procedures*). Xi'an: Shaanxi renmin jiaoyu chubanshe.

Lipton, P. (1991). *Inference to the best explanation.* London: Routledge.

Livingston, E. (1986). *The ethnomethodological foundations of mathematics.* London: Routledge & Kegan Paul.

Lombard, L. B. (1995). Event. In R. Audi (Ed.), *Cambridge Dictionary of Philosophy* (pp. 292–293). Cambridge, UK: Cambridge University Press.

Löwe, B., & Müller, T. (Eds.). (2010). *Philosophy of mathematics: sociological aspects and mathematical practice.* London: College Publications.

Lucent Technologies. (2002, September). *Report of the investigation committee on the possibility of scientific misconduct in the work of Hendrik Schön and coauthors.* Retrieved from http://www.alcatel-lucent.com/wps/Document StreamerServlet?LMSG_CABINET=Docs_and_Resource_Ctr&aLMSG_CONTENT_FILE=Corp_Governance_Docs/researchreview.pdf

Lynch, M. (1982). Technical work and critical inquiry: investigations in scientific laboratory. *Social Studies of Science, 12*, 499–533.

Lynch, M. (1985). *Art and artifact in laboratory science: A study of shop work and shop talk in a research laboratory*. London: Routledge and Kegan Paul.

Lynch, M. (1993). *Scientific practice and ordinary action: Ethnomethodology and social studies of science*. Cambridge & New York: Cambridge University Press.

Lynch, M. (Ed.). (2012). *Science and technology studies* (Vols. 1–4). London & New York: Routledge.

Lynch, M., Livingston, E., & Garfinkel, H. (1983). Temporal order in laboratory work. In K. Knorr-Cetina & M. Mulkay (Eds.), *Science observed: Perspectives on the social study of science* (pp. 205–238). London & Beverly Hills: Sage.

Lynch, M., & Woolgar, S. (1988). Introduction: Sociological orientations to representational practice in science. *Human studies, 11*(2), 99–116.

MacKenzie, D. (1981). *Statistics in Britain 1865–1930*. Edinburgh, UK: Edinburgh University Press.

MacKenzie, D. A., & Spinardi, G. (1995) Tacit knowledge, weapons design, and the uninvention of nuclear weapons. *American Journal of Sociology, 101*, 44–99.

Maddy, P. (1997). *Naturalism in mathematics*. Oxford: Clarendon Press.

Maddy, P. (2007). *Second philosophy. A naturalistic method*. Oxford: Oxford University Press.

Mahan, B. H. (1975). *University chemistry*. Reading, MA: Addison-Wesley Publishing Company.

Maher, P. (1988). Prediction, accommodation, and the logic of discovery. In A. Fine & J. Leplin (Eds.), *PSA Proceedings 1988* (Vol. 1, pp. 273–285). East Lansing, MI: Philosophy of Science Association.

Mancosu, P. (Ed.) (2008). *The philosophy of mathematical practice*. Oxford & New York: Oxford University Press.

Mannheim, K. (1929/1995). *Ideologie und Utopie*. Frankfurt am Main: Klostermann.

Martinson, B. C., Anderson, M. S., & De Vries, R. (2005). Scientists behaving badly. *Nature, 435*, 737–738.

Massachusetts Institute of Technology. (2005, May). *From useful abstractions to useful designs—Thoughts on the foundations of the engineering method, Part I* (Draft). Engineering Council for Undergraduate Education, MIT.

Mayo, D. G. (1994). The new experimentalism, topical hypotheses, and learning from error. In *PSA: Proceedings of the Biennial Meeting of the Philosophy of Science Association* (pp. 270–279). Philosophy of Science Association.

Mayo-Wilson, C. (*forthcoming*). Reliability of testimonial norms in scientific communities. In C. Imbert, R. Muldoon, J. Sprenger, & K. Zollman, (Eds.). *The Collective Dimension of Science, special issue of Synthese*. Retrieved from http://www.contrib.andrew.cmu.edu/~conormw/Papers/Testimonial_Norms.pdf

Mazurs, E. G. (1974). *Graphic representations of the periodic system during one hundred years*. Tuscaloosa, AL: University of Alabama Press.

McDowell, J. (1994). *Mind and world*. Cambridge: Harvard University Press.

Medawar, P. (1964, August 1) Is the scientific paper fraudulent? Yes; it misrepresents scientific thought. *Saturday Review*, 42–43.

Medawar, P. B. (1976/1996). The strange case of the spotted mice. In *The strange case of the spotted Mice: And other classic essays on science* (pp. 132–143). Oxford, UK: Oxford University Press. (Reprinted from the *New York Review of Books, 23*(6), 6–11)

Mendeleev, D. I. (1869). On the relationship of the properties of the elements to their atomic weights. *Zeitschrift für Chemie, 12*, 405–406.

Mendeleev, D. I. (1871). Die periodische Gesetzmassigkeit der chemischen Elemente. *Annalen der Chemie und Pharmacie* (Supplement 8, pp. 133–229) Leipzig & Heidelberg: Wintersche Verlaghandlung.

Mendeleev, D. I. (1891). *The principles of chemistry* (1st English translation of 5th Russian Edition). New York: G. Kemensky, Collier.

Merton, R. K. (1942). Science and technology in a democratic order, *Journal of Legal and Political Science, 1,* 115–126.

Merton, R. K. (1957). Manifest and latent functions. In R. K. Merton (Ed.), *Social theory and social structure* (pp. 73–138). New York: The Free Press.

Merton, R. K. (1968). *Social theory and social structure.* New York: Free Press.

Merton, R. K. (1993). *On the shoulders of giants. A Shandean postscript.* Chicago: University of Chicago Press.

Meyer, J. L. (1870). Die Natur der chemischen Elemente als Funktion ihrer Atomgewichte. *Annalen der Chemie und Pharmarcie,* Supplement 7, 354–364.

Millikan, R. (1984). *Language, thought and other biological categories.* Cambridge: MIT Press.

Mills, C. W. (1940). Situated actions and vocabularies of motive. *American Sociological Review, 5,* 904–913.

Mitcham, C. (1994): *Thinking through technology: The path between engineering and philosophy.* Chicago & London: University of Chicago Press.

Mitchell, S. D. (2000) Dimensions of scientific law. *Philosophy of Science, 67,* 242–265.

Mooney, C., & Sokal, A. (2007, February 4). Taking the spin out of science. *Los Angeles Times.* Retrieved from http://articles.latimes.com/2007/feb/04/opinion/op-mooney4

Morgan, M. S., & Morrison, M. (Eds.). (1999). *Models as mediators: Perspectives on natural and social science* (Vol. 52). Cambridge: UK, Cambridge University Press.

Mumma, J. (*forthcoming*). Constructive geometric reasoning and diagrams. *Synthese.*

Nahmias E. (2007). Autonomous agency and social psychology. In M. Marraffa, M. De Caro, & F. Ferretti (Eds.) *Cartographies of the mind: Philosophy and psychology in intersection* (pp. 169–185). Dordrecht: Springer.

Nersessian, N. J. (1988). Reasoning from imagery and analogy in scientific concept formation. *PSA: Proceedings of the Biennial Meeting of the Philosophy of Science Association: Vol. 1* (pp. 41–47). Chicago: University of Chicago Press.

Nersessian, N. J. (2005). Interpreting scientific and engineering practices: Integrating the cognitive, social, and cultural dimensions. In M. Gorman, R. D. Tweney, D. Gooding, & A. Kincannon (Eds.), *Scientific and technological thinking* (pp. 17–56). Hillsdale, NJ: Lawrence Erlbaum.

Nersessian, N. J. (2012). Modeling practices in conceptual innovation: An ethnographic study of a neural engineering research laboratory. In U. Feest & F. Steinle (Eds.), *Scientific concepts and investigative practice* (pp. 245–269). Berlin: DeGruyter.

Nersessian, N. J., & Patton, C. (2009). Model-based reasoning in interdisciplinary engineering. In A. W. M. Meijers (Ed.), *The handbook of the philosophy of technology & engineering sciences* (pp. 678–718). Amsterdam: Elsevier.

Nisbett, R. E., Peng, K., Choi, I., & Norenzayan, A. (2001). Culture and systems of thought: Holistic versus analytic cognition. *Psychological review, 108,* 291–310.

Noë, A. (2005). *Action in perception.* Cambridge, MA.: MIT Press.

Norenzayan, A., Smith, E. E., Kim, B. J., & Nisbett, R. E. (2002). Cultural preferences for formal versus intuitive reasoning. *Cognitive sciences, 26,* 653–684.

Odling-Smee, F. J., Laland, K., & Feldman, M. (2003). *Niche construction.* Princeton: Princeton University Press.

Oreskes, N., & Conway, E. M. (2010). *Merchants of doubt: How a handful of scientists obscured the truth on issues from tobacco smoke to global warming.* New York, NY: Bloomsbury Publishing.

Orlikowski, W. J. (2000). Using technology and constituting structures: A practice lens for studying technology in organizations. *Organization Science, 11*(4): 404–428.

Pease, A., & Aberdein, A. (2011). Five theories of reasoning: Interconnections and applications to mathematics. *Logic and Logical Philosophy, 20*(1–2), 7–57.

Peng, Hao (2001). *Zhangjiashan hanjian «Suanshushu» zhushi* 張家山漢簡《算數書》注釋 (Commentary on the *Book of Mathematical Procedures,* a writing on bamboo strips dating from the Han and discovered at Zhangjiashan). Beijing: Science Press (Kexue chubanshe).

Perini, L. (2005). Explanation in two dimensions: Diagrams and biological explanation. *Biology & Philosophy, 20,* 257–269.

Perini, L. (2010). Scientific representation and the semiotics of pictures. In P. D. Magnus & J. Busch (Eds.), *New waves in philosophy of science* (pp. 131–154). New York: Palgrave Macmillan.

Pickering, A. (1981a). Constraints on controversy: The case of the magnetic monopole. *Social Studies of Science, 11,* 63–93.

Pickering, A. (1981b). The hunting of the quark. *Isis, 72,* 216–236.

Pickering, A. (1984a). *Constructing quarks: A sociological history of particle physics.* Chicago: University of Chicago Press.

Pickering, A. (1984b). Against putting the phenomena first: The discovery of the weak neutral current. *Studies in the History and Philosophy of Science, 15,* 85–117.

Pickering, A. (1987). Against correspondence: A constructivist view of experiment and the real. A. Fine & P. Machamer (Eds.), *PSA 1986, Vol. 2* (pp. 196–206). Pittsburgh: Philosophy of Science Association.

Pickering, A. (1988). Review. Peter Galison. How experiments end. *Isis, 79*(3), 298.

Pickering, A. (1989a). Editing and epistemology: Three accounts of the discovery of the weak neutral current. *Knowledge and society: Studies in the sociology of science past and present, 8,* pp. 217–232.

Pickering, A., (1989b). Living in the material world: On realism and experimental practice. In D. Gooding, T. Pinch, & S. Schaffer (Eds.), The uses of experiments: Studies in the natural sciences (pp. 275–297). Cambridge, UK: Cambridge University Press.

Pickering, A. (Ed.). (1992). *Science as practice and culture.* Chicago: University of Chicago Press.

Pickering, A. (1995a). *The mangle of practice. Time, agency and science.* Chicago: University of Chicago Press.

Pickering, A. (1995b). Beyond constraint: The temporality of practice and the historicity of knowledge. In J. D. Buchwald (Ed.), *Scientific practice: Theories and stories of doing physics* (pp. 42–55). Chicago: University of Chicago Press.

Pickering, A. (2012). The robustness of science and the dance of agency. In L. Soler, E. Trizio, T. Nickles, & W. C. Wimsatt (Eds.), *Characterizing the robustness of science after the practice turn in philosophy of science* (pp. 317–327). Boston Studies in the Philosophy of Science. Dordrecht, Heidelberg, London, & New York: Springer.

Pickering A. & Stephanides A. (1992). Constructing quaternions: On the analysis of conceptual practice. In A. Pickering (Ed.), *Science as practice and culture* (pp. 139–167). Chicago: University of Chicago Press.

Pickstone, J. V. (2000). *Ways of knowing: A new history of science, technology and medicine.* Manchester: Manchester University Press.

Pickstone, J. V. (2007). Working Knowledges before and after circa 1800: Practices and disciplines in the history of science, technology and medicine. *Isis, 98*, 489–516.

Pinch, T. J. (1977). What does a proof do if it does not prove? In E. Mendelsohn, P. Weingart, & R. Whitley (Eds.), *The social production of scientific knowledge* (pp. 171–215). Dordrecht, the Netherlands: Reidel.

Pinch, T. J. (1986). *Confronting nature: The sociology of solar neutrino detection.* Dordrecht, the Netherlands: D. Reidel.

Pinch, T. J. (2011). Karen Barad, quantum mechanics, and the paradox of mutual exclusivity. *Social Studies of Science, 41*(3), 431–441.

Pinch, T. J., & Bijker, W. E. (1984). The social construction of facts and artifacts. *Social Studies of Science, 14*(3), 399–441. (Reprinted in Bijker et al., 1987)

Pinch, T. J., Collins, H. M., & Carbone, L. (1996). Inside knowledge: Second order measures of skill. *The Sociological Review, 44*(2), 163–186.

Polanyi, M. (1958/1962) *Personal knowledge: Towards a post-critical philosophy* (corrected edition). Chicago: University of Chicago Press. (Originally published in 1958)

Polanyi, M. (1967). *The tacit dimension.* London: Routledge & Kegan Paul.

Pólya, G. (1945). *How to solve it: A new aspect of mathematical method.* Princeton: Princeton University Press.

Pollard, B. (2005). The rationality of habitual actions. *Proceedings of the Durham-Bergen Postgraduate Philosophy Seminar* (Vol. 1, pp. 39–50). Bergen, NOR: Philosophy Departments of the University of Bergen; Durham, UK: University of Durham.

Popper, K. (1963). *Conjectures and refutations.* London: Routledge & Kegan Paul.

Price, D. J. D. S. & Beaver, D. D. B. (1966). Collaboration in an invisible college. *American Psychologist, 21*, 1011–1018.

Putnam, H. (1975). The meaning of meaning. In *Mind, Language and Reality* (pp. 215–271). Cambridge: Cambridge University Press.

Qian, B. (1963). *Suanjing shishu* 算經十書 *(Qian Baocong jiaodian* 錢寶琮校點*)* (Critical punctuated edition of *The Ten Classics of Mathematics*, 2 Vols.) Beijing 北京: Zhonghua shuju 中華書局.

Rayfield, D. (1968). Action. *Nous, 2*(2), 131–145.

Rennie, D., Yank, V., & Emanuel, L. (1997). When authorship fails: A proposal to make contributors accountable. *JAMA, 278*, 579–585.

Restivo, S., Van Bendegem, J. P., & Fischer, R. (Eds.). (1993). *Math worlds: Philosophical and social studies of mathematics and mathematics education.* Albany: SUNY Press.

Restivo, S. (1985). *The social relations of physics, mysticism, and mathematics.* Dordrecht, Boston: D. Reidel.

Restivo, S. (1992). *Mathematics in society and history. Sociological inquiries* (Series Episteme, Vol. 20). Dordrecht: Kluwer Academic Publishers.

Restivo, S. (2011). *Red, black, and objective. Science, sociology, and anarchism.* Farnham: Ashgate.

Reuchlin, M. (1978). Processus vicariants et différences individuelles. *Journal de psychologie normale et pathologique, 2*, 133–145.

Rheinberger, H.-J. (1997). *Toward a history of epistemic things: Synthesizing proteins in the test tube.* Stanford: Stanford University Press.

Rheinberger, H.-J. (2005). A Reply to David Bloor: "Toward a Sociology of Epistemic Things". *Perspectives on Science, 13*, 406–410.

Ricoeur, P. (1986). *Facts and values.* Dordrecht, the Netherlands: Springer.

Rip, A. (2003). Constructing expertise: In a third wave of science studies? *Social Studies of Science*, 33, 419–434.

Risinger, D. M. (2000). Preliminary thoughts on a functional taxonomy of expertise in a post-Kumho world. *Seton Hall Law Review, 31*, 508–536.

Rorty, R. (1979). *Philosophy and the mirror of nature*. Princeton, NJ: Princeton University Press.

Rorty, R. (1998). Pragmatism as romantic polytheism. In M. Dickstein (Ed.), *The revival of pragmatism* (pp. 21–36). Durham, London: Duke University Press.

Rosental, C. (2008). *Weaving self-evidence: A sociology of logic*. Princeton: Princeton University Press.

Roth, P., & Barrett, R. (1990). Deconstructing quarks. *Social Studies of Science, 20*(4), 579–632.

Rouse, J. (1987). *Knowledge and power*. Ithaca: Cornell University Press.

Rouse, J. (1996). *Engaging science: How to understand its practices philosophically*. Ithaca & London: Cornell University Press.

Rouse, J. (1999). Understanding Scientific Practices: Cultural Studies of Science as a Philosophical Program. *Science Studies Reader* (pp. 442–456), New York: Routledge.

Rouse, J. (2002). *How scientific practices matter*. Chicago: University of Chicago Press.

Rouse J. (2007). Practice theory. In S. Turner, M. Risjord, D. Gabbay, P. Thagard, & J. Woods (Eds.), *Handbook of the philosophy of science. Vol. 15: Philosophy and anthropology and sociology* (pp. 639–682). Amsterdam & Oxford: Elsevier.

Rouse, J. (2009). Laboratory fictions. In M. Suarez (Ed.), *Fictions in science* (pp. 37–55). New York: Routledge.

Rouse, J. (2011). Articulating the world: Experimental systems and conceptual understanding. *International Studies in Philosophy of Science, 25*, 243–254.

Rouse, J. (*forthcoming*). *Articulating the world*.

Ruse, M. (1992). Response to the commentary: *Pro judice*. *Science, Technology, & Human Values, 7*(41), 19–23.

Ruse, M. (1996). Commentary: The academic as expert witness. *Science, Technology, & Human Values, 11*, 68–73.

Sacks, H. (1972). On the analyzability of stories by children. In J. Gumperz & D. Hymes (Eds.), *Directions in sociolinguistics: The ethnography of communication* (pp. 325–345). New York: Rinehart & Winston.

Salanskis, J.-M. (2000). *Modèles et pensées de l'action*. Paris: L'Harmattan.

Salanskis, J. M. (2001). Kant, la science et l'attitude philosophique. *Cahiers d'histoire et de philosophie des sciences, 50*, 199–235.

Salanskis J.-M. (2003). Entretien avec Olivier Picavet. *Le Philosophoire, 20*, 21–39.

Salanskis J.-M. (2010) Action et décision. *Intellectica, 53–54*, 163–179.

Salmon, W. C. (1984). *Scientific explanation and the causal structure of the world*. Princeton, NJ: Princeton University Press.

Sartre, J.-P. (1943). *L'être et le néant*. Paris: Gallimard.

Scerri, E. R. (1997). Has the periodic table been successfully axiomatized? *Erkenntnis, 47*, 229–243.

Scerri, E. R. (2007). *The periodic table: Its story and its significance*. Oxford: Oxford University Press.

Scerri, E. R., & Worrall, J. (2001). Prediction and the periodic table. *Studies in History and Philosophy of Science, 32*, 407–452.

Schatzki, T. (1996). *Social practices: A Wittgensteinian approach to human activity and the social*. New York: Cambridge University Press.

Schatzki, T. R. (2001). Practice mind-ed orders. In T. R. Schatzki, K. Knorr-Cetina, & E. Von Savigny (Eds.), *The practice turn in contemporary theory* (pp. 42–55). London: Routledge.

Schatzki, T.R., Knorr-Cetina, K., & Von Savigny, E. (Eds.). (2001). *The practice turn in contemporary theory.* London: Routledge.

Scheffler, I. (1967). *Science and subjectivity.* Cambridge: Harvard University Press.

Schön, D.A. (1983). *The reflective practitioner: How professionals think in action.* New York: Basic books.

Schön, D.A. (1992). The crisis of professional knowledge and the pursuit of an epistemology of practice. *Journal of Interprofessional Care, 6*(1), 49–63.

Scriven, M. (1962). Explanations, predictions, and laws. In H. Feigl & G. Maxwell (Eds.), *Scientific explanation, space, and time* (pp. 170–230). Minneapolis: University of Minnesota Press.

Sellars, W. (1985). Toward a theory of predication. In J. Bogen & J. McGuire (Eds.), *How things are* (pp. 285–322). Dordrecht: D. Reidel.

Sellars, W. (1997). *Empiricism and the philosophy of mind.* Cambridge: Harvard University Press.

Sellars, W. (2007). Philosophy and the scientific image of man. In R. Brandom & K. Scharp (Eds.), *In the space of reasons* (pp. 369–408). Cambridge: Harvard University Press.

Service, R.F. (2000). Breakdown of the year: Physics fraud. *Science, 298*, 2302.

Shapere, D. (1977). Scientific theories and their domains. In F. Suppe (Ed.), *The Structure of Scientific Theories* (2nd ed., pp. 518–565). Urbana: University of Illinois Press.

Shapin, S. (1979). The politics of observation: Cerebral anatomy and social interests in the Edinburgh Phrenology Disputes. In R. Wallis (Ed.), *On the margins of science: The social construction of rejected knowledge* (Sociological Review Monograph, 27, pp. 139–178). Keele, UK: Keele University Press.

Shapin, S. (1984). Talking history: Reflections on discourse analysis. *Isis, 75*(1), 125–130.

Shapin, S. (1986). History of science and its sociological reconstructions. In R.S. Cohen and T. Schnelle (Eds.), *Cognition and Fact: Materials on Ludwik Fleck* (pp. 325–386). Dordrecht, Boston, Lancaster, & Tokyo: Reidel Publishing Company.

Shapin, S. (2007). Expertise, common sense, and the Atkins diet. In J. N. Porter & P. W. B. Phillips (Eds.), *Public science in liberal democracy* (pp. 174–193). Toronto: University of Toronto Press.

Shapin, S., & Schaffer, S. (1985). *Leviathan and the air pump.* Princeton: Princeton University Press.

Shin, S.-J. (1995). *The logical status of diagrams.* Cambridge: Cambridge University Press.

Simon, H. (1969). *The sciences of the artificial.* Cambridge: MIT Press.

Siu, M.-K., & Volkov, A. (1999). Official curriculum in traditional Chinese mathematics: How did candidates pass the examinations? *Historia Scientiarum, 9,* 85–99.

Sneed, J.D. (1971). *The logical structure of mathematical physics.* Dordrecht: Reidel.

Soler, L. (2008a). Are the results of our science contingent or inevitable? *Studies in History and Philosophy of Science. Part A, 39*(2), 221–229.

Soler, L. (2008b). Revealing the analytical structure and some intrinsic major difficulties of the contingentist/inevitabilist issue. *Studies In History and Philosophy of Science. Part A, 39*(2), 230–241.

Soler, L. (2009). *Introduction à l'épistémologie* (2nd ed., rev.). Paris: Ellipses.

Soler, L. (2011). Tacit aspects of experimental practices: Analytical tools and epistemological consequences. *European Journal for the Philosophy of Science, 1*(3), 394–433.

Soler, L. (2012a). The solidity of scientific achievements: structure of the problem, difficulties, philosophical implications. In L. Soler, E. Trizio, T. Nickles, &

W. C. Wimsatt (Eds.), *Characterizing the robustness of science after the practice turn in philosophy of science* (pp. 1–60). Springer, Boston Studies in the Philosophy of Science. Dordrecht, Heidelberg, London, & New York: Springer.

Soler, L. (2012b). Robustness of results and robustness of derivations: The internal architecture of a solid experimental proof. In L. Soler, E. Trizio, T. Nickles, & W. C. Wimsatt (Eds.), *Characterizing the robustness of science after the practice turn in philosophy of science* (pp. 227–266). Springer, Boston Studies in the Philosophy of Science. Dordrecht, Heidelberg, London, & New York: Springer.

Soler, L. (2012c). *Étudier les pratiques scientifiques: Étudier quoi? A la place de quoi?* [website]. Retrieved from http://www.sphere.univ-paris-diderot.fr/IMG/pdf/LSoler_Practices_8Feb12.pdf

Soler, L., Trizio, E., Nickles, T., & Wimsatt, W. C. (Eds.). (2012). *Characterizing the robustness of science after the practice turn in philosophy of science*. Springer, Boston Studies in the Philosophy of Science. Dordrecht, Heidelberg, London, & New York: Springer.

Soler, L., Trizio, E., & Pickering, A. (in progress). *Science as it could have been. Discussing the contingent/inevitable aspects of scientific practices*.

Soler, L., Zwart, S. D., & Catinaud, R. (Eds.). (2013). Tacit and explicit knowledge: Harry Collins' framework. *Philosophia Scientiæ* 17(3). Special issue.

Sørensen, K. H. (2009). The role of social science in engineering. In A. W. M. Meijers (Ed.), *The handbook of the philosophy of technology and engineering sciences* (pp. 93–115), Amsterdam: Elsevier.

Steen, R. G. (2011). Retractions in the medical literature: How many patients are put at risk by flawed research? *Journal of medical ethics, 37*, 688–692.

Steneck, N. H. (1994). Research universities and scientific misconduct: History, policies, and the future. *Journal of Higher Education Policy and Management, 65*, 310–330.

Steneck, N. H. (1999). Confronting misconduct in science in the 1980s and 1990s: What has and has not been accomplished? *Science and Engineering Ethics, 5*, 161–176.

Stern, D. (2003). The practical turn. In S. P. Turner & P. Roth (Eds.), *The Blackwell guidebook to the philosophy of the social sciences* (pp. 185–206). Oxford: Blackwell.

Stewart, P. J. (2007). A century on from Dmitrii Mendeleev: Tables and spirals, noble gases and Nobel prizes. *Foundations in Chemistry, 9*, 235–245.

Suárez, M. (2005). An inferential conception of scientific representation. *Philosophy of Science, 71*, 767–779.

Swiddler, A. (2001). What anchors cultural practices. In T. R. Schatzki, K. Knorr-Cetina, & E. Von Savigny (Eds.), *The practice turn in contemporary theory* (pp. 83–101). London: Routledge.

Taylor, C. (1971). Interpretation and the sciences of man. *The Review of Metaphysics, 25*(1): 3–51.

Taylor, C. (1985). *Philosophical papers* (2 volumes). Cambridge: Cambridge University Press.

Teller, P. (2001). Twilight of the Perfect Model Model. *Erkenntnis, 55*, 393–415.

Thagard, P. (1993). Societies of minds: Science as distributed computing. *Studies in History and Philosophy of Science, 24*, 49–67.

Tiles, J. E. (1992). Experimental evidence vs. experimental practice? *British Journal for the Philosophy of Science, 43*, 99–109.

Thom, R. (1972/1976). *Structural stability and morphogenesis* (D. H. Fowler, Trans.). Reading, MA.: W. A. Benjamin. (Translation of *Stabilité structurelle et morphogenèse*, 1972, Paris: Interéditions)

Thom, R. (1974). *Modèles mathématiques de la morphogenèse*. Paris: Union générale d'Editions.

Travis, G. D. L. (1980). On the importance of being earnest. In K. Knorr, R. Krohn, & R. Whitley (Eds.), *The social process of scientific investigation: Sociology of the sciences yearbook, 4* (pp. 11–32). Dordrecht, the Netherlands: Reidel.

Travis, G. D. L. (1981). Replicating replication? Aspects of the social construction of learning in planarian worms. *Social Studies of Science, 11,* 11–32.

Traweek, S. J. (1988). *Beamtimes and lifetimes: The world of high energy physicists.* Cambridge: Harvard University Press.

Trevelyan, J. P. (2007). Technical coordination in engineering practice. *Journal of Engineering Education, 96*(3), 191–204.

Trevelyan, J. P. (2009). Engineering education requires a better model of engineering practice. In *Proceedings of the research in engineering education symposium* (pp. 20–23). Retrieved from http://rees2009.pbworks.com/f/rees2009_submission_52.pdf

Trevelyan, J. P. (2010). Reconstructing engineering from practice. *Engineering Studies, 2*(3), 175–195.

Turner, S. (1994). *The social theory of practices: Tradition, tacit knowledge and presuppositions.* Cambridge: Polity Press.

Turner, S. (2001). What is the problem with experts? *Social Studies of Science, 31*(1), 123–149.

Turner, S. (2003). The third science war. *Social Studies of Science, 33*(4), 581–611.

Tymoczko, T. (Ed.) (1985). *New directions in the philosophy of mathematics,* Boston: Birkhäuser.

University of Pittsburgh. (2006). *Summary investigative report on allegations of possible scientific misconduct on the part of Gerald P. Schatten, Ph.D.* Pittsburgh, PA. Retrieved from http://ecommons.library.cornell.edu/bitstream/1813/11589/1/Gerald_Schatten_Final_Report_2.08.pdf

University of Tilburg. (2011). *Interim report regarding the breach of scientific integrity committed by Prof. D. A. Stapel.* Retrieved from http://www.tilburguniversity.edu/upload/547aa461-6cd1-48cd-801b-61c434a73f79_interim-report.pdf

U.S. Congress Committee on Governmental Operations, Subcommittee on Investigations and Oversight. (1988). *Scientific fraud and misconduct and the federal response.* Washington, DC: GPO.

U.S. Congress Committee on Science and Technology, Subcommittee on Investigations and Oversight. (1981). *Fraud in biomedical research.* Washington, DC: GPO.

Van Bendegem J. P. (1993). Foundations of mathematics or mathematical practice: Is one forced to choose? In: R. Fischer, S. Restivo, & J. P. Van Bendegem (Eds.), *Math worlds: New directions in the social studies and philosophy of mathematics* (pp. 21–38). New York: State University New York Press.

Van Bendegem, J. P. (2004). The creative growth of mathematics. In D. Gabbay, S. Rahman, J. Symons, & J. P. Van Bendegem (Eds.), *Logic, epistemology and the unity of science (LEUS)* (pp. 229–255). Dordrecht: Kluwer Academic Press.

Van Bendegem, J. P., & Van Kerkhove, B. (2008). Pi on earth, or mathematics in the real world. *Erkenntnis, 68*(3), 421–435.

Van Bendegem, J. P., De Vuyst, J., & Van Kerkhove, B. (Eds.). (2010). *Philosophical perspectives on mathematical practice* (Texts in Philosophy 12). London: College Publications.

Van Benthem, J. (2011). *Logical dynamics of information and interaction.* Cambridge: Cambridge University Press.

Van Brakel, J. (2000). *Philosophy of chemistry: Between the manifest and the scientific image.* Leuven, Belgium: Leuven University Press.

Van Dijk, T. A. (1974). Philosophy of action and narrative. *Poetics, 5,* 287–332.

Van de Poel, I., & Royakkers, L. (2011). *Ethics, technology, and engineering: An introduction.* Chicester: Wiley-Blackwell.

Van Dongen, J. (2007). Emil Rupp, Albert Einstein, and the canal ray experiments on wave-particle duality: Scientific fraud and theoretical bias. *Historical Studies in the Physical Sciences, 37,* 73–120.

Van Fraassen, B. C. (1980). *The scientific image.* Oxford: Clarendon Press.

Van Kerkhove, B. (Ed.). (2009). New perspectives on mathematical practices. *Essays in philosophy and history of mathematics.* Singapore: World Scientific.

Van Kerkhove, B., & Van Bendegem, J. P. (2004). The unreasonable richness of mathematics. *Journal of Cognition and Culture, 4*(3–4), 525–549.

Van Kerkhove, B. & Van Bendegem, J. P. (Eds.). (2007). *Perspectives on mathematical practices: Bringing together philosophy of mathematics, sociology of mathematics, and mathematics education.* Dordrecht: Springer.

Van Kerkhove, B., & Van Bendegem, J. P. (to appear). Another look at mathematical style, as inspired by Le Lionnais and the OuLiPo. In J. Meheus, E. Weber, & D. Wouters (Eds.), *Logic, reasoning and rationality.* New York: Springer.

Van Spronsen, J. W. (1969). *The periodic system of the chemical elements: A history of the first hundred years.* Amsterdam: Elsevier.

Venable, F. P. (1896). *The development of the periodic law.* Easton, PA: Chemical Publishing Company.

Vermersch, P. (1994). *L'entretien d'explicitation en formation continue et initiale.* Paris: ESF.

Vincenti, W. (1979). The air-propeller tests of W. F. Durand and E. P. Lesley: A case study in technological methodology. *Technology and culture, 20*(4), 712–751.

Vincenti, W. (1990). *What engineers know and how they know it: Analytical studies from aeronautical history* (Johns Hopkins Studies in the History of Technology [New. Ser., No. 11]). Baltimore: Johns Hopkins University Press.

Vinck, D. (1992). *Du laboratoire aux réseaux. Le travail scientifique en mutation.* Luxembourg: Office des publications officielles des Communautes europeennes.

Vinck, D. (Ed). (2003). *Everyday engineering. Ethnography of design and innovation.* Cambridge, MA: MIT Press.

Vorms, M. (2010). The theoretician's gambits: scientific representations, their formats and content. In L. Magnani, W. Carnielli, & C. Pizzi (Eds.), *Model-based reasoning in science and technology: abduction, logic, and computational discovery* (pp. 533–558). Studies in Computational Intelligence, Vol. 314. Berlin & Heidelberg: Springer-Verlag.

Wagenknecht, S. (*forthcoming*). Facing the incompleteness of epistemic trust: Managing dependence in scientific practice. *Social Epistemology.*

Warwick, A. (1992). Cambridge mathematics and Cavendish physics: Cunningham, Campbell, and Einstein's relativity, 1905–1911. Pt. 1, The uses of theory. *Studies in the History and Philosophy of Science Part A, 23,* 625–56.

Warwick, A. (1993). Cambridge mathematics and Cavendish physics: Cunningham, Campbell, and Einstein's relativity, 1905–1911. Pt. 2, Comparing traditions in Cambridge physics. *Studies in the History and Philosophy of Science Part A, 24,* 1–25.

Warwick, A. (2003). *Masters of theory: Cambridge and the rise of mathematical physics.* Chicago: University of Chicago Press.

Williams, B., & Figueiredo, J. (2011). Engineering practice-inputs for course design. In *IEEE, Global Engineering Education Conference (EDUCON), 2011 IEEE* (pp. 809–815). Piscataway, NJ: IEEE.

Wilson, A., & Ashplant, T. G. (1988). Whig history and present centered history. *The Historical Journal, 31*(1), 1–16.

Wilson, G., & Shpall, S. (2012, Summer). Action. In E. N. Zalta (Ed.), *The Stanford Encyclopedia of Philosophy.* Retrieved from http://plato.stanford.edu/archives/sum2012/entries/action/

Wimsatt, W. (1981). Robustness, reliability, and overdetermination. In D. T. Campbell, M. B. Brewer, and B. E. Collins (Eds.), *Scientific inquiry and the social sciences* (pp. 125–163). San Francisco: Jossey-Bass.

Winch, P. (1958). *The idea of a social science*. London: Routledge & Kegan Paul.

Winner, L. (1993). Upon opening the black box and finding it empty: Social constructivism and the philosophy of technology. *Science, Technology, & Human Values, 18*(3), 362–378.

Wittgenstein, L. (1958). *Philosophical investigations* (Vol. 255). Oxford: Blackwell.

Woodhouse, E., Hess, D., Breyman S., & Martin, B. (2002). Science studies and activism: Possibilities and problems for reconstructivist agendas. *Social Studies of Science, 32*(2), 297–319.

Woodward, J. (2003). *Making things happen: A theory of causal explanation.* Oxford: Oxford University Press.

Woody, A. I. (2000). Putting quantum mechanics to work in chemistry: The power of diagrammatic representation. *Philosophy of Science, 67,* S612–S627.

Woody, A. I. (2003). On explanatory practice and disciplinary identity. *Chemical Explanation: Characteristics, Development, Autonomy. Annals of the New York Academy of Sciences, 988,* 22–29.

Woody, A. I. (2004a). More telltale signs: What attention to representation reveals about scientific explanation. *Philosophy of Science, 71,* 780–793.

Woody, A. I. (2004b). Telltale signs: What common explanatory strategies in chemistry reveal about explanation itself. *Foundations of Chemistry, 6,* 13–43.

Woody, A. I. (2011). How is the ideal gas law explanatory? *Science & Education.* doi:10.1007/s11191–011–9424–6

Woody, A. I. (2012). Concept amalgamation and representation in quantum chemistry. In A. I. Woody, R. F. Hendry, & P. Needham (Eds.), *Handbook of the philosophy of science, volume 6: Philosophy of chemistry* (pp. 427–466). Amsterdam: Elsevier Press.

Woolf, P. (1981). Fraud in science: How much, how serious? *Hastings Center Report, 11,* 9–14.

Woolgar, S. (1988). *Knowledge and reflexivity. New frontiers in the sociology of knowledge.* London: Sage.

Workman, M., Ford, R., & Allen, W. (2008). A structuration agency approach to security policy enforcement in mobile ad hoc networks. *Information Security Journal, 17,* 267–277.

Wray, K. B. (2002). The epistemic significance of collaborative research. *Philosophy of Science, 69,* 150–168.

Wuchty, S., Jones, B. F., & Uzzi, B. (2007). The increasing dominance of teams in production of knowledge. *Science, 316,* 1036–1039.

Wynne, B. (1996). May the sheep safely graze? A reflexive view of the expert-lay knowledge divide. In S. Lash, B. Szerszynski, & B. Wynne (Eds.), *Risk, environment and modernity: Toward a new ecology* (pp. 44–83). London: Sage.

Wynne, B. (2003). Seasick on the third wave? Subverting the hegemony of propositionalism: Response to Collins & Evans (2002). *Social Studies of Science, 33*(3), 401–417.

Xiao, C. (2011). Research on the Qin strips "Mathematics" kept at the Academy Yuelu 嶽麓書院藏秦簡《數》研究, Ph. D. in History, Hunan University 湖南大學.

Zhu, H., & Chen, S. (Eds.). (2011). *Yuelu shuyuan cang Qin jian (er)* 嶽麓書院藏秦簡（貳）(Qin Bamboo slips kept at the Academy Yuelu (2)). Shanghai 上海: Shanghai cishu chubanshe 上海辞出版.

Zuckerman, H. (1967). Nobel Laureates in science: Patterns of productivity, collaboration, and authorship. *American Sociological Review, 32,* 391–403.

Zwart, S.D., Jacobs, J., & Van de Poel, I. (2013). Values in engineering models: Social ramifications of modeling in engineering design. *Engineering Studies, 5*(2), 1–24. doi:10.1080/19378629.2013.809349

Zwart, S.D., & Kroes, P. (to appear). Substantive and procedural contexts of engineering design. In C. Hyldgaard, B. Newberry, M. Meganck, & C. Didier (Eds.) *Engineering identities, values, contexts and epistemologies* (Vol. 2 of Issues in Engineering Studies, IiES). Dordrecht & London: Springer.

Index